Advances in Industrial Control

Other titles published in this series:

Digital Controller Implementation and Fragility
Robert S.H. Istepanian and James F. Whidborne (Eds.)

Optimisation of Industrial Processes at Supervisory Level
Doris Sáez, Aldo Cipriano and Andrzej W. Ordys

Robust Control of Diesel Ship Propulsion
Nikolaos Xiros

Hydraulic Servo-systems
Mohieddine Jelali and Andreas Kroll

Model-based Fault Diagnosis in Dynamic Systems Using Identification Techniques
Silvio Simani, Cesare Fantuzzi and Ron J. Patton

Strategies for Feedback Linearisation
Freddy Garces, Victor M. Becerra, Chandrasekhar Kambhampati and Kevin Warwick

Robust Autonomous Guidance
Alberto Isidori, Lorenzo Marconi and Andrea Serrani

Dynamic Modelling of Gas Turbines
Gennady G. Kulikov and Haydn A. Thompson (Eds.)

Control of Fuel Cell Power Systems
Jay T. Pukrushpan, Anna G. Stefanopoulou and Huei Peng

Fuzzy Logic, Identification and Predictive Control
Jairo Espinosa, Joos Vandewalle and Vincent Wertz

Optimal Real-time Control of Sewer Networks
Magdalene Marinaki and Markos Papageorgiou

Process Modelling for Control
Benoît Codrons

Computational Intelligence in Time Series Forecasting
Ajoy K. Palit and Dobrivoje Popovic

Modelling and Control of Mini-Flying Machines
Pedro Castillo, Rogelio Lozano and Alejandro Dzul

Ship Motion Control
Tristan Perez

Hard Disk Drive Servo Systems (2nd Ed.)
Ben M. Chen, Tong H. Lee, Kemao Peng and Venkatakrishnan Venkataramanan

Measurement, Control, and Communication Using IEEE 1588
John C. Eidson

Piezoelectric Transducers for Vibration Control and Damping
S.O. Reza Moheimani and Andrew J. Fleming

Manufacturing Systems Control Design
Stjepan Bogdan, Frank L. Lewis, Zdenko Kovačić and José Mireles Jr.

Windup in Control
Peter Hippe

Nonlinear H_2/H_∞ Constrained Feedback Control
Murad Abu-Khalaf, Jie Huang and Frank L. Lewis

Practical Grey-box Process Identification
Torsten Bohlin

Control of Traffic Systems in Buildings
Sandor Markon, Hajime Kita, Hiroshi Kise and Thomas Bartz-Beielstein

Wind Turbine Control Systems
Fernando D. Bianchi, Hernán De Battista and Ricardo J. Mantz

Advanced Fuzzy Logic Technologies in Industrial Applications
Ying Bai, Hanqi Zhuang and Dali Wang (Eds.)

Practical PID Control
Antonio Visioli

(continued after Index)

Daniel Sbárbaro · René del Villar
Editors

Advanced Control and Supervision of Mineral Processing Plants

Springer

Daniel Sbárbaro, Prof.
Universidad de Concepción
Departamento de Ingeniería Eléctrica
Casilla 160-C, Correo 3
Concepción
Chile
dsbarbar@udec.cl

René del Villar, Prof.
Université Laval
Département de Génie des Mines,
de la Métallurgie et des Matériaux
Pavillon Adrien-Pouliot 1065
Avenue de la Médecine
Québec, QC G1V 0A6
Canada
rene.delvillar@gmn.ulaval.ca

ISSN 1430-9491
ISBN 978-1-4471-2585-3 ISBN 978-1-84996-106-6 (eBook)
DOI 10.1007/978-1-84996-106-6
Springer London Dordrecht Heidelberg New York

British Library Cataloguing in Publication Data
A catalogue record for this book is available from the British Library

Cover design: eStudioCalamar, Figueres/Berlin

Printed on acid-free paper

Springer is part of Springer Science+Business Media (www.springer.com)

Advances in Industrial Control

Series Editors' Foreword

The series *Advances in Industrial Control* aims to report and encourage technology transfer in control engineering. The rapid development of control technology has an impact on all areas of the control discipline. New theory, new controllers, actuators, sensors, new industrial processes, computer methods, new applications, new philosophies..., new challenges. Much of this development work resides in industrial reports, feasibility study papers and the reports of advanced collaborative projects. The series offers an opportunity for researchers to present an extended exposition of such new work in all aspects of industrial control for wider and rapid dissemination.

In control conferences around the globe, there is much activity today on all aspects of control in networks; computer networks, communication networks, wireless networks, sensor networks, and so on. Another highly visible topic is control in the field of vehicles; automobile control, engine control, control of autonomous vehicles on land, at sea and in the air, and cooperative control and formation control of convoys, fleets and swarms of autonomous vehicles are all areas of current interest. Although such topics are exciting state-of-the-art technological developments, it is too easy to overlook the continuing need for control engineering to optimize or at least to improve the processes of heavy or primary industries. Some of these industries, like energy production and petroleum refining, are quite mature users of advanced control technology, but others in this economically important category have only recently automated their process units and are now ready to exploit the additional benefits that more advanced control systems can bring. A case in point is the mining and mineral processing industry. Automation of this industry's processes is widespread but the control used is simple, open-loop, or closed-loop using rule-based and classical PID control. The process lines are large-scale and mechanically vigorous, so the control emphasis is on achieving continuous operation (with as little downtime as possible) and efficient extraction performance. However, some control engineers are now beginning to realize that it is possible to do better, and when the end product is a commodity like gold, copper, zinc, cobalt or another valuable mineral, it is easy to see that enhanced extraction performance quickly brings an economic return but there are some important communication and training barriers to overcome:

- a need for a general awareness in the industry of the benefits that more advanced control systems can bring;
- a need for engineering specialists in the industry to be conversant with control concepts and ideas; and
- a reciprocal need for engineers, researchers and academics from the control community and related research groupings to understand the processes, the technology, the process objectives, and the technical language of the mining and mineral processing industries.

This *Advances in Industrial Control* monograph, *Advanced Control and Supervision of Mineral Processing Plants*, edited by Professors René del Villar (Canada) and Daniel Sbárbaro (Chile) should be seminal in initiating the dialogue needed to understand the measurement and control of processes in the industry. The monograph covers the types of process variables to be measured, novel measurement methods, the status of industrial control products currently used, the objectives of control and what the control community can contribute to enhance the performance of mineral processing units. The Editors (who also contribute to the book as authors) have brought together a strong international authorial team who illuminate vividly the engineering needed for these control system developments.

The monograph comprises seven chapters. The opening Chapter 1 gives the reader an introductory overview of the general process structure for mineral processing, namely: mining, size reduction, classification, concentration, and final product handling leading to transportation or smelting. The next three chapters offer a thorough look at the process parameters and the process qualities that can be measured, or that the industry would like to measure. Three novel state-of-the-art technologies for measurement are described: models and data reconciliation, image processing from machine vision, and soft sensors. Measurement is followed by process simulation, and in Chapter 5 a MATLAB®-based library of simulation modules for the units and equipment of the mineral processing line is described. This chapter also has case study examples showing how the library is used. While the case study of Chapter 5 is a grinding circuit, that used in Chapter 6 for an in-depth control study is a flotation column. In this chapter there is an interesting table showing 18 previously-published control solutions for one aspect of the column control. The methods used were as follows: PID (7), MPC (5), fuzzy logic (3), expert system (2) and nonlinear model reference (1), but the table also shows that the *industrial* controllers used in practice were either an expert system, fuzzy logic or PID controller. The final chapter surveys current industrial control products used in the mineral processing industry and these are mainly general process industry control packages used on mineral processing applications with just a few products specifically developed for the industry.

The Editors and contributing authors are to be congratulated on producing such a useful contribution to the industrial control literature. The monograph will be required reading for any control engineer working in the mineral processing industry. The chapters are self-contained – state-of-the-art but with good survey and historical perspective sections. Other readers likely to enjoy and benefit from this book include mineral processing engineering specialists, related process industry engineers and industrial control system engineers. Academics, researchers and research students working in the control and process control communities, especially those seeking new and challenging topics to apply advanced measurement and control methods will also find the monograph an inspirational read. This is an exemplary addition to the *Advances in Industrial Control* series.

Industrial Control Centre
Glasgow
Scotland, UK
2009

M.J. Grimble
M.A. Johnson

Preface

In 2008 the idea of writing a book on issues of automatic control of mineral processes was suggested by Professor Michael Johnson and discussed between three Chilean academics working in this field, each belonging to a different University (Daniel Sbarbaro from Universidad de Concepción, Guillermo Gonzalez from Universidad de Chile and Aldo Cipriano from Pontificia Universidad Católica de Chile). During the 12th IFAC International Symposium on automation in Mining, Mineral and Metal Processing organized by Université Laval in Québec City (August 2007) the decision to integrate this automation group of researchers was adopted. As such, four new professors, Daniel Hodouin, René del Villar, André Desbiens and Carl Duchesne, each contributing with his particular field of expertise, joined the endeavor.

This book provides an up-to-date review of modern automation developments at industrial level, highlighting the main role played by modeling, control theory and measuring systems in the design and operation of effective supervisory strategies for mineral processing processes. It describes how to use dynamic models of major equipments in the design of automatic control, data reconciliation and soft sensing schemes. Through examples, it illustrates modern design tools for integrating simulation and control system design for comminution (crushing, grinding) circuits and flotation columns. It fully covers the design of soft sensors based on standard single point measurements and those based on more complex measurements, such as digital images. It surveys the main issues concerning steady-state and dynamic data reconciliation and their application to the design of instrumentation architectures and fault diagnosis systems for the mineral processing processes. Considering that most mineral processing plants have distributed control systems and information management systems, this book also describes the current platforms and toolkits available for implementing these advanced data processing and control systems. Some applications to real mineral processing plants or laboratory/pilot scale set-ups highlight the benefits obtained with the techniques described in the book.

The book will benefit engineers working in the mineral processing industry by providing valuable tools and information about the use of modern software platforms and techniques. It is also of interest to senior students of chemical, metallurgical and electrical engineering looking for applications of control technology to mineral processing plants. Control engineers and academics will find the industrial application areas for the new control techniques of general interest.

The editors would like to thank the book's contributors for the work accomplished, and the numerous students that, at each extreme of the American continent, have contributed to the results presented by their professors (naming all of them would be almost impossible). A special word of thanks is addressed to Prof. Michael Johnson and Aislinn Bunning from Springer, whose encouragement and technical support was instrumental to the conclusion of the book.

Québec and Concepción,
July 2009

René del Villar
Daniel Sbarbaro

Contents

List of Contributors

Dr Jocelyn Bouchard
XStrata Process Support, Falconbridge, Ontario, Canada, e-mail: jbouchard@xstrataprocesssupport.ca

Prof. Aldo Cipriano
Department of Electrical Engineering, Pontificia Universidad Católica de Chile, Santiago, Chile, e-mail: aciprian@ing.puc.cl

Prof. René del Villar
Department of Mining, Metallurgical, and Materials Engineering, Université Laval, Québec, Canada, e-mail: rene.delvillar@gmn.ulaval.ca

Prof. André Desbiens
Department of Electrical and Computer Engineering, Université Laval, Québec, Canada, e-mail: desbiens@gel.ulaval.ca

Prof. Carl Duchesne
Department of Chemical Engineering, Université Laval, Québec, Canada, e-mail: carl.duchesne@gch.ulaval.ca

Prof. Guillermo D. González
Department of Electrical Engineering, Universidad de Chile, Santiago, Chile, e-mail: gugonzal@ing.uchile.cl

Prof. Daniel Hodouin
Department of Mining, Metallurgical, and Materials Engineering, Université Laval, Québec, Canada, e-mail: Daniel.Hodouin@gmn.ulaval.ca

Dr Miguel Maldonado
Department of Electrical and Computer Engineering, Université Laval, Québec, Canada, e-mail: miguel.maldonado.1@ulaval.ca

Prof. Daniel Sbarbaro
Department of Electrical Engineering Universidad de Concepción, Concepción, Chile, e-mail: daniel.sbarbaro@udec.cl

Chapter 1
Introduction

Daniel Sbarbaro and René del Villar

Abstract A general overview of the processes for extracting and upgrading valuable minerals is given. A concentration plant is then described in terms of the main processes used and process control challenges. The main functions of an automation system are defined in terms of their use to address these challenges. The benefits normally obtained through the implementation of these functions are identified, and a historical perspective of the evolution of automation systems in the mineral industry is highlighted.

1.1 Introduction

Valuable minerals normally locked within the ore body are extracted using different exploitation methods. These methods are determined by the characteristics of the mineral deposit and the limits imposed by safety, technology, environmental concerns and economics. Geological conditions such as the dip, shape and strength of the rocks play a key role in selecting the method. Mining unit operations are the basic steps to extract the valuable minerals from the deposit. They are grouped into two main categories: rock breakage and material handling. The first consists of two basic unit operations: drilling and blasting, whereas material handling encompasses loading or excavation and haulage (horizontal transport) [1].

Rocks are in general brittle heterogeneous materials; this heterogeneity is caused by the mineral composition, grain size, grain shape, grain strength, porosity and micro-cracks from blasting, among other factors. As a consequence, the raw ma-

Daniel Sbarbaro
Department of Electrical Engineering, University of Concepción, Concepción, Chile, e-mail: daniel.sbarbaro@udec.cl

René del Villar
Département Génie des Mines, de la Métallurgie et des Matériaux, Université Laval, Québec City, Canada, e-mail: rene.delvillar@gmn.ulaval.ca

terial extracted from a mine is usually highly variable, which imposes serious and continuous challenges for downstream engineers.

Rock that can be processed at a profit is called ore. To extract the minerals carrying the valuable metal from the ore, a combination of separation operations, typically known as mineral processing or ore beneficiation is required. The objective of the mineral processing plant is to increase, in an economically viable manner, the concentration of valuable minerals so that the subsequent metal extraction process becomes profitable.

This book will focus on sulfides ore processing, although the same processing stages can be applied to treat many other minerals. The main operations of a mineral processing plant are depicted in Figure 1.1 along with the products obtained in each stage, in terms of the processed mineral.

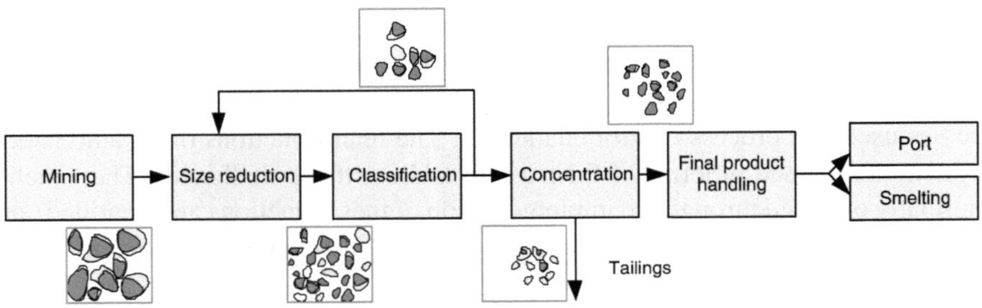

Figure 1.1 Main operations of a mineral processing plant

Size reduction. Rock fragmentation (comminution) liberates valuable minerals from the ore matrix (in mineral processing), thereby increasing the surface area for high reactivity (in hydrometallurgy) and facilitating the particle transport between unit operations. Comminution is carried out in crushers and grinding mills. Whereas crushers operate on dry ores, grinding mills mostly work on slurries, which means that a certain amount of water must be added to the ore before it enters the milling circuit.

Size separation. Since the comminution processes tend to produce broad size distribution particles, some might not be the right size for adequate grain liberation, thus requiring further comminution. The separation of adequate-sized particles (liberated) from still large (locked) particles is done through a size classification process. Screens are used for coarse particulate sizing whereas cyclones or mechanical classifiers are used for fine particles.

Mineral concentration. It is the process of ore enrichment that gets rid of most of the useless species (gangue) keeping as much as possible of the valuable minerals (concentration process). Froth flotation is the most commonly used method for sulfide separations, whereas gravimetric separation is used for heavy minerals (such as precious metals, iron oxides, *etc.*), with magnetic and electrostatic separation for industrial minerals and iron oxides. In froth flotation, air bubbles are injected into an agitated pulp. By adjusting some pulp properties (pH, pulp potential (Eh), *etc.*)

and/or using adequate reagents, it is possible to render the valuable minerals air-avid (aerophilic) and the gangue minerals water-avid (aerophobic). This results in separation by transferring the valuable minerals to the air bubbles which form a froth floating on the surface of the pulp, from where it is removed as valuable product (concentrate).

Final product handling. It normally involves two steps:

1. *Dewatering processes*, by which the solid particles, either in the concentrate or the tailings streams, are concentrated by elimination of part of the contained water, which is recovered for re-use. This is normally done by using thickeners, although in some cases (concentrates) filters and driers might be necessary to further decrease the water content.
2. *Material transport*: concentrate is then transported as thicker pulp to the final handling facility for further transportation to the port or smelting facilities, whereas tails are dumped into ponds in an environmentally friendly manner.

1.2 A Concentration Plant

A mineral concentration plant, called a concentrator or simply a mill, comprises several units where the above-mentioned operations take place, as depicted in Figure 1.2. In this section, a brief description of these unit operations is provided as an overview of the processes involved in a concentrator and their interactions. Further description can be found elsewhere [2–9].

1.2.1 Crushing Circuits

This is the first unit operation after the initial fragmentation provided by underground or open-pit blasting. The mechanisms of size reduction are based on the slow compression of large particles against rigid surfaces. Crushing is performed in one or more stages with smaller reduction ratios; *i.e.*, the ratio feed top-size to product top-size. The first stage (primary crushing) is usually done at the mine, reducing the size of the blasted ore to a level adequate for direct feeding semi-autogenous mills or convenient for further crushing, when the grinding mill is of a conventional type. In addition, screens are used before the crusher to minimize packing the crushing chamber by fines or downstream to separate already final-product size material. The discharge of the secondary and tertiary crushers, normally on the surface, is usually classified on a screen to produce an oversize stream which is recirculated to the tertiary crushers and an undersize product that goes to fine ore bins for feeding the concentrator. This final screening eliminates packing of fines in the tertiary crusher, reduces the production of finely sized material, and thus helps maximize circuit throughput.

1.2.2 Grinding Circuits

The crushed material is then processed by wet grinding circuits. In the case of the most commonly used equipment (tumbling mills and tower mills) size reduction is accomplished through a combination of impact, chipping, and abrasion events caused by energy transferred from grinding media, such as balls, rods or large particles. The primary grinding stage can be carried out by rod mills, in which the grinding media consists of iron rods, by autogenous mills, which use large ore particles, and by semi-autogenous mills, which use a small load of balls. The final stages of grinding (secondary and rarely tertiary) are normally accomplished by ball mills, where the grinding media is made of balls or slugs. In a very few cases, ball mills are also used as a primary grinding stage. The discharge product of the ball mills requires size classification via hydrocyclones or mechanical (rake, spiral) classifiers, to produce a fine stream (cyclone overflow), which feeds the separation circuit, and a coarse stream (cyclone underflow) which is sent back to the ball mill for further grinding (liberation).

1.2.3 Separation Circuits

Only flotation separation processes will be considered in this book. A flotation circuit is usually composed of various sub-circuits (sections) with each having a specific objective. Each section is composed of a series of interconnected tank cells or banks of conventional cells, where flotation takes place. Tailings from the first cell become the feed of the next cell of the bank and so on; the tailing stream of the last cell in the bank is sent to a different bank or section for further processing, or discarded from the circuit (to the tailings pond). Usually, the concentrate stream of each cell in a bank or section are combined into an overall concentrate, which is then directed to the next section for further processing, unless it is the final concentrate of the circuit in which case it is sent to the dewatering stage. Sometimes, regrinding or thickening units exist within the flotation circuit. Regrinding is necessary when floating two minerals with very different optimal particle size distribution or when final concentrate grade cannot be achieved because of gangue contamination in non-liberated particles. Thickeners may be needed to increase the solid content of slurry. Before the actual flotation process, slurries need to be conditioned in an agitated tank, called a conditioner, where appropriate chemical reagents are added. More additions might be needed further down the circuit. Three types of sections are normally encountered in a separation circuit: rougher, scavenger and cleaner. The rougher section, the first after the grinding circuit, is fed with the classifier fine stream, the objective being the recovery of as much valuable material as possible with some disregard for the product quality (grade). Here, most of the fast floating valuable minerals are separated by flotation from the slurry and recovered from the froth of the concentrate. Tailings from the rougher stage are usually exhausted in the scavenger section, where maximum recovery is the target. The tailings of the scavenger section usually become the final circuit tailings, whereas the concentrate is normally recirculated to the rougher feed for reprocessing. The rougher concentrate is processed in one or more cleaner sections (re-cleaners) where an improved (optimal) concentrate grade is the aim. Since this typically implies getting rid of entrained gangue particles, cells having froth-washing capabilities (flotation columns) are preferred. Columns usually outperform conventional mechanical cells in cleaning stages (better product grade) due to their particular froth-washing features and thus, improved product grade. Whereas flotation columns do not have any mechanical agitator for slurry suspension (air bubbles injected at the bottom keep the particles in suspension), conventional cells achieve pulp dispersion through agitation by mechanically driven impellers. These cells employ one of two types of pulp flow and aeration systems: cell-to-cell flow with adjustable weirs between cells or open-flow without weirs and air intake via suction resulting from the operation of an impeller. When dealing with multi-valuable species ores, the processing scheme can be of two types: selective flotation, where all valuable minerals are separated sequentially (one at a time in its own circuit, the tailings of the first mineral circuit becoming the feed for the next mineral circuit, gangue is mostly found in the last circuit tailings), or bulk flotation, where the separation of gangue from the all valuable mineral is accomplished in a single circuit, whose concentrate (bulk concentrate) is selectively processed in a separate circuit. Bulk flotation is typically used for Cu-Mo ores, whereas selective flotation is used for Cu-Zn ores.

1.2.4 Dewatering Circuits

Depending on the degree of water elimination required, dewatering circuits can be composed of one or more of the following processes: thickening (up to 65% solids), filtration (up to 93% solids) and drying (up to 99% solids). Thickeners are a specific type of sedimentation equipment which operates continuously, performing a wide variety of tasks in the mineral industry. Their most common application is the enhancing of the solid content of mineral concentrates (for transportation purposes and in respect of net smelter return contract clauses), as well as recovering water from flotation tailings (reclaim of the excess of water for process re-use). The objective of these units is to produce clean overflow water and maximum solid concentration in the underflow. Flocculants are normally used to agglomerate the solids to increase the settling rate and improve the overflow clarity. The use of synthetic flocculants has increased the efficiency of sedimentation rates allowing increased production. Filtration is used when even smaller amounts of water are permitted (typically the case of concentrates); filters are then fed by the thickener underflow. Filters typically operate by passing the mineral slurry through a membrane (cloth, ceramics, *etc.*) whose pores retain the solid particles letting the liquid pass through. In some cases, the solid discharge is continuous (vacuum filters such as belt-, drum- and disk-filters) and in others the solid is discharged every few minutes (pressure plate-and-frame filters). Further drying of the solid product (filter cake) can be achieved in rotary driers using a heated (oil or gas) air stream. This operation is rarely used in base-metal processing. Niobium concentrates are rotary dried whereas graphite concentrates are filtered on a heated belt-filter.

These unit operations are tightly coupled as is depicted in the generic flowsheet diagram shown in Figure 1.2.

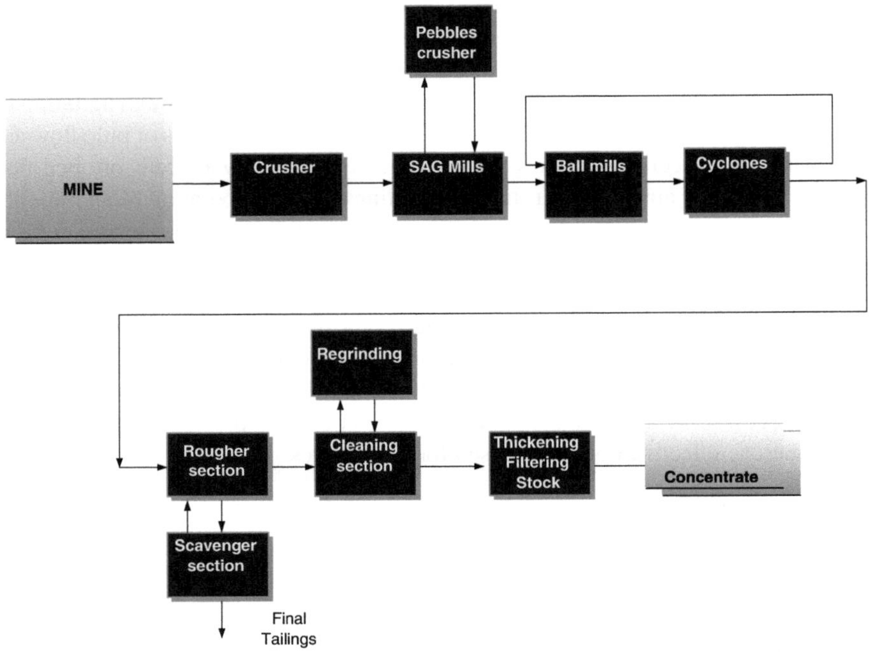

Figure 1.2 Generic concentrator flowsheet and its unit operations

Clearly the operation of this processes chain will be affected by important geological factors such as those listed in Table 1.1. Unknown and unexpected changes in these factors could translate into large variations in the main process variables. The use of any available information about them could be used to reduce their effect by taking some feed-forward actions. Thus, any improvement in the predictability of the main disturbances and/or input variables will have an impact on the operation of the whole process. Currently, there is an increasing interest in obtaining and using information related to these factors which can be directly used in control strategies. Some important advances are expected in the next decade.

Besides the effect of the external factors, the effect of the upstream processes on downstream ones is also very important in this serial process structure. There are some well-documented examples illustrating this statement. For instance, blasting and crushing can have a significant impact on SAG mill performance, since their throughput rate is dependent on the ore hardness and feed size distribution [11, 12]. Hardness is dependent on rock strength and mill feed size is dependent on *in situ*

Table 1.1 Geological factors and their effect on metallurgical operations [10]

Geological/mineralogical factor	Area of linkage	Metallurgical operation
Primary rock type and size distribution	Hardness	Crushing, Grinding
Ore assemblage and ore formation	Solubility, Hardness, Presence of talc	Grinding, Flotation
Alteration	Clays, hardness	Grinding, S/L separation
Down temperature (Hypogene)	Clay, oxidation	Flotation
Faulting	Clay, presence of talc	Grinding, S/L separation
Metamorphism	Hardness	Flotation

rock structure, rock strength, blast intensity and primary crushing. In the area of mineral separation, some examples dealing with the coordination of grinding and flotation stages have also been presented [13, 14].

As pointed out by Hodouin [15] the task of devising a global coordinating strategy has been hampered by a lack of liberation measurement and reliable models. However, the idea of integrating the operation of all the processes remains the ultimate goal of a mineral processing automation system.

1.3 Main Functions of Automation Systems in Mineral Processing Plants

The main goal of an automation system is to provide timely information and to take prompt actions to ensure a safe and stable operation in spite of the external and internal disturbances affecting the process. In order to accomplish these objectives, the automation system requires the carrying out of a set of functions, which can be divided into two categories: basic and advanced functions. The former are fundamental functions essential to any automation system for controlling or monitoring a process. The advanced functions, which require the implementation of the previous functions, are oriented to the analysis of information and to the higher decision-making process.

1.3.1 Basic Functions

1. Human-machine interface (HMI). In modern automation systems, operators are located in a room away from the process. Consequently, the interaction between the automation system and the operator plays an important role in ensuring an effective operation. Timely presentation to the operators of relevant information regarding both normal and abnormal process states is of prime importance for the success of the automation system, and for its acceptance by them. Compli-

cated or confusing presentation of any information can lead to serious operational problems. The displays can take different forms and in general include mimic diagrams of the process displaying the status and continuous measurement values. In many modern installations these displays are supplemented by closed circuit television cameras, which give an even more enhanced level of plant control. The operator can override the process through specialized keyboards or pointing devices such as a mouse or trackballs.

2. Data acquisition and processing. This function processes the data obtained from the sensors and sends suitable commands to the actuators. Sensors are devices that obtain information in the form of one or more physical quantities, and convert it into an electrical output signal. They consist of two main basic components: (i) a primary measuring element, referred to as a transducer; and (ii) a transmitter unit, responsible for producing electrical output that has some known relationship to the physical measurement. Normally, the performed processing actions are signal filtering, signal sampling and data conversion from raw values to engineering units. Most of these functions are now performed within the instrument itself. The advent of more sophisticated sensors, such as visual sensors, has increased the complexity of this function; some related challenges are described in Chapter 4. The real-time storage of important variables is also carried out by this function. This information can then be used for further process analysis.
3. Communication. This function deals with communication between the control system function and other computer systems. In many situations, control actions need to be supplemented with data located in other systems and communication between different control or information systems is required. This is also a key function associated with data acquisition, since all the measurement data can be digitally transmitted through dedicated fieldbuses such as Foundation fieldbus or Profibus.
4. Regulatory control. This comprises a combination of discrete and continuous control functions, which aims at stabilizing the process and keeping its operation within safe margins for the process, equipment and the workforce.

1.3.2 Advanced Functions

1. Process analysis. This includes tools for modeling and analyzing the process, as well as incoming flows. These functions can be used to analyze the whole process starting with the raw materials through all the process steps down to the end-products. The operational variables can be analyzed to assess inter-dependencies, to identify possible improvements based on these findings, and to devise new operational scenarios. In addition, a valuable process tool is signal-characterization which can reveal how often and by how much disturbances change.
2. Optimization. This function seeks to improve the operational level by taking actions according to a given performance index. These improvements can be ap-

plied to a single piece of equipment, a unit operation or a plant section. In general, this function is always associated with process analysis functions, since it requires information concerning relevant constraints and variables affecting the operational performance. Before implementing this function, it is always advisable to perform holistic process analysis, so that improvements throughout the process can be identified.

3. Fault detection. It aims at promptly detecting failures in the equipment and automation hardware and/or software. Data reconciliation plays an important role in implementing fault detection functions.

The relationship between all these functions is depicted in Figure 1.3. As described in previous sections, mineral processing plants contain integrated operations involving multiple processes in which all automation functions are required to operate in a safe and economical way. The possible benefits offered by the implementation of these advanced functions can only be achieved if the functional specifications of the automation system are clearly defined.

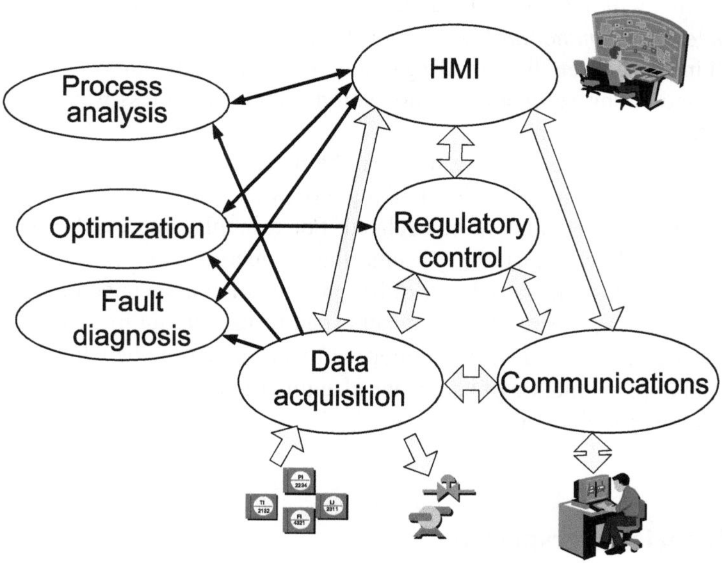

Figure 1.3 Functions of an automation system

1.4 Benefits

The benefits of an automation system can be classified into tangible and intangible benefits. Tangible ones can easily be assigned a monetary value [16]. Although it is difficult to assign a monetary value to intangible benefits, they are real and valuable.

Automation systems can be beneficial in many different ways. An assessment of their impact on a mineral processing plant should consider the following:

Increased production. A throughput increase has a clear monetary value. However, it must be kept in mind that a production increase in some units must be consistent with the processing capacity of downstream units. For instance, it is not sensible to produce more flotation concentrate if the filtering plant is not capable of processing it.

Process stability improvements. Process stability in the context of a mineral processing plant refers to the capacity of the process to run smoothly and to consistently meet specifications. These specifications can be imposed by either a downstream unit or an external customer.

Better use of raw materials. Automation systems help prevent the use of raw materials that will not contribute directly to the process objectives.

Reduced maintenance and improved safety. A well-implemented automation system will increase safety by providing timely information about dangerous situations (alarms) and taking automatic prompt actions. This means less wear and tear, and fewer catastrophic failures.

Improved process knowledge. This can be regarded as an intangible benefit if it is not directly linked to a decision-making process.

Many of these benefits are interrelated. For example, improved process stability may reduce downtime, thereby increasing production. A number of examples in Chapter 6 will illustrate these kinds of benefits and their associated monetary value. The economic assessment of these benefits can be carried out by standard methodologies as described in [17]. It is important to stress that the benefits, arising from a potential implementation of an advanced automation solution, must be the core of its functional specification.

1.5 Historical Perspective

The mineral processing industry has been very slow to adopt automation technologies in contrast to the petrochemical and pulp and paper industries. In the last decades though, there has been an increased interest in intensifying their use [18]. Figure 1.4 illustrates the main developments since 1960 and the range of their effects.

During the 1960s, some elementary control systems involving measurements of solid concentration and pulp level were developed and some control loops were implemented aiming at stabilizing the process operation. In the 1970s, new modern measurement devices, specifically designed for particulate systems, came into existence and mathematical models were developed. The availability of such models and the increased capability of digital computers provided new impetus for the development of control strategies aiming at simultaneously controlling several pieces of equipment to pursue an overall control objective.

In the 1980s, almost all large- and medium-sized mineral processing plants worldwide implemented some form of automatic control involving the use of a number of stabilizing control loops. The technology was a mixture of analog and digital controllers. During this decade, plants adopted many newly developed sensors to measure the main process variables, which were processed through distributed control systems [19, 20].

The 1990s brought a wider industrial acceptance of advanced technologies, for instance, the use of advanced control strategies based on expert systems applied to SAG mill circuits. Data management systems were also implemented for collecting and storing operating data from almost all plant-units in a unique database. The enhanced data availability and the existence of friendlier systems for accessing it improved process accountability. In addition, this led to the development of data reconciliation algorithms and multivariable data analysis techniques for process monitoring. The use of vision-based sensors for particle size distribution measurement and froth image analysis also extended to almost all medium- and large-size installations [15].

Finally, since 2000 the industry has witnessed an increase in the number of activities using image-based measurements, artificial-intelligence applications to unit supervision and the use of multivariable non-linear control of flotation systems.

Future developments will likely include an integrated plant operation, coordinating the operation of the different units aimed at increasing the overall benefit level. The coordination of these units will also have an impact on how the plant is operated.

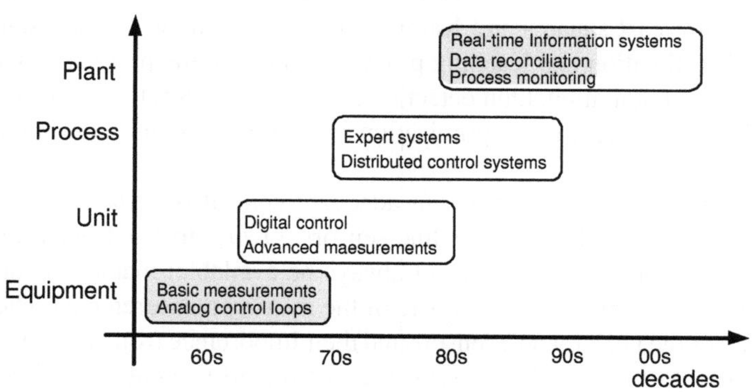

Figure 1.4 Evolution of the automation system in mineral processing

1.6 Synopsis

This book is organized into seven chapters highlighting the central role played by measuring systems, modeling and control theory in the design of effective supervisory and control strategies for mineral processing plants.

This first chapter has provided a summary of the main processes involved in a concentrator plant. Their main characteristics have been described along with the external and internal variables affecting their operation. A set of basic and advanced functions have been defined to address the challenges imposed by operational problems. Possible benefits obtained from their implementation have been briefly described. Finally, a historical review of developments in automation systems in mineral processing plants, emphasizing their technical and economical achievements, has explored possible future developments. The following chapters present the main techniques required to implement advanced functions in modern automation systems.

The second chapter is devoted to the process of data acquisition for the purposes of process performance audits, diagnosis and supervision, advanced automatic control, and real-time optimization. This is very important, since control and supervisory systems rely on data acquired by numerous sensors scattered around in a very hostile environment. The consistence and availability of this information plays a key role in building robust systems. Emphasis is put on the constraints of mass and energy conservation, which are used as a foundation for the design of measurement strategies, the upgrading of measured values by measurement-error filtering techniques and the estimation of unmeasured process variables. Since key variables in any mineral processing unit are usually flow rates and solid and/or metal concentrations, their reconciliation with the laws of mass conservation is central to the discussed techniques. Tools are proposed for three different kinds of operating regimes: steady-state, stationary and dynamic. These reconciliation methods are based on standard least squares and Kalman filtering techniques. Case studies involving grinding, flotation and leaching processes illustrate the problems of sensor placement, data reconciliation, fault detection and diagnosis. Strategies for coupling data reconciliation with real-time optimization and automatic control techniques are also proposed.

As described in previous chapters an adequate control of mineral processing plants relies on the availability of on-line sensors. Owing to harsh environments or operating conditions, sensors may not always be available; mathematical models can then be used to provide an estimate of the missing measurements. The third chapter addresses the need for soft sensors and their importance from a technical and economical point of view. It also presents state-of-the-art technology and possible future developments. Different classes of soft sensors, soft sensor models and their design methodologies are presented along with soft sensor management systems (*e.g.* monitoring, automatic updating and selection) which attest to the robustness of their operation. Several examples illustrate their implementation in industrial applications.

A new set of sensors based on digital images has been widely applied in mineral processing. Owing to the multidimensional nature of the information provided in the images, they pose new challenges for their integration with automation systems. Chapter four addresses these issues by first providing a brief description of the following concepts and/or methods: (i) what a digital image is and what information and/or features (color, textural, geometrical) can be extracted and/or quantified from it; (ii) the multivariate statistical methods used in image analysis, such as principal component analysis (PCA) and partial least squares regression, and finally (iii) current image analysis techniques: multivariable image analysis (MIA), multivariate image regression (MIR), and multi-resolution multivariate image analysis (MR-MIA). After presenting various concepts and methods, a number of case studies and applications to the mineral processing industry will be presented: (i) flotation froths: concentrate grade prediction, froth health monitoring and froth appearance control based on multivariate projection methods; and (ii) rock classification on conveyor belts based on lithological compositions or grindability/hardness.

The fifth chapter deals with dynamic modeling, simulation and control of comminution circuits (crushing and grinding), the first stage of every mineral processing plant. Short case studies highlight the use of various techniques in the control of such circuits.

Since their first commercial application for mineral separation in the early1980s, flotation columns have become a standard piece of equipment in mineral concentrators particularly for cleaning operations. The sixth chapter presents and discusses the most recent advances in dynamic modeling, instrumentation and automatic control of flotation columns. It also examines how current industrial practice could benefit from recent academic developments in these areas. A particular emphasis is placed on the development of specific sensors for the continuous monitoring of process operations and their regulation.

As a result of their success, many of these technological advances have been marketed. Chapter 7 surveys these successes by reviewing commercial advanced control systems, new sensors for grinding and flotation and system development environments for mineral processing plants. It presents solutions based on expert systems, fuzzy logic and neural networks currently available commercially. Tools for linear and non-linear modeling, simulation, advanced control, monitoring, diagnosis and operator training are also reviewed. The chapter ends with a discussion of benefits produced by advanced control in mineral processing.

References

[1] Hartman HL, Mutamsky JM (2002) Introductory Mining Engineering. John Wiley and Sons, New Jersey

[2] Weiss NL (1985) SME Mineral Processing Handbook. Society for Mining Metallurgy & Exploration, New York

[3] Lynch AJ (1979) Mineral Crushing and Grinding Circuits: Their Simulation, Design, and Control (Developments in Mineral Processing Series, Vol.1). Elsevier Science Ltd., Amsterdam

[4] Lynch AJ (1981) Mineral and Coal Flotation Circuits: Their Simulation and Control (Developments in Mineral Processing). Elsevier Science Ltd., Amsterdam
[5] King RP (2001) Modeling Simulation of Mineral Processing Systems. Butterworth-Heinemann
[6] Gupta A, Yan D (2006) Mineral Processing Design and Operation: An Introduction. Elsevier Science Ltd., Amsterdam
[7] Mular AL, Barratt DJ, Halbe DN (2002) Mineral Processing Plant Design, Practice, and Control (2-volume set). Society for Mining Metallurgy & Exploration, New York
[8] Fuerstenau MC (Editor), Han KN (Editor) (2003) Principles of Mineral Processing. Society for Mining Metallurgy & Exploration, New York
[9] Napier-Munn T, Wills BA (2006) Wills' Mineral Processing Technology, Seventh Edition: An Introduction to the Practical Aspects of Ore Treatment and Mineral Recovery. Butterworth-Heinemann, New York
[10] Williams SR, Richardson JM (2004) Geometallurgical mapping: a new approach that reduces technical risks. In: Proceedings of 36th Annual Meeting of the Canadian Mineral Processors
[11] McKee DJ, Chitombo GP, Morrell S (1995) The relationship between fragmentation in mining and comminution circuits throughput. Mineral Engineering 8:1265–1274
[12] Titichoca G, Magne L, Pereira G, Andrades G, Molinet P (2007) Effect of blasting modifications in the SAG grinding plant of Codelco Chile Andina. In: Cu2007 Mineral Processing II:265–276
[13] Bascur OA (1991) Integrated grinding/flotation controls and managment. In: Proceedings of Copper 91 II:411–427
[14] Sosa-Blanco C, Hodouin D, Bazin C, Lara-Valenzuela C, Salazar J (2000) Economic optimisation of a flotation plant through grinding circuit tuning. Minerals Engineering 13:999–1018
[15] Hodouin D, Jamsa-Jounela SL, Carvalho MT, Bergh L (2001) State-of-the-art and challenges in mineral processing control. Control Engineering Practice 9:995–1005
[16] Stout TM (1973) Economic justification of computer control systems. Automatica 9:9–19
[17] Bauer M, Craig IK (2008) Economic assessment of advanced process control. Journal of Process Control 18:2–18
[18] Thwaites P (2007) Process control in metallurgical plants from a Xstrata perspective. Annual Reviews in Control 31:221–239
[19] Herbst JA, Bascur OA (1984) Mineral-processing control in the 1980s-realities and dreams. In: Herbst JA (ed) In Control'84 Mineral/Metallurgical Processing, SME, New York
[20] Lynch AJ (1984) The automatic control of mineral preparation and concentration circuits. In: Herbst JA (ed) In Control'84 Mineral/Metallurgical Processing, SME, New York

Chapter 2
Process Observers and Data Reconciliation Using Mass and Energy Balance Equations

Daniel Hodouin

Abstract This chapter is devoted to data reconciliation for process audit, diagnosis, monitoring, modeling, advanced automatic control, and real-time optimization purposes. The emphasis is put on the constraints of mass and energy conservation, which are used as a foundation for measurement strategy design, measured value upgrading by measurement error filtering techniques, and unmeasured process variables estimation. Since the key variables in a mineral processing unit are usually flowrates and concentrations, their reconciliation with the laws of mass conservation is central to the discussed techniques. Tools are proposed for three different kinds of operating regimes: steady-state, stationary and dynamic. These reconciliation methods are based on the usual least squares and Kalman filtering techniques. Short examples involving grinding, flotation, leaching and thermal processes are presented to illustrate the problems of data reconciliation, sensor placement, fault detection and diagnosis. Strategies for coupling data reconciliation with real-time optimization and automatic control techniques are also proposed. A nomenclature section is included at the end of the chapter

2.1 Introduction

The production goal of a mineral or metallurgical plant (MMP) is ideally to maintain complex unit operating conditions at values where some plant performance index is optimized. The performance index could be expressed either by technical factors, such as the tonnage of valuable material produced, or by the quality of the material produced (*e.g.*, concentrate grade or metal purity). More globally, since a trade-off between the productivity, the material quality, and the production costs is required,

Daniel Hodouin
Département de Génie des Mines, de la Métallurgie et des Matériaux, Université Laval, Québec City, Canada, e-mail: daniel.hodouin@gmn.ulaval.ca

the performance is frequently expressed as an economical index which embeds all these aspects. When the operating conditions drift away from the range of optimal performance, the plant experiences economic losses. The ability of a plant to remain in the vicinity of its optimum operation is related to real-time decision making processes, *i.e.*, to production supervisory systems, real-time optimization systems, and automatic control strategies. Regardless of the strategy used for maintaining a plant close to an optimum performance, the variability around the optimum value relies in the first place upon an efficient evaluation of the performance index–in other words upon the variance of its estimate. The lower the variance, the better is the plant performance.

A plant performance index observer uses the available measurements of the process variables. Typically, in a metallurgical plant, these variables are overall material, phase, and metal flowrates, material chemical compositions, energy flowrates, temperatures, consumed power, *etc.* As any other observer, the plant performance observer simultaneously uses measured values and process models. These models are required to cope with common data processing problems such as measurement uncertainties – which are quite large in a metallurgical operation – lack of measurement availability for critical variables (obviously a performance index is usually not directly measurable), limited knowledge of the process behavior (a difficult problem, particularly in extractive metallurgy), and information redundancy in the available measurements and prior process knowledge. As the process model uncertainties are very large in metallurgical industries, it is common practice to use only constraints, *i.e.*, sub-models – in the sense that they are not causal models as assumed in traditional model-based control and observation. Since the level of confidence in these sub-models must be high to prevent distorting the data set information content by uncertain models, the selected constraints are essentially laws of mass and energy conservation. In the metallurgical, and more generally chemical industries, these observation methods are called reconciliation methods, in the sense that they reconcile the measurement data with the laws of mass and/or energy conservation.

Estimation of process states is required for process performance audit, process modeling, monitoring, supervision, control, and real-time optimization. Whatever the process scale, laboratory, pilot, or full industrial scale, the first step of state estimation is to collect experimental data. Unfortunately, and this is particularly true at the industrial scale, measurements are extremely difficult and inaccurate in the metallurgical engineering field. Production units treat materials that are multi-phase and usually contain extremely heterogeneous particulate phases [1]. The data is highly inaccurate and incomplete, and requires to be improved before being used in the above mentioned applications. The usual statement "Garbage in, garbage out" is particularly true in this context, use of poor data leading invariably to poor models, poor decisions, and improperly designed and operated systems. Therefore, using additional information to the experimental data through process prior knowledge leads to better state estimates. Mathematical models are usually the most efficient way to encapsulate process behavior knowledge. Unfortunately, in metallurgical processes, the knowledge is frequently fuzzy and less accurate than in mechanical and chemical industries.

Figure 2.1 summarizes the concept under discussion in this chapter. The core of a data reconciliation procedure is a mathematical algorithm that can be called either an observer, or an estimator, or a filter. It is an observer in the general sense that it allows the process state observation, *i.e.*, the observation of the variables upon which the process behavior and performance are qualified. It is an estimator in the sense that it estimates numerical values of state variables which may not necessarily be measured or measurable. It is a filter in the sense that, if a state variable is measured, it will correct the experimental value of this process state variable. In this chapter the words observer, estimator and filter, as well as data reconciliation, will be commonly used, without strict meaning differences. Generically, Y is the measurement vector, X the state vector, and $\hat{X}$ its reconciled (or estimated, or filtered, or observed) value. The constraint equations $f(X) = 0$, normally a sub-model as mentioned earlier, are here mainly mass and energy conservation equations. Figure 2.1 presents a steady-state reconciliation (SSR) procedure, but it will be seen later on that stationary and dynamic reconciliation methods can also be considered when the process is not operating in steady-state conditions.

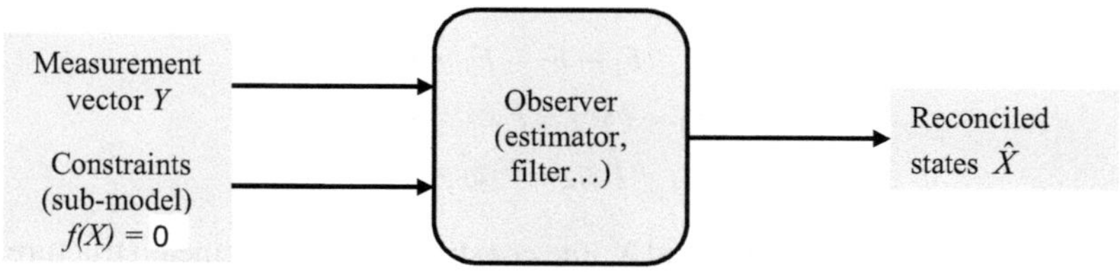

Figure 2.1 Scheme of a data reconciliation procedure using mass and energy conservation constraints

An introductory example. Before going into deeper and more rigorous definitions of the concepts used in data reconciliation, let us give a simple example for qualitatively introducing the key words used in this chapter. The considered plant is the flotation unit of Figure 2.2, and the corresponding data is given in Table 2.1: measured value, measured value standard deviation, as well as reconciled values as explained in Section 2.8.2.

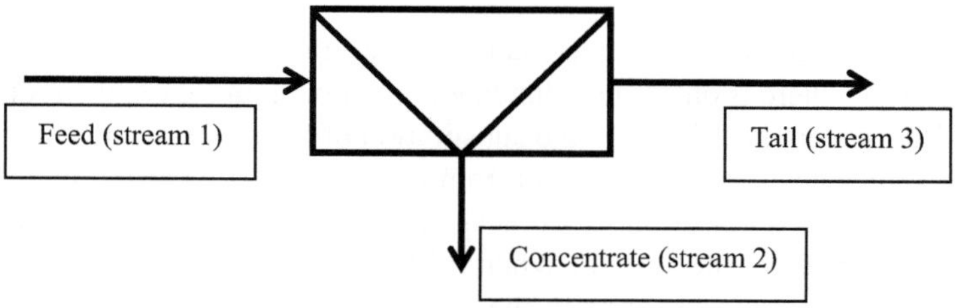

Figure 2.2 A flotation unit

Table 2.1 Measured and reconciled values for the flotation unit of Figure 2.2

Process variable	Flowrates $F(t/h)$			% Cu (100c)			% Zn (100z)		
Stream	Meas. values Y	error s.d.	Reconc. values $\hat{X}$	Meas. values Y	error s.d.	Reconc. values $\hat{X}$	Meas. values Y	error s.d.	Reconc. values $\hat{X}$
Feed	100	1	100.0	2.39	0.05	2.36	8.53	0.19	8.57
Concentrate		0	9.25	23.2	1,11	24.36	7.54	0.34	7.53
Tail		0	90.75	0.12	0,01	0.12	9.65	0.98	8.68

The nine state variables X are the three ore mass flowrates F and the six copper and zinc mass fractions c and z. The measured process variables are a subset of the state variables. The measurement vector Y contains the measured values of the ore feedrate and of the six metal mass fractions. The measurement values are assumed to give an image of the process steady-state behavior. Hence, the constraints of mass conservation $f(X) = 0$ are

$$F_1 - F_2 - F_3 = 0, \tag{2.1}$$

$$F_1 c_1 - F_2 c_2 - F_3 c_3 = 0, \tag{2.2}$$

$$F_1 z_1 - F_2 z_2 - F_3 z_3 = 0. \tag{2.3}$$

Considering this selection of Y and X, the constraints have a bilinear structure. The information content (measurements + constraints) is said to be redundant since it contains two unknown state variables (F_2 and F_3) and three equations. The system is said to be observable since F_2 and F_3 could be estimated by resolving Equations 2.1 and 2.2 or 2.1 and 2.3. The first case leads to

$F_2 = 9.83t/h$; $F_3 = 90.17t/h$

and the second one to

$F_2 = 53.1t/h$; $F_3 = 47.9t/h$.

The conflict between these two possible solutions explains why there is a need for reconciling the measurements with the constraint equations. Eliminating unmeasured process states from Equations 2.1 to 2.3 gives

$$c_1 z_3 - c_1 z_2 + c_2 z_1 - c_2 z_3 - c_3 z_1 + c_2 z_2 = 0. \tag{2.4}$$

This equation is called a redundancy equation because it contains only measured quantities. Since there is only one redundancy equation in this case, the redundancy degree of the system is 1. In (2.4), the substitution of the process variables by their measured values gives a value different from zero (2.1×10^{-3}) because of the conflict existing between constraints and measurements. This residual variable generates the parity space of the system. When using reconciled values, Equation 2.4 is exactly verified, as obviously are also the constraints (2.1) to (2.3). The application of a reconciliation procedure to this system generates the following advantages:

- Unmeasured process variables F_2 and F_3 are estimated.
- Process variable measured values are corrected.
- Reconciled values are consistent with the law of mass conservation.
- Plant performance, such as copper recovery, calculated with reconciled data becomes independent of the calculation path (using copper data to estimate the flowrates gives a recovery of 95.4%, and, using zinc data, gives an absurd value of 515%, whereas the reconciled value gives 95.9%). Furthermore results are more accurate than values calculated by any other methods.
- Models that are subsequently estimated from reconciled data (for instance flotation kinetic models) are much more reliable.
- Decisions made from reconciled data are necessarily more efficient than decisions made from raw data.

Reconciliation of mass and energy balance with raw data is a technique which has been known for a long time in mineral processing (see, for instance, [2–4]), but it was already mentioned in chemical engineering as early as 1961 [5]. Crowe wrote a good survey paper in 1996 [6], but the first organized books addressing the chemical process reconciliation topic appeared only at the end of the 20th century [7, 8]. However, the contributions of the MMP community to the field of data reconciliation is mentioned only in the first one. Off-line reconciliation methods for steady-state processes are now quite mature and various computer packages are available. For example, in the MMP field, although they are not at all limited to these applications, one can mention: Bilmat™ and Metallurgical Accountan™[1] (Algosys) [9], Bilco™ and Inventeo™[2] (Caspeo) [10], JKMultibal™ [3](JKTech) [11], Movazen™ (Banisi) [12] and more chemical process oriented: Sigmafine™[4] (OSIsoft) [13], Datacon (IPS)[14], Advisor™[5] (AspenTech) [15], and VALI™ [6](Belsim) [16].

The most usual reconciliation techniques are based on the minimization of quadratic criteria, therefore assuming that uncertainties mainly belong to Gaussian distributions. This chapter focuses only on this type of approach. However, alternative reconciliation methods based on artificial neural networks have also been proposed. But those do not offer either the same rigorous statistical and physical background or the same result reliability analytical evaluation tools as in the approach presented here (see, for instance, [17–20]). Linear matrix inequality (LMI) methods have also been proposed by Mandel *et al.* [21].

Steady-state methods are applied off-line to mineral processes such as comminution [22, 23], flotation [59], gold extraction [25–28], hydrometallurgy [29] and [30], pyrometallurgy [31–33], and cement preparation [34]. On-line applications to steady-state processes are actively used, while stationary-state methods that make a

---

[1] Bilmat and Metallurgical Accountant are registered trademarks of Algosys, www.algosys.com

[2] Bilco and Inventeo are registered trademarks of Caspeo, www.caspeo.net

[3] JKMultibal is a registered trademark of JKTech, www.jktech.com.au

[4] Sigmafine is a registered trademark of OSIsoft, www.osisoft.com

[5] Advisor is a registered trademark of AspenTech, www.aspentech.com

[6] VALI is a registered trademark of Belsim, 174k rue De Bruxelles, 4340 Awans, Belgium, www.belsim.com

trade-off between model uncertainties and measurement errors are well developed but still not frequently used [24, 35]. Dynamic methods have been proposed and are being developed [36–40], but applications of those are not documented.

Other topics are related to data reconciliation, such as sampling error and reconciliation criterion weighting factor evaluation [41, 79, 96], reconciled value accuracy evaluation [44], use of reconciled values to calculate and display plant performance indices, such as concentrate grade and recovery. Owing to their better reliability, these indices may improve manual or automatic process performance optimization [45, 46]. Coupling of dynamic reconciliation with control has also been investigated [47, 48], as well as gross error detection, fault isolation and diagnosis [49, 50]. Finally, instrumentation design can be performed on the basis of data reconciliation methods used as process observers [51, 52].

The chapter includes the following parts. Section 2.2 begins with definitions of plant process variables and operating regimes that may be considered in reconciliation methods. Then, in Section 2.3, mass and energy conservation equations are written for different plant operating modes. As is the case for all the sections of this chapter, strong emphasis is placed on mass balance problems, rather than on energy. Then, since the basic incentive for reconciling data is the presence of measurement errors, Section 2.4 covers measurement problems, while Section 2.5 presents the observation equations. Section 2.6 introduces the general principles of data reconciliation algorithms based on least-squares procedures, while Section 2.7 gives the steady-state and stationary operating regime solutions for the linear constraints case. Section 2.8 briefly discusses the non-linear reconciliation cases. Section 2.9 is devoted to the reliability of reconciled data analysis. Section 2.10 presents some reconciliation methods for plants operating in the dynamic regime, while Section 2.11 briefly addresses the issue of how reconciliation methods can help to improve metallurgical plant instrumentation design strategies. Section 2.12 explores how mass and energy conservation constraints can also be used for detecting abnormal process behaviors or measurement problems. Finally Section 2.13 makes way for the integration of reconciliation techniques into optimization or control loops.

2.2 Process Variables and Operating Regimes

Variables involved in mineral processing units characterize process states relatively to material quantities (extensive properties) or qualities (intensive properties). Mass or volume flowrates and hold-ups of solid, liquid, slurry and gas phases (or of given species within these phases) belong to the first category, as well as other related variables such as levels and flowrates. The second category frequently includes the following variables:

- concentrations of chemical species or minerals in solid, liquid and gaseous phases;
- solid percentage in slurries;
- particle size and density distributions;

- concentrations of chemical species or minerals in particle size or density classes;
- pressures, temperatures.

Process variables may or may not be directly measured by sensors or analytical devices. For instance, ore flowrates may be obtained through the information given by a slurry volume flowmeter, a density gauge, and ore specific gravity measurement. Metal flowrates are other examples of process variables that are obtained through a combination of measurements (flowrates and concentrations).

There is no unique way to select the set of variables X that characterize process states. The structure of the models and constraints describing a process behavior depends upon variable selection. This will have an impact on the observation or data reconciliation method, although the resulting values of reconciled states should not rest upon the problem formulation, if consistent information processing methods are used. Similarly, the measured values Y of the Z process variables used as input to the reconciliation procedure may or may not be raw measurements of the process states X. Furthermore, they can be obtained by combining several sources of raw measurements. The structure of the database Y and of its uncertainties may have a significant impact on the reconciliation method and sometimes on the reconciled results.

The variation of the process states X as a function of time depends on the intrinsic dynamics of the process, on the variations of the operating conditions applied to the process, and on the disturbance dynamics. The process operating regimes can be classified into six types:

- the steady-state regime, when all the process input and state variables are constant;
- the stationary regime, when the process dynamics are limited to random variations around a steady-state regime;
- the transient regime, where the process evolves from one steady-state to another one;
- the quasi-stationary regime, which corresponds to stationary random variations around persistent mean value changes;
- the cyclic regime, when the process operates cycles of production, such as in the carbon-in-pulp process where the carbon transfer is cyclic, or in smelting processes where the material is cyclically cast;
- the batch regime, when the states evolved according to a trajectory from an initial state to a final state.

Figure 2.3 illustrates process state variations for four different operating regimes. Local stochastic state variations are mainly due to input disturbances (ore grade variations for instance), while the trends are mainly the results of deterministic changes of the manipulated variables. Since in real processes it is impossible to maintain strictly constant conditions, the steady-state regime corresponds to virtual operating conditions.

The statistical properties of any stationary process variable, a scalar or a vector x (either input disturbances and/or process states and/or process outputs), can be

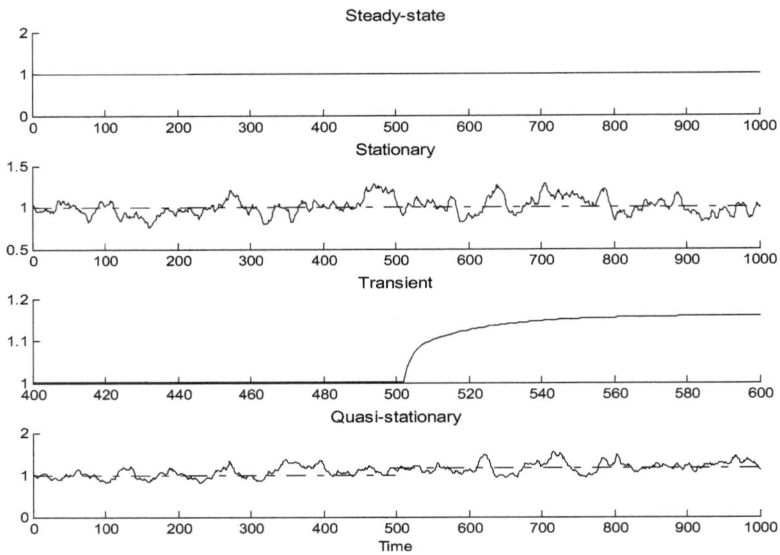

Figure 2.3 Typical variations of process variables for various operating conditions

described by the usual statistical features: the mean μ_x, the variance $V_x(0)$ and the autocovariance $V_x(k)$ (or autocorrelation $\rho(k)$ if normalized by the variance):

$$\begin{aligned} E(x) &= \mu_x; \\ E((x(t)-\mu_x)(x(t-k)-\mu_x)^T) &= V_x(k). \end{aligned} \tag{2.5}$$

The nature of the reconciliation procedure to be applied to filter industrial data must be adapted to the operating regime that was prevailing during data gathering, and to the measurement strategy then applied. There is no systematic method to decide whether a steady-state, stationary or dynamic filter must be applied, but some hints can be helpful. Some of the conditions for the application of a steady-state observer are:

- The process deviation from a theoretical steady-state is of low magnitude compared with measurement error amplitudes (variance matrix V). In other words, the diagonal terms of V are large in comparison with the diagonal terms of $V_x(0)$.
- The process deviation from a theoretical steady-state is significant with respect to the measurement errors, but the dynamic variations are produced by stationary disturbances of high frequency spectra in comparison with the natural process dynamics – in other words, the process variable autocorrelogram time widths are small in comparison with the width of the cross-correlation between process inputs and outputs. However, instantaneous measurements data set could be processed by the steady-state method providing that the high frequency disturbance variances are added to the measurement error variances.

- The process is operating in a stationary regime while the frequency of the variations around the steady-state is of the same range as the process natural dynamics and of significant amplitude. The steady-state method could be applied to averaged measurements in a time window sufficiently large to significantly dampen the dynamics of the variations around the steady-state values. The time window must be at least two times larger than the process time constants. Measurement variances should be augmented to take account of the residual dynamics of the variations around the steady-sate values.

Obviously, stationary methods could be applied to instantaneous measurements when the process is stationary, *i.e.*, when the process is not operating during significant changes of the underlying steady-state values. In other words, the process must be sampled at a time sufficiently far from a deterministic change of the average operating conditions. When the process is clearly in a transient regime or cyclic or batch, a dynamic reconciliation method should be used to take account of the lags occurring between the various process states.

2.3 Models and Constraints

The component stream networks, where the material components (or enthalpy) are flowing through the industrial unit, are described as oriented graphs, made of p branches representing streams, and n_n nodes representing accumulation equipments. In the following, emphasis is placed on mass balance to alleviate the presentation. Energy balance equations have the same structure as mass balance equations, enthalpy (or heat) being added to the list of the $n+1$ components (including total mass) that must be conserved. Since reconciliation techniques are quite similar when energy balances are considered, there is no need to repeat the expression "mass and/or energy balance" throughout the text. The selected components used for writing balance equations can be either phases, or species as minerals, metals, atoms, molecules, ions, or classes of physical properties (size, density), and enthalpy. Each component may have its own network characterized by an incidence matrix M_i, whose entries 1, -1, and 0 represent either a node input, or output, or a not connected stream. The process states used to write the mass conservation constraints are usually mass flowrates and mass fractions, therefore leading to bilinear equations. When considering component flowrates – instead of total flowrates and component mass fractions – the state equations can be kept linear. This case will be considered first.

2.3.1 The Dynamic Linear Mass Balance Equation

The dynamic mass conservation equations for any i component (including phases, species, properties and the total material flowing in the various streams) are

$$\frac{dm_i}{dt} = M_i f_i + P_i - \varepsilon_i, \quad \text{for } i = 0 \text{ to } n+1, \tag{2.6}$$

where m_i is the component i mass vector (including the total mass) accumulated at the network nodes, f_i the stream mass flowrate vector, P_i the component i production rate vector at the various nodes (for the total material $P_0 = 0$), and finally ε_i an uncertainty vector providing for structural errors such as forgotten secondary streams or intermittent streams, or errors in the production rates evaluation.

Usually rates of production are unknown and cannot be measured independently from other states. There are three possible situations:

1. the component i is transformed, and the conservation equation must not be written at the corresponding node;
2. the transformation is of very low magnitude and is simply incorporated into ε_i:

$$\frac{dm_i}{dt} = M_i f_i - \varepsilon_i, \quad \text{for } i = 0 \text{ to } n; \tag{2.7}$$

3. the component i is not transformed at a given node, and $P_i = 0$. Therefore, in the absence of structural uncertainties, this leads to the exact dynamic conservation constraint:

$$\frac{dm_i}{dt} = M_i f_i, \quad \text{for } i = 0 \text{ to } n. \tag{2.8}$$

Defining the state vector as $x_i = (m_i^T, f_i^T)^T$, Equation 2.8 can be written in the generic form:

$$E_i \frac{dx_i}{dt} = D_i x_i, \quad \text{for } i = 0 \text{ to } n+1. \tag{2.9}$$

Rather than a model allowing process simulation, this is a singular model, *i.e.*, a set of constraints linking the state variables. Its discrete version is

$$E_i x_i(t+1) = F_i x_i(t), \quad \text{for } i = 0 \text{ to } n+1. \tag{2.10}$$

Example. In a complex ore comminution or separation plant, the conservation constraints could, for instance, be written for the following components: slurry, water, ore, copper, lead, zinc, gold, particle size classes, and gold in particle size classes. If ten size classes are defined, the number of component conservation equations is 27 ($n+1 = 27$). As will be discussed in Section 2.3.5, the component definition selected here will create additional constraints since, among others, the gold species is selected at two different levels of the mass balance equations.

2.3.2 The Linear Stationary and Steady-state Cases

When the process is in a stationary operating regime, *i.e.*, a regime randomly fluctuating around a steady-state, the rate of accumulation dm_i/dt can be omitted and incorporated into the uncertainties ε_i:

$$M_i f_i = \varepsilon_i, \quad \text{for } i = 0 \text{ to } n+1, \tag{2.11}$$

where ε_i is interpreted as a stochastic term containing the neglected accumulation rate, production rate and structural uncertainties. When the process is strictly operating in a stationary mode, as defined in Section 2.2, ε_i can be interpreted as a stationary random signal with the following statistical properties deduced from the statistical variations of the component flowrates that are usually assumed to behave as normal random variables. If V_f is the covariance matrix of f, the stacked f_i vectors, then the stacked vector ε of the ε_is has the following properties:

$$\varepsilon \sim N(0, V_\varepsilon), \quad \text{with } V_\varepsilon = M V_f M^T, \tag{2.12}$$

where M is the block diagonal matrix of the M_is.

The steady-state case is then a particular case of the stationary equations, when V_ε has a zero value.

2.3.3 The Bilinear Case

Instead of using, as state variables, the component flowrates (including the total mass flowrate), one can use the total mass flowrates and the phase or species mass fractions. Obviously the models are strictly identical, but the selection of these more usual variables changes the structure of the equations with respect to the state variables. For the total mass conservation, the equations are now

$$\frac{dm_0}{dt} = M_0 f_0 - \varepsilon_0, \tag{2.13}$$

where M_0 is the total mass incidence matrix. For the phases or species mass fractions c_i in the streams, and h_i in the node loads, the equations are

$$\frac{dm_0 \bullet h_i}{dt} = M_i (f_0 \bullet c_i) + P_i - \varepsilon_i, \quad \text{for } i = 0 \text{ to } n, \tag{2.14}$$

where $\bullet$ is Hadamard's product. One can also incorporate Equation 2.13 into (2.14) which becomes

$$m_0 \bullet \frac{dh_i}{dt} = M_i (f_0 \bullet c_i) - h_i \bullet (M_0 f_0 - \varepsilon_0) + P_i - \varepsilon_i \quad \text{for } i = 0 \text{ to } n. \tag{2.15}$$

Dynamic equations similar to (2.7) and (2.8) can also be obtained. More particularly the stationary case becomes

$$M_0 f_0 = \varepsilon_0, \tag{2.16}$$

$$M_i (f_0 \bullet c_i) = \varepsilon_i, \quad \text{for } i = 1 \text{ to } n, \tag{2.17}$$

where it is clear that these equations contain flowrates and mass fractions cross-products, thus giving a bilinear to the equation set.
Example. For the same example as in Section 2.3.1, f_0 would be the ore flowrate, and c_i would correspond to copper, lead, zinc, gold and the 10 size class mass fractions. To keep the bilinear structure, it is necessary, in this particular case, to ignore the data levels corresponding to slurry and gold content in size classes.

2.3.4 Multi-linear Constraints

Process measurements usually concern flowrates and material compositions. In that case, process states are defined with two levels of properties as depicted in the bilinear case above. Unfortunately, when the performances of a mineral or metallurgical processing plant must be deeply assessed, more than two levels of material properties need to be handled [53]. Material streams may contain various phases (ore, carbon, aqueous, organic, and gas phases) which are characterized by many properties such as flowrates, particle density and size distributions, as well as mineral and chemical compositions of each phase and particle class. The mass conservation equations complexity can rapidly increase with the level of detail needed for process analysis.
Example. Gold ore processing plants involve different phases (slurry, water, carbon and ore) that are characterized by various properties (size distributions, along with chemical and mineral compositions), as well as mineral and chemical compositions of ore and carbon size classes. Only a few studies on data reconciliation in the gold ore processing industry are available [25–27],[54]. Figure 2.4 shows a possible multi-level representation of the stream materials. First, the slurry phase is divided into liquid, ore and carbon phases. Then, each phase is subdivided into its specific components:

- the liquid phase into chemical reagents (CN^-, O_2) and leached species (Au, Ag, Cu);
- the ore phase into populations of particles (such as coarse and fine) which are subsequently split into classes of particles (such as -38μm, $+38/-53\mu$m, $+53/-75\mu$m), each particle size class being characterized by its mineral (native gold, pyrite, hematite) content, and, subsequently, the mineral metal contents (Au, Ag, Cu);
- the carbon phase into size classes, each class being characterized by its metal content.

In such a complex system with six different levels of information, the mass conservation system becomes 6-linear. In addition, various components are simultaneously considered at various levels, thus creating a complex set of additional constraints that are necessary to ensure gold conservation is consistent between the different levels. This case study is discussed in [28], and the next section gives some more information on the additional constraints that are usual in MMP plants.

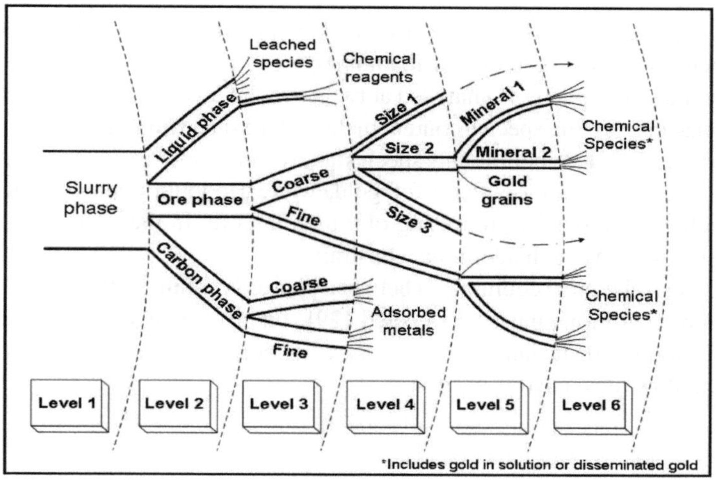

Figure 2.4 Multi-level representation of stream matter for a generic gold ore processing plant

2.3.5 Additional Constraints

In addition to the above basic mass conservation constraints, one may have to add equations that represent either additional structural constraints, or stoichiometric constraints, or process behavior modeling assumptions (mass or energy transfer and rate equations). In the first case, the most frequent constraint arises when an exhaustive species analysis is available. Then the species mass fractions must sum to 1,

$$\sum_i c_i = 1 \tag{2.18}$$

for i values corresponding to exhaustive analyses. Alternatively, the constraint may be expressed by writing that the sum of some component flowrates must be equal to the total flowrate:

$$f_0 = \sum_i f_i. \tag{2.19}$$

In the above example, these constraints are valid for carbon, water and ore components, and also for particle size classes.

A stoichiometric constraint may arise for instance when minerals are assumed to have fixed composition, such as FeS_2 or $CuFeS_2$. When both sulfides are present and are the only sulfur, copper, and iron carriers, there is a relationship between sulfur, copper, and iron contents that must be verified. Another possible constraint arises when there are stream splitting systems where an equal share of the streams is assumed. Another possibility is that a species production rate has a prior estimated value that must be obeyed with some level of uncertainty.

Other constraints may arise in multi-phase systems, when a species transfers from one phase to another one. This is obviously the case in leaching, elution, adsorption and solvent extraction processes when a metal transfers from a solid to a liquid phase or from a liquid phase to another one.

When the same species is analyzed at two different levels, for instance in ore and particle size classes, the species content in the ore must be simultaneously consistent with particle size distribution and species assays in size classes. The example of Section 2.3.1 shows a case study for gold, while Hodouin and Vaz Coelho [53] show another example for processing of a uranium ore. In the example of Section 2.3.4 there are many such additional constraints.

Also, when chemical equilibrium between phases is assumed, the thermodynamic equilibrium conditions must be satisfied [29]. Finally, inequality constraints may also be in force. For instance, species mass fractions must have values between 0 and 1, and flowrates must be positive.

2.3.6 Summary of Stationary Conservation Equations

Equations 2.11 or 2.16 and 2.17 coupled to specific additional constraints can be gathered into the following form:

$$f(X) = \varepsilon, \tag{2.20}$$

where X is a vector which gathers all the n_X state variables and f the q conservation equations. In the linear case all the equations are gathered into

$$MX = \varepsilon, \tag{2.21}$$

where M is a matrix of coefficients containing blocks such as M_0 and M_i and possibly other terms depending upon the additional constraints. It may happen that constant terms are also present in the constraints; Equation 2.21 would then become

$$MX = K + \varepsilon. \tag{2.22}$$

To simplify the presentation, this case will not be covered in the following developments. If this is required, the modification of the formulae to take account of K would not be complex.

2.4 Sensors, Measurement Errors and Observation Equations

Measurement errors play a central role in data reconciliation, since one important feature of the technique is to correct data that is contaminated by measurement inaccuracies. This section proposes a description of the statistical properties of measurement errors, with some emphasis on particulate material sampling and analysis.

2.4.1 Statistical Properties of Measurements and Measurement Errors

The measurement error statistical properties, as well as the methods used to process them, depend upon the category they belong to. Measurements can be classified into the following categories:

- On-line or off-line measurement. For instance, flowmeters or density gauges are on-line while chemical or physical analysis of a sample in a laboratory is off-line. Automatic sampling followed by centralized X-ray fluorescence analysis is considered an on-line analysis.
- Measurement either on part of the material, for instance, chemical analysis of an ore sample, or on the whole material, as performed by a flowmeter installed around a pipe.
- Continuous or discrete measurement. A flowmeter or a particle size analyzer delivers a continuous signal, while the analysis of a sample of material taken at constant time intervals is delivered at a given frequency.
- Averaged measurement value or instantaneous measurement. The reading of a flowmeter can be averaged in a given time window, or a sample made of composite increments can be analyzed for its physical or chemical properties. On the contrary a sampled flowmeter signal gives an instantaneous value.
- For discrete measurements, averaged or instantaneous samples can be taken at constant time periods (systematic sampling), or randomly (random sampling), or randomly within constant time intervals (stratified sampling).

Moreover, one can distinguish three main types of measurement errors that must be processed differently by the process observers:

- Systematic errors, or biases, are the consequences of sensor calibration drifts, interaction effects – such as the interaction of foreign chemical elements on the analyzed species – or biases in sampling procedures.
- Random centered errors. They results from many independent sources of noise due to the heterogeneous nature of the material to be analyzed and to the inherent fluctuations of the analytical devices. They have a zero mean value and are usually considered as obeying normal distributions, unless the error variance is large compared with the nominal value of the process variable, in which case the normal distribution might mean that a process variable could have negative values, an unacceptable property for inherently positive variables such as concentrations or flowrates.
- Accidental gross errors, due, for instance, to contamination of samples, tagging mistakes, transmission faults.

For correct use of measurements, systematic errors must be detected, corrected, and their sources eliminated, for instance by maintenance and calibration of sensors, or by redesigning sampling procedures. Accidental gross errors must also be detected, using fault detection and isolation (FDI) techniques, and the corresponding data eliminated or replaced by interpolation techniques. Random errors are filtered by observers.

The quality of a measured value is characterized by its accuracy, *i.e.*, the absence of bias, and by its reliability, quantified by random error variances. In the following, the statistical properties of any measurement error e will be defined by

$$e \sim N(b, v), \tag{2.23}$$

where N stands for the normal law, b for the bias and v for the variance. For multiple unbiased measurements (vector e), the statistical properties are

$$e \sim N(0, V), \tag{2.24}$$

where, most frequently, V is a diagonal matrix. However, it may happen that the measurement errors of different process variables are correlated. In these conditions V contains non-zero off-diagonal terms. This may arise when a common sample or a common measuring device is used to measure different process variables. The following situations are examples of such correlated errors:

- Particle size analysis by sieving. Particles that are not on the right sieve are necessarily present on another one, thus creating a negative correlation between the errors of fractions belonging to different size intervals [55].
- Measurement systems involving a common sensor, such as an X-ray fluorescence analyzer used at different sampling locations, create correlation between the errors.
- Matrix effects in X-ray fluorescence analysis might correlate concentration measurement errors of different metals.
- Synchronous incremental sampling of different streams may induce error correlation due to intercorrelation of the streams dynamics, which is created by the process itself (see integration error in the following section).

It is important to point out common misinterpretation of error correlation. Even if two process variables are correlated, *e.g.*, concentrations of two different metals in an ore, their measurement errors are not necessarily correlated. If the sampling and analysis steps are uncorrelated, the measurement errors are usually uncorrelated, even when the process variables are correlated.

2.4.2 Measurement Errors for Particulate Materials

Since they are made of randomly organized grains of various minerals, ores are heterogeneous materials. Also, as these grains take random shapes and sizes, local properties of ores may not be representative of their overall average properties. Breakage of ores into smaller particles may help ore homogenization. However, if the particles are not perfectly mixed, local properties of a batch of particles may not be representative of the average whole batch properties. Based on Gy [56] and Pitard [57], the main error sources for particulate materials are now briefly presented.

Fundamental error. Unless the material is perfectly homogeneous at a micro-scale level, sampling a limited number of particles to characterize the whole properties of an ore batch inherently leads to measurement errors. According to Gy, the errors induced in a sample by ore constitution heterogeneity will be called a fundamental error. It cannot be eliminated. However, it can be evaluated. It has a zero mean and a variance that can be estimated by

$$\sigma_F^2 = E\left[\left(\frac{x_s - x^*}{x^*}\right)^2\right] = \frac{1}{M_s} c \sqrt{\frac{d_{lib}}{d}} f g d^3, \tag{2.25}$$

where x_s and x^* are the mineral contents, respectively, of the ore sample and the ore batch, M_s the sample mass, d_{lib} the liberation particle size, d the sieve opening retaining 25% of the particles, f the shape factor, g the size distribution factor, and c the composition factor defined as

$$c = ((1 - x^*)/x^*)((1 - x^*)\rho_{min} + x^* \rho_{gan}), \tag{2.26}$$

where ρ denotes the mineral or gangue density.

Integration error. This error is induced by the particle distribution heterogeneity, either spatially distributed for a fixed batch of ore, or as a function of time for a flowing material. Hybrid integration errors may also occur when a sample is taken from a given location of a flowing stream, rather than from all the possible locations in the stream (using, for instance, a cross-stream cutter). This error can be evaluated if the signal heterogeneity statistics, such as mineral content autocovariance, or geostatistical variograms of the particle batch, are known. For a time related concentration of a stream sampled in a time window of width T, the definition of the integration error is simply

$$e_I = \left\{\begin{matrix}\text{composition}\\ \text{of the incre-}\\ \text{ment reunion}\end{matrix}\right\} - \left\{\begin{matrix}\text{average stream}\\ \text{composition over}\\ \text{time window T}\end{matrix}\right\}. \tag{2.27}$$

The variance of e_I can be calculated from the sampling strategy parameters and the autocovariance of e_I. This calculation can be extended to a composition vector of several streams [79].

Materialization error. This arises when extracting the sample from the ore batch to be characterized, *i.e.*, when executing the designed sampling scheme. For example, when cutting a stream flow with a moving sampler, errors may occur if all the particles do not have the same probability of entering into the sampler, either because the sampler opening is too tight or its speed too high or not uniform. Other materialization errors may arise if the sampler is overflowing or when gathering the various sampling increments to generate a composite sample. Spillage and contamination are also materialization error sources. Careful design of sampling devices and sampling execution allows elimination of this type of error, which is usually almost impossible to quantify.

Preparation errors. They are related to the operations that condition the final sample to be analyzed by the measuring equipment (X-ray fluorescence device for instance). The preparation involves secondary sampling, and thus induces new fundamental and integration errors. Drying, grinding are also operations that may induce contamination or transformation of the phases, by oxidation for example.

Analysis error. Finally, as the analytical method is based either on mechanical methods (sieving for instance) or chemical principles (titration) or spectral analysis (X-ray fluorescence, atomic adsorption), the analyzing device itself necessarily induces errors. Usually, these errors can be quantified but cannot be avoided.

Total measurement error. The variance of the resulting measurement error is obtained by summing the variances of all the contributing errors.

2.5 Observation Equations

The general form of a process observation equation is

$$\begin{aligned} Z &= g(X), \\ Y &= g(X) + e, \end{aligned} \tag{2.28}$$

where X is the vector of state variables, Z the measured process variables and Y the measurement value of Z. There are two different ways to define the measurements. Logically, the measured variables should be selected as the process variables that are directly measured by on-line sensors or laboratory analytical instruments. However, there is frequently some raw measurement preliminary processing. For instance the particle masses retained on sieves are converted to mass fractions [58], or slurry volume flowrates and densities converted to ore flowrates, thus assuming that the ore specific mass is a known parameter. The propagation of the measurement errors through the preliminary processing must be evaluated properly, since the calculation process may not only increase the variance, but also create covariance terms in the V matrix. These covariance terms structure the reconciliation results and should not be ignored, as shown for instance by Hodouin *et al.* [103]and Bazin and Hodouin [55] for particle size distributions.

Moreover, when the measured variables are not state variables, one may find it easier to combine measured variables in such a way to obtain state variable measurements even if those have not been directly measured. An advantage of this procedure is that the observation equation becomes linear and can be written as

$$\begin{aligned} Z &= CX, \\ Y &= CX + e, \end{aligned} \tag{2.29}$$

where C is a matrix of coefficients with values pointing at measured state variables. A possible drawback of the method is that the covariance of the pseudo-measured states should be calculated from the variance of the measurement errors of the source measured variables.

Alternatively, as the state variable selection is not unique, one may also select the state variables in such a way that they are confounded with directly measured variables. The drawback of this procedure is that it usually does not allow state equations to be linear. In summary, selection of the X and Y sets is not unique, but it must be made such that the functions f and g and the X and Y uncertainties covariance matrices be the least complex possible. Obviously, there is no perfect selection, since simplifying one function necessarily implies that the other one becomes more complex.

2.6 General Principles of Stationary and Steady-state Data Reconciliation Methods

The core of a data reconciliation procedure is a model-based observer that makes use simultaneously of process models (stationary regime constraints as defined in Section 2.3.6) and measurements as defined in Section 2.5. It optimally estimates unmeasured process variables in such a way that the data is reconciled with the process model, while respecting the measurement and model uncertainties (see Figure 2.5). The observer is based on the minimization of a reconciliation criterion, which usually consists of a quadratic sum of residuals $J(X)$ containing both the node imbalances ε and the measurement errors e. The reconciliation problem is formulated as

$$\begin{aligned} \hat{X} &= \arg\min_X [(Y-Z)^T V^{-1}(Y-Z) + \varepsilon^T V_\varepsilon^{-1} \varepsilon \\ Z &= g(X) \\ Y &= Z + e; \;\; e \sim N(0,V) \\ \varepsilon &= f(X); \;\; \varepsilon \sim N(0,V_\varepsilon) \end{aligned} \tag{2.30}$$

subject to

$$X_{min} \leq X \leq X_{max}, \tag{2.31}$$

where X are the plant states, Y the measured values of Z, e the measurement errors assumed to have zero mean values and known variance matrix V, and ε the constraint uncertainty values assumed also to have zero mean values and known variance matrix V_ε. State equation $f(.)$ consists of mass conservation constraints and additional constraints (as well as energy constraints when needed), while observation equation $g(.)$ relates measured variables to state variables. Finally, estimated plant states X have to be within physically meaningful intervals. For instance, mass fractions have to be between 0 and 1 and flowrates should have positive values.

The criterion in Equation 2.30 can be viewed as an empirical least-squares procedure or as the maximum likelihood solution of the state estimation problem if measurement errors and model uncertainties are Gaussian. The latter might be not strictly verified since model structure, parameters, and neglected dynamics may not be Gaussian, as well as measurement error uncertainties, which, obviously, can-

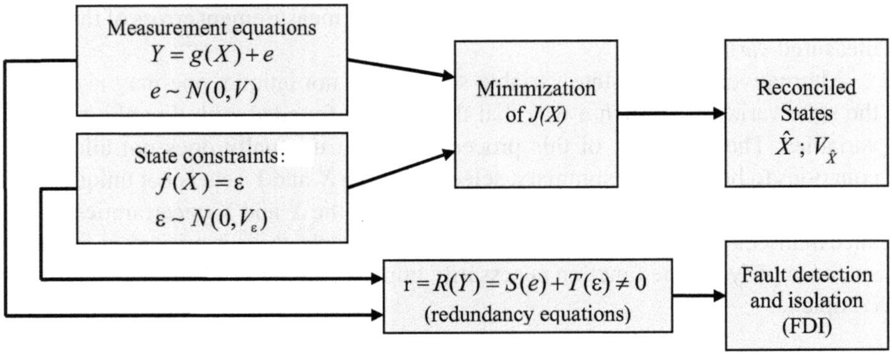

Figure 2.5 General scheme for stationary data reconciliation and fault detection and isolation (FDI)

not be anything between minus and plus infinity, because of physical inequality constraints. Without pretending that this is a statistically correct statement, it is in practice observed that this criterion is sufficiently powerful for improving data at industrial or even lab scales. Alternative techniques such as LMI [21], robust estimators [60], and artificial neural networks have also been proposed [17–20],[61].

In parallel with the reconciliation procedure, Figure 2.5 also shows optional FDI procedures. Redundancy equations R are obtained by elimination of X between the measurement equations and state constraints. Because of the conflict generated by the uncertainties e and ε, these equations generate residual values that are not zero but functions of e and ε. These residuals can be used to detect measurement biases or abnormal deviations to mass and energy conservation laws. These concepts will be discussed later on in Section 2.12. In Figure 2.5, the term $V_{\hat{X}}$ is the variance matrix of the reconciled values that will be discussed in Section 2.9.

The stationary optimization problem defined by Equation 2.30 degenerates into two limit reconciliation problems when e or ε are assumed to have null values: the steady-state case and the node imbalance case.

Steady-state data reconciliation. The SSR case is obtained by setting V_ε to zero, thus removing the second term of the criterion. As already said the steady-state case is an ideal situation. There is a continuum between the stationary and the steady-sate case, and one can superficially say that SSR is legitimate when V_ε is small in comparison with V.

Node imbalance data reconciliation. At the other end of the relative values of V and V_ε spectra, one may consider the case where the measurement values are much more accurate than the conservation constraints. In other words V is small in comparison with V_ε. The first term of the reconciliation criterion disappears and only the residuals ε are estimated. These estimates are called the reconciled node imbalances.

The feasibility of reconciliation procedures is governed by process observability [62, 63] and, its corollary, information redundancy. These two concepts are now discussed briefly.

2.6.1 Observability and Redundancy

A process state x_i is said to be observable or estimable if it can be calculated using simultaneously the measurement values and the conservation constraints, or part of them. Therefore the problem is to find a unique estimate of the variables x_i of the vector X that satisfies the following system:

$$\begin{aligned} f(X) &= 0, \\ g(X) &= Z, \end{aligned} \tag{2.32}$$

where Z is known and is the exact value of the measured variables. Equation 2.32 represents the constraint and observation equations of (2.30) where the uncertainty variables are set to their most probable values, *i.e.*, zero. A state variable, such as a metal concentration, which is directly measured, is obviously observable since one possible estimate is its measured value. Hence, the concept of observability is only important for state variables that are not directly measured.

When, in system (2.32), there is at least one of the equations (state or measurement equation) that cannot be removed without losing x_i observability, x_i is said to be non-redundant. When there is more than one possible way to estimate the value of a state, using different equations of the system (2.32), this state variable is said to be redundant. When the state is directly measured, it is redundant when it is still possible to estimate its value in the case the measurement is unavailable. Because of the inherent uncertainties of the Y values, the estimate value obtained from the following system for a redundant state x_i:

$$\begin{aligned} f(X) &= 0, \\ g(X) &= Y, \end{aligned} \tag{2.33}$$

depends on the subset of equations that is kept for calculating the state variable.

When all the state variables are estimable, the process is said to be observable. When at least one state variable is redundant the process information is said to be redundant. When all the state variables are observable and non-redundant, the system is said to be of minimal observability.

One can define the process information overall degree of redundacy as the largest number of equations that can be eliminated from the system without losing process observability. Usually it is related to the number of equations minus the minimum number of required equations to obtain minimal observability. Redundancy degrees for individual states can also be defined [64].

The redundancy degree is strongly coupled to the data reconciliation performance: the higher the redundancy, the higher the reconciled value reliability (for state estimate reliability, see Section 2.9). Moreover, the higher the redundancy, the higher the robustness of the observer. This means roughly that the number of possible sensor failures that do not hinder process observability increases with the degree of redundancy [64]. Assuming that all the variables are observable, a redundancy degree can be defined as

$$\rho_r = \frac{q+m-n_X}{q}, \tag{2.34}$$

where q and m are, respectively, the number of state equations and observation equations, and n_X the number of state variables. This index varies between 0 and 1 from the overall minimal observability to the case where all the states are measured directly or indirectly. The determination of an unmeasured state observability may be a difficult task for non-linear f and g functions. Let's have a look at the linear case.

The linear case. The system of equations containing the plant information is

$$\begin{aligned} MX &= 0, \\ CX &= Z, \end{aligned} \tag{2.35}$$

It can be globally rewritten as

$$\Psi X = \Phi Z. \tag{2.36}$$

A state variable x_i is observable if there is at least a subset of equations in (2.36) that allows the calculation of x_i when Z is known. The state vector would be observable if the rank of the matrix Ψ is n_X, the number of state variables. If the process is globally at minimal observability, then Ψ is an invertible matrix (regular matrix).

Redundancy equations. Another way of looking at redundacy is to eliminate the state variable X from the system (2.32) (or (2.35)) in the linear case). The remaining set of equations is

$$R(Z_r) = 0 \text{ or } RZ_r = 0 \text{ in the linear case.} \tag{2.37}$$

$R(.)$ is the set of redundant equations, and Z_r is here the generic vector of the measured process variables which are redundant (Z or a subset of Z). The number of equations it contains is usually $m+q-n_X$, where $n_X = (n+1) \times p$ in the linear case with $(n+1)$ components and p streams. The number of redundant equations is thus directly related to the redundancy degree of the reconciliation problem . When replacing Z_r by the measurement values Y_r, the system (2.37) is no longer verified because of the unavoidable measurement errors. The resulting vector is a residual, that is a vector of a space called parity space, physically related to nodes or joint nodes imbalances. It is a function of the measurement errors and model uncertainties insofar as stationary conditions are assumed. It vanishes when the uncertainties have zero values:

$$R(Y_r) = r = S(e_r) + T(\varepsilon) \text{ or } RY_r = Re_r + T\varepsilon \text{ in the linear case.} \tag{2.38}$$

The linear stationary case with $Z = X_m$. Let us consider, as an illustrative case, the situation where the measured variables are state variables (X_m), and the stationary conservation constraints are linear, and gathered into

$$MX = \varepsilon. \tag{2.39}$$

The vector X can be reorganized as

$$X = \begin{pmatrix} X_m \\ X_{um} \end{pmatrix}, \tag{2.40}$$

where X_m is the set of measured variables, *i.e.*, the vector Z of this particular case, and X_{um} the set of unmeasured variables. The matrix M can also be decomposed into two sub-matrices, one for X_m, and the other one for X_{um}:

$$M = (M_m, M_{um}). \tag{2.41}$$

This allows Equation 2.39 to be written as

$$M_m X_m + M_{um} X_{um} = \varepsilon. \tag{2.42}$$

Providing that the system is at minimal observability, *i.e.*, that the measured values are properly placed on the state vector, M_{um} is invertible and X_{um} calculable as

$$X_{um} = (M_{um})^{-1}(\varepsilon - M_m X_m). \tag{2.43}$$

This last situation corresponds to $n_X - m = q$, and X_{um} is estimated by setting $\varepsilon = 0$ and $X_m = Y$.

When $n_X - m > q$, there are more unmeasured variables than conservation constraints. The process is not fully observable (some states or even all states are non-estimable).

When $n_X - m < q$, there are more equations than unmeasured states. It is highly improbable that, due to the measurement uncertainties, a value of X_{um} that would simultaneously satisfy all the conservation constraints could exist. Since the number of equations in Ψ, $m+q$, is larger than the number of states to be estimated n_X, the observation system is redundant. The elimination of X_{um} from (2.42) leads to the linear redundancy equations:

$$RX_{mr} = T\varepsilon, \tag{2.44}$$

where X_{mr} is the vector of the redundant measured states. By replacing X_{mr} by the measurement Y_r, one obtains the values of the parity vector:

$$r = RY_r = Re_r + T\varepsilon. \tag{2.45}$$

This equation clearly shows that the redundancy equations residuals are not zero because of the uncertainties prevailing in the measurement and mass conservation constraints. To manage this problem, one possibility could be to remove redundant measurements. But this is the wrong approach, since experimental information is lost and, furthermore, the estimate values would depend upon the data that would have been removed. The right approach is to reconcile the information by the procedures discussed in this chapter.

Classification of the process variables:.The condition $n_x - m \leq q$ is not sufficient to ensure process observability. It may happen that the process observability is only partial, *i.e.* that some states are non-observable. It may also happen that, though the system is not redundant, some variables are observable redundant, while, as a consequence, others are non-observable. In fact any system of observation as depicted by Equation 2.32 or 2.35 can be decomposed into unobservable and observable parts. Moreover, the observable states can also be classified into minimal observability and redundant observability classes. Figure 2.6 shows the classification tree that can be drawn for the measured Z variables and the process states. The particular case where measured process variables are state variables ($Z = X_m$) is given within brackets.

The classification of the process variables and the decomposition of the equation system (2.32), or (2.35) in the linear case, into its redundancy part of Equation 2.37 are two related problems. The initial Equation 2.32 can be rewritten as

$$\begin{aligned} R(Z_r) &= 0, \\ Q(X_o) &= Q'(Z_r) + Q''(Z_{nr}), \\ X_{no}, & \end{aligned} \tag{2.46}$$

where the second equation corresponds to the deductible part of the system, allowing calculation of the observable states X_o. Various methods are available for this decomposition of the initial system and the classification of the process variables. They have been developed mainly for linear systems, but also extended to multi-linear systems. The graph theory is used in [65] and [66]. The Gauss Jordan elimination technique is used in [67], projection matrices in [68] and [69], a mixture of graph theory and linear algebra in [70], and the QR factorization in [71]. Using adapted process variable combinations, these methods have been extended to bilinear systems and even to multi-linear systems [72].

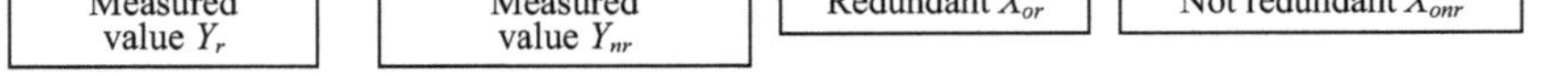

Figure 2.6 Scheme showing the status of the various process variables

2.6.2 General Principles for State Estimate Calculation

The data reconciliation problem to be solved is schematically represented in Figure 2.5, where the known information is given at the reconciliation procedure input and the process states to be estimated at the procedure output. When f and g are linear functions and inequality constraints are inactive, the solution to this problem is analytical. This occurs when the measured process variables are plant states or linear combinations of them, and represent component flowrates. From the state estimates, one can subsequently correct the values of the measured variables and estimate the unmeasured process variables.

The optimization method that can be used to solve the unconstrained problem defined by (2.30) is the usual approach where the reconciled states are the $\hat{X}$ values that cancel the derivatives of J with respect to X. The resolution methods are slightly different for the degenerate cases when either ε or e is zero.

When $\varepsilon = 0$, the optimization problem

$$\min_X J(X) \text{ subject to} f(X) = 0 \tag{2.47}$$

can be processed by one of two main approaches. The substitution method consists in exploiting the equality constraints to decrease the number of variables the criterion has to be minimized with respect to. The Lagrange method, on the contrary, increases the number of search variables to optimize the criterion by incorporating the equality constraints into it. New variables λ, called the Lagrange multipliers, are associated to these constraints.

- The substitution method consists of the following steps:

  1. Select a set of independent variables among X. Such a set is composed of the smallest number of variables X_{ind} that, if they were measured, would allow the remaining variables from the system $f(X) = 0$ (the dependent ones X_{dep}) to take unique values. In other words, the $n_X - q$ independent variables X_{ind} are selected in such a way that the q variables X_{dep} are at minimal observability.
  2. Express X_{dep} as a function of X_{ind} by solving

  $$f(X_{ind}, X_{dep}) = 0 \tag{2.48}$$

  with respect to X_{dep}:

  $$X_{dep} = h(X_{ind}). \tag{2.49}$$

  3. Replace X in the criterion J by its expression as a function of X_{ind}. The initial constrained minimization problem with respect to X is thus transformed into an unconstrained minimization problem with respect to X_{ind}:

  $$\min_{X_{ind}} J(X_{ind}). \tag{2.50}$$

  4. Find the $\hat{X}_{ind}$ value and calculate $\hat{X}_{dep}$ from (2.49).

- The Lagrange method consists in integrating the constraint $f(X) = 0$ into the J criterion to form a new function, the Lagrangian $\mathscr{L}$, which has to be optimum with respect to the variables X and λ , the latter being the Lagrange multipliers:

$$\mathscr{L} = J(X) + \sum_i \lambda_i f_i(X), \tag{2.51}$$

where f_i stands for the ith conservation constraint in the system $f(X)$, for $i = 1$ to q. The initial constrained problem has been replaced by a new optimization problem with $n_X + q$ unknown parameters.

These optimization problems can be solved by classical numerical techniques. When J is quadratic and the constraints are linear, an analytical solution can be obtained by resolving the system expressing that the derivatives of the reconciliation criterion, or the Lagrange function, have zero values. In other cases, either non-linear programming methods can be used to minimize the criterion, or numerical methods used for solving the non-linear equations expressing that the derivatives of the criterion are zero.

When $e = 0$, the optimization problem

$$\min_X J(X) = f(X)^T V_\varepsilon^{-1} f(X) \text{ subject to } g(X) = Y \tag{2.52}$$

can again be processed by one of the two above approaches: substitution and Lagrange methods. This reconciliation method is common practice in the bilinear case. It allows to estimate unmeasured flowrates through the use of species concentrations in the ore streams. It is called the node imbalance method.

2.7 The Linear Cases: Steady-state, Stationary and Node Imbalance Data Reconciliation Methods

2.7.1 The Steady-state Case

The unconstrained SSR problem can be formulated as the following particular case of (2.30):

$$\begin{aligned} \hat{X} &= \arg\min_X [(Y - CX)^T V^{-1} (Y - CX)], \\ Y &= CX + e; \;\; e \sim N(0, V). \end{aligned} \tag{2.53}$$

In the substitution method, $n_X - q$ independent variables are selected using either heuristic or systematic methods, and the X vector is restructured as (X_{ind}, X_{dep}), while the matrix M is partitioned into compatible blocks, such that the conservation constraints can be rewritten as

$$\begin{pmatrix} M_{ind} & M_{dep} \end{pmatrix} \begin{pmatrix} X_{ind} \\ X_{dep} \end{pmatrix} = 0. \tag{2.54}$$

or

$$M_{ind} X_{ind} + M_{dep} X_{dep} = 0. \tag{2.55}$$

The dependent variables X_{dep} can then be expressed as functions of X_{ind}:

$$X_{dep} = -(M_{dep})^{-1} M_{ind} X_{ind} \tag{2.56}$$

assuming that the independent variables have been selected to ensure M_{dep} invertibility. Therefore X can be expressed as

$$X = \begin{pmatrix} X_{ind} \\ X_{dep} \end{pmatrix} = \begin{pmatrix} X_{ind} \\ -M_{dep}^{-1} X_{ind} \end{pmatrix} = LX_{ind}, \tag{2.57}$$

with $L = \begin{pmatrix} I & -M_{dep}^{-1} \end{pmatrix}^T$.

The linear reconciliation criterion:

$$J = (CX - Y)^T V^{-1} (CX - Y) \tag{2.58}$$

can then be written as a quadratic expression of X_{ind}:

$$J(X_{ind}) = X_{ind}^T L^T C^T V^{-1} CLX_{ind} - 2Y^T V^{-1} CLX_{ind} + Y^T V^{-1} Y. \tag{2.59}$$

J is minimized with respect to X_{ind} by writing that the derivatives of J with respect to X_{ind} have zero values. One obtains a system of $n_X - q$ equations system with $n_X - q$ unknown states, which has the following solution:

$$\hat{X}_{ind} = (L^T C^T V^{-1} CL)^{-1} L^T C^T V^{-1} Y. \tag{2.60}$$

Knowing X_{ind}, the dependent variable estimates can be calculated from Equation 2.56:

$$\hat{X}_{dep} = -(M_{dep})^{-1} M_{ind} \hat{X}_{ind}. \tag{2.61}$$

The Lagrange method can also be applied to solve the linear SSR problem. The Lagrangian function is

$$\mathscr{L} = (CX - Y)^T V^{-1} (CX - Y) + \lambda^T MX. \tag{2.62}$$

The $\mathscr{L}$ stationarity conditions are

$$\frac{d\mathscr{L}}{dX} = 2C^T V^{-1} CX - 2C^T V^{-1} Y + M^T \lambda = 0, \tag{2.63}$$

$$\frac{d\mathscr{L}}{d\lambda} = MX = 0. \tag{2.64}$$

The X solution of this $n_X + q$ equation system with $n_X + q$ unknown variables is

$$\hat{X} = \Pi^{-1} (I - M^T (M \Pi^{-1} M^T)^{-1} M \Pi^{-1}) C^T V^{-1} Y, \tag{2.65}$$

where

$$\Pi = C^T V^{-1} C. \tag{2.66}$$

There are other alternatives to the solution of the reconciliation problem. Two of them should be mentioned, in the particular case where the measured variables are states ($Z = X_m$):

- The simplest one, for software development, consists in selecting $C = I$, *i.e.*, assuming that all the states are measured. For states that are not actually measured, corresponding variances of measurement errors are simply given very large values. The solution is thus directly obtained from (2.65) as

$$\hat{X} = (I - VM^T(MVM^T)^{-1}M)Y. \tag{2.67}$$

This method is used in the BILMAT™ algorithm [22], and has never numerically failed.

- The system of conservation constraints can be partitioned using the redundancy equations as follows (linear version of Equation 2.46):

$$\begin{cases} RX_{mr} = 0, \\ QX_o = Q'X_{mr} + Q''X_{mnr}. \end{cases} \tag{2.68}$$

The solution is then obtained in two steps. First, the reconciled values of the measured redundant states are estimated by

$$\hat{X}_{mr} = (I - V_{mr}R^T(RV_{mr}R^T)^{-1}R)Y_{mr}, \tag{2.69}$$

where V_r and Y_r are formatted to match with X_{mr}. Second, the following equation is solved for X_o :

$$Q\hat{X}_o = Q'\hat{X}_{mr} + Q''Y_{nr}. \tag{2.70}$$

Tuning of the measurement error matrix covariance. Before using the steady-state method, it is always necessary to test: (1) that the main contribution to the variability of the process variables is related to the measurement and sampling errors; and (2) that there is an underlying steady-state regime during the period of time corresponding to the measurement and sampling process. In this case, the measurement errors due to the process time variations, such as the integration error, should be incorporated into the measurement error. In other words data obtained during a transient regime of large magnitude in comparison with the measurement errors should not be used for SSR. Methods to find the measurement variance matrix have been discussed by [41],[73–78].

Four methods are proposed to estimate the V covariance matrix:

- Estimation from the properties of the equipment involved in the measurement process and the properties of the material flowing in the stream. For instance the theory of particulate material sampling [56, 79, 80] can be applied as seen in Section 2.4, in conjunction with a statistical analysis of the reliability of the measuring devices for the measured specific material property.
- Direct estimation of the variance of the measurement Y from a large set of data for the same plant operated under the same operating conditions. In this case, the estimate of V is obtained from the Y records by standard statistical estimation techniques, and is noted $Var(Y)$. This estimate of V does not strictly consist of measurement errors, since it also includes some process dynamic disturbances, such as the integration error. However, this is the right approach, since the recon-

ciled values are expected to give an estimate of the underlying plant steady-state (*i.e.*, average) behavior.

- Indirect estimation from the constraint residuals (the node imbalances) in the particular case where $C = I$ (the identity matrix). The constraint residuals r are calculated for each Y value of the data set and their variance estimated:

$$\begin{cases} r = MY = MX + Me, \\ Var(r) = MVM^T. \end{cases} \tag{2.71}$$

Techniques have been proposed [77] for extracting V from this last equation. The main advantage of this technique is that it takes account of the mass conservation constraints.

- Simultaneous estimation of V and of a rough model of the plant. For instance, a mineral separation plant can be simply modeled by mineral separation coefficients at each separation node of the flowsheet. The plant model is then expressed as

$$X = B(s)X_f \tag{2.72}$$

where X_f is the state vector of the feed streams and s the separation coefficient vector. The variance of Y estimated from a measurement data set is then

$$Var(Y) = B(s)Var(X_f)B(s)^T + V. \tag{2.73}$$

From Equation 2.73, variance V can be extracted simultaneously to s and $Var(X_f)$ by a least-squares procedure [78].

2.7.2 The Stationary Case

The unconstrained stationary reconciliation problem is formulated as the following particular case of (2.30):

$$\begin{cases} \hat{X} = \arg\min_X [(Y - CX)^T V^{-1}(Y - CX) + \varepsilon^T V_\varepsilon^{-1} \varepsilon], \\ Y = CX + e; \; e \sim N(0, V), \\ \varepsilon = MX; \; \varepsilon \sim N(0, V_\varepsilon). \end{cases} \tag{2.74}$$

The solution is obtained by canceling the derivatives of J:

$$\frac{dJ}{dX} = 2C^T V^{-1} CX - 2C^T V^{-1} Y + 2M^T V_\varepsilon^{-1} MX = 0, \tag{2.75}$$

which leads to

$$\hat{X} = (C^T V^{-1} C + M^T V_\varepsilon^{-1} M)^{-1} C^T V^{-1} Y \tag{2.76}$$

or alternatively, using a matrix inversion lemma,

$$\hat{X} = \Pi^{-1}[I - M^T(V_\varepsilon + M\Pi^{-1}M^T)^{-1}M\Pi^{-1}]C^T V^{-1} Y, \tag{2.77}$$

where Π is defined in Equation 2.66. Equation 2.77 gives Equation 2.65 when V_ε is set to 0.

Tuning of the accumulation rate covariance matrix. Three methods are proposed to estimate the V_ε covariance matrix:

- Measurement of the state variables by a suitable combination of instrumentation for a sufficiently long period of time. This may require transient installation of instruments on the streams. Since it may happen that not enough sensors are available to simultaneously measure all the state variables, the different parts of the circuit may be sequentially studied, assuming that the statistical properties of the operation do not change much from a test on a plant sub-network to a test on another sub-network. For such a subset of X, the measurement equation is

$$Y = X + e. \tag{2.78}$$

  Multiplying Equation 2.78 by the incidence matrix M of the sub-network corresponding to the fully instrumented part of the plant, one obtains

$$MY = MX + Me = \varepsilon + Me. \tag{2.79}$$

  Assuming that there is no correlation between the state variables and their measurement errors, the variance-covariance matrix is given by:

$$MVar(Y)M^T = V_\varepsilon + MVM^T, \tag{2.80}$$

  where $Var(Y)$ is the measurement variance–covariance matrix of Y, the elements of which can be evaluated by the usual statistical estimation methods. V_ε can then be calculated by

$$V_\varepsilon = MVar(Y)M^T - MVM^T. \tag{2.81}$$

  The total V_ε matrix can therefore be constructed by consistently assembling the various terms of the partial variance matrices. Obviously some covariance terms will not be evaluated, but since they necessarily correspond to distant streams in the plant, they can be neglected.
- Alternatively, one may evaluate the V_ε matrix by an approximation of the plant dynamics, using for instance first order transfer functions with rough evaluation of their time constants. The process dynamics are subsequently gathered into a plant state-space model (see Section 2.10 for more detailed explanations). This model allows an estimation of V_x, the state variable variance characterizing the underlying unknown random dynamic variations of the process. V_ε can finally be estimated by

$$V_\varepsilon = MV_xM^T. \tag{2.82}$$

- A third option is to use V_ε as a tuning factor that can be adjusted heuristically following an evaluation of the data reconciliation performance in comparison with a desired behavior. In fact, this is the subjective method that is mostly adopted, for example, when tuning model uncertainty relative variances and measurement uncertainties in Kalman filtering techniques.

Example. Figure 2.7 shows a standard flotation circuit arrangement.

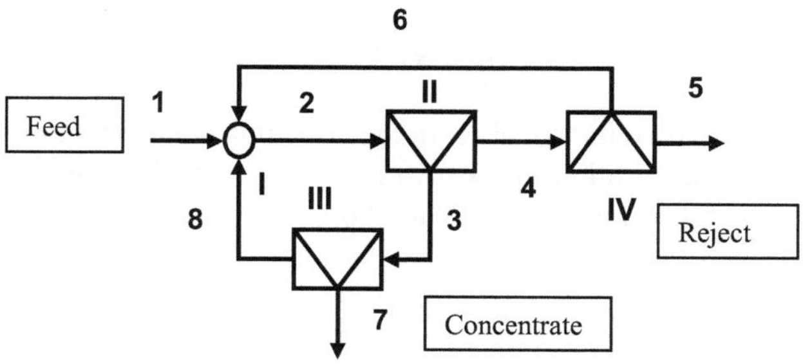

Figure 2.7 A flotation mineral separation circuit

The state vector X consists of the valuable mineral flowrates in the eight streams. There are four nodes, two recycling streams and two plant outputs, the concentrate, which contains the valuable mineral to be sold for metal extraction, and the reject. Table 2.2 gives the measured values Y of the flowrates of the mineral to be concentrated, as well as their reliability σ_e (standard deviations of e). Table 2.2 shows also the values of the reconciled values $\hat{X}$. The reconciled values do not exactly satisfy the steady-state mass conservation equations since the plant is assumed to be operated in a stationary dynamic regime where V_ε is different from zero.

Table 2.2 Data for the circuit of Figure 2.7 (fifth column will be explained in Section 2.9)

Stream	$Y(t/h)$	σ_e	$\hat{X}$	$\sigma_{\hat{X}}$
1	15,8	0,6	14,86	0,44
2	22,2	0,8	23,08	0,45
3	-	∞	14,19	0,47
4	8,3	0,3	8,18	0,27
5	1,4	0,2	1,47	0,19
6	6,3	0,3	6,23	0,23
7	11,3	0,5	11,62	0,46
8	2,5	0,2	2,45	0,2

2.7.3 *The Node Imbalance and Two-step Methods for Bilinear Systems*

In bilinear constraint cases, when either flowrates or species concentrations are assumed to be exactly measured, the reconciliation procedure degenerates into a LQ (linear-quadratic) optimization problem. The node imbalance for the component i is defined as the mass flowrate residuals at the network nodes that are obtained when using the measured values of the component i concentrations:

$$\varepsilon_i = M_i(f_0 \bullet c_i) = M_i C_i^d f_0, \tag{2.83}$$

where C_i^d is the diagonal matrix of c_i. In the node imbalance method, it is assumed that component concentrations are known, and that flowrates only have to be estimated. The optimal flowrate estimates are those that minimize

$$J(f_0) = \sum_{i=1}^{n} \varepsilon_i^T V_{\varepsilon_i}^{-1} \varepsilon_i + (Y_f - C_f f_0)^T V_f^{-1} (Y_f - C_f f_0), \tag{2.84}$$

where the first term of J corresponds to the node imbalances and the second term to the flowrate measured values, if any:

$$Y_f = C_f f_0 + e_f; \; e_f \sim N(0, V_f). \tag{2.85}$$

The node imbalance estimates of the flowrates are therefore

$$f = (\Sigma_i \gamma_i + C_f^T V_f^{-1} C_f)^{-1} C_f^T V_f^{-1} Y_f, \tag{2.86}$$

where

$$\gamma_i = C_i^{dT} M_i^T V_{\varepsilon_i}^{-1} M_i C_i^d. \tag{2.87}$$

When the flowrates have been estimated by the node imbalance method, species concentrations can be estimated in a second step. When the flowrates are assumed to be known, the species concentration constraints become linear, as shown in this expression:

$$M_i(f_0 \bullet c_i) = M_i F_0^d c_i = 0, \tag{2.88}$$

where F_0^d is the diagonal matrix of the known f_0 values. The quadratic criterion to be minimized, subject to (2.88), is

$$J(\ldots, c_i, \ldots) = \sum_{i=1}^{n} (Y_i - C_i c_i)^T V_i^{-1} (Y_i - C_i c_i), \tag{2.89}$$

where C_i is the observation matrix of the component i concentration, and V_i its corresponding measurement error variance matrix. The solution is given by the same equations as (2.65) and (2.66), providing that there are no additional constraints that couple the mass balance equations of the various components:

$$\hat{c}_i = \alpha_i [I - F_0^d M_i^T (M_i F_0^d \alpha_i F_0^d M_i^T)^{-1} M_i F_0^T \alpha_i] C_i^T V_i^{-1} Y_i, \tag{2.90}$$

where

$$\alpha_i = (C_i^T V_i^{-1} C_i)^{-1}. \tag{2.91}$$

When the node imbalance estimation of the flowrates is followed by the reconciliation of the species concentrations for these flowrate estimates, the method is a sub-optimal two-step LQ method [34]. The next section shows that a similar hierarchical technique can be used to find the true optimal solution to the bilinear reconciliation problem.

2.8 The Non-linear Cases

As discussed above, when the constraints and the measurement equations are linear, the solution of the reconciliation problem can be developed analytically. However, in non-linear cases, it is normally impossible to derive an explicit expression for the reconciled states. Several methods (for instance [26], [81–86]) are possible depending upon the approach selected to handle the constraints (substitution or Lagrange multipliers techniques), and the optimization technique that is used to minimize the criterion. The following optimization techniques are possible options:

- Any Nnon-linear programming (NLP) method, involving a search algorithm to iteratively approach the optimal values. The substitution method is well adapted to these procedures since it decreases the number of search variables to be manipulated by the NLP algorithm.
- A hierarchical minimization method where the criterion and the search variables are split into blocks. These approaches allow hybrid minimization methods, in which some parts are optimized by analytical methods, others by NLP algorithms.
- Iterative numerical resolution of the system of equations expressing that the Lagrange criterion derivatives have zero values.

It would be too long to detail all the possible options for solving non-linear reconciliation problems, but examples may help to illustrate some possible approaches.

2.8.1 An Example of Substitution Methods: Mass and Heat Balance of a Thermal Exchanger

Let us consider the mixer-exchanger of Figure 2.8 [87], which heats an electrolyte solution containing dissolved copper and nickel subsequently fed to an electrorefinery plant. Part of the solution is directly recycled to the mixer-exchanger, and another part is cleaned from impurities before being recycled. Data collected and averaged for a time period of 3 h are presented in Table 2.3, together with an estimation of their standard deviations. The process variables are the solution flowrates F, densities ρ, temperatures T, copper and nickel concentrations c and n for the streams 1, 2 and 3, and heating power P. The system parameters are the specific heats of the solutions C, respectively 4400, 4210, 4350 J/(kg.K) for streams 1, 2 and 3. They are assumed to be constant and perfectly known. The data are assumed to reasonably represent a steady-state operating regime of the plant. The thermal losses are neglected.

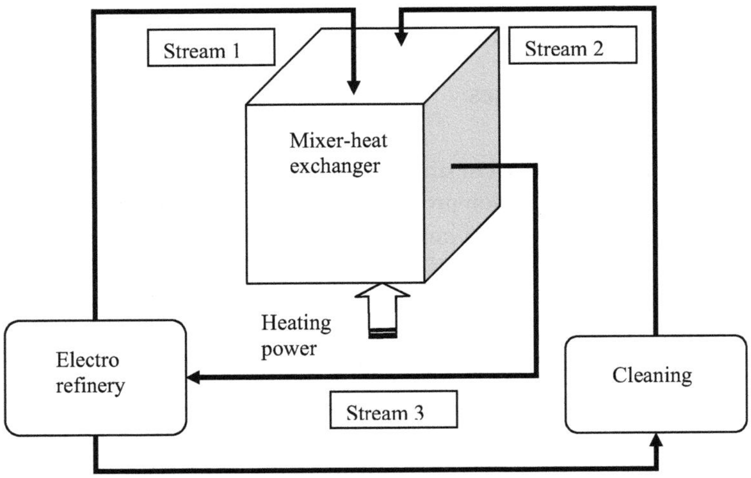

Figure 2.8 Mixer-exchanger for conditioning an electrolyte

Table 2.3 Data for the plant of Figure 2.8

Variable	T (^{O}C)			F (m^3/min)			ρ (kg/l)			c(g Cu/l)			n(g Ni/l)		
Stream	Y	σ	$\hat{X}$	Y	σ	$\hat{X}$	Y	σ	$\hat{X}$	Y	σ	$\hat{X}$	Y	σ	$\hat{X}$
1	46	2	45.3	20	4	20.4	1.31	0.12	1.25	40	3	41.1	10	2.0	10.1
2	27	2	26.7			9.6	1.15	0.06	1.15	10	2	10.2	1	0.3	1.0
3	42	3	42.1			27.2	1.28	0.12	1.34	37	4	34.4	8	2.0	7.9
Power data (kJ/s) $P: Y = 6000; \sigma = 500; \hat{X} = 5978$															

The mass balance constraints around the mixer-exchanger are written as

$$\begin{cases} F_1\rho_1 + F_2\rho_2 - F_3\rho_3 = 0 \\ F_1c_1 + F_2c_2 - F_3c_3 = 0 \\ F_1n_1 + F_2n_2 - F_3n_3 = 0 \end{cases} \tag{2.92}$$

and the heat balance as

$$F_1\rho_1C_1T_1 + F_2\rho_2C_2T_2 - F_3\rho_3C_3T_3 + P = 0 \tag{2.93}$$

The least-squares criterion is

$$J(X) = \Sigma_{i=1}^{3}\left[\frac{(T_i-T_i^m)^2}{\sigma_{T_i}^2}+\frac{(c_i-c_i^m)^2}{\sigma_{c_i}^2}+\frac{(n_i-n_i^m)^2}{\sigma_{n_i}^2}+\frac{(\rho_i-\rho_i^m)^2}{\sigma_{\rho_i}^2}\right] + \frac{(F_1-F_1^m)^2}{\sigma_{F_1}^2}+\frac{(P-P^m)^2}{\sigma_P^2}, \tag{2.94}$$

where the superscript m stands for the measured value of the process variable (Y) and the subscript i for the stream number. The minimization of $J(X)$ subject to the bilinear and trilinear constraints (2.92) and (2.93) can be performed with a PNL algorithm applied to the substitution method. The independent variables are selected as

$$X_{ind} = \left(F_1\ F_2\ F_3\ T_1\ T_2\ T_3\ \rho_1\ \rho_2\ c_1\ c_2\ n_1\ n_2\right)^T. \tag{2.95}$$

The dependent variables are then

$$X_{dep} = \left(\rho_3\ c_3\ n_3\ P\right)^T. \tag{2.96}$$

The minimization algorithm is depicted in Figure 2.9. It requires an initialization of the independent variables, which can be estimated from the conservation constraints and the measured values, using, for instance, the node imbalance method.

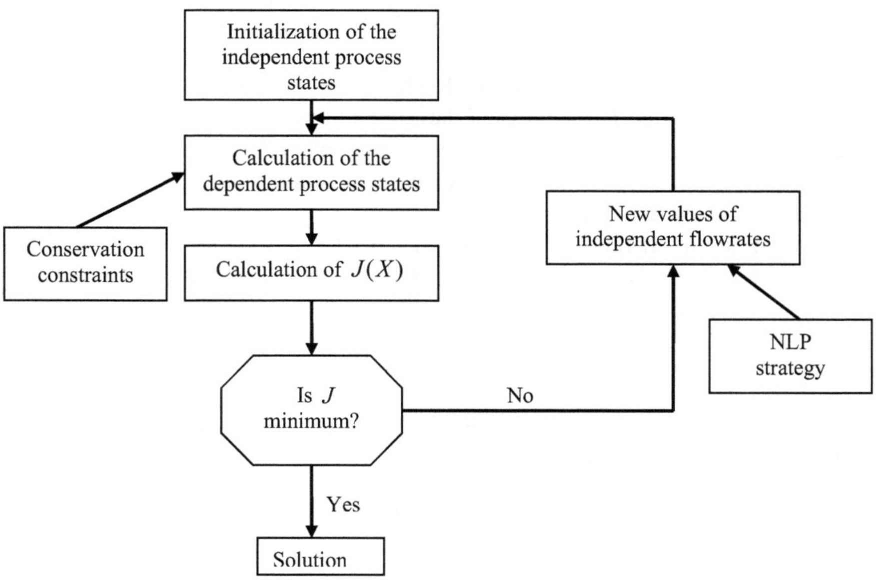

Figure 2.9 Algorithm for the calculation of the reconciled values by the substitution/PNL method

The results of the reconciliation procedure appear in Table 2.3. The simultaneous negative and positive corrections of, respectively, power and stream 3 temperature indicate that thermal losses should be taken into account. Repetition of the measurement campaign as well as application of FDI techniques (see Section 2.12) could confirm this assumption. In that case a model for estimating the losses could be used to generate a virtual measure of the heat loss that can be added to the reconciliation problem.

2.8.2 An Example of Hierarchical Methods: BILMAT™ Algorithm

The BILMAT™ algorithm has been developed mainly for bilinear problems [22] but it can be extended to trilinear systems [53, 59]. The basic principle is that, when the main phase flowrates are known (upper level of the hierarchy), the species mass conservation constraints become linear functions of the species mass fractions (see Section 2.7.3). Therefore the idea is to define a lower level reconciliation problem where the criterion contains the mass fraction part of the data, and the constraints are written only for the species conservation. This problem, being LQ, has a direct analytical solution (see Section 2.7.3). At the upper level, only the flowrate variables are manipulated to minimize the overall criterion. Formally the problem is thus split into two minimization problems defined by

$$\min_{X_f,X_m} J(X_f,X_m) = \min_{X_f}\left(\min_{X_m} J_m(X_m|X_f) + J_f\right), \tag{2.97}$$

where X_f and X_m are, respectively, the flowrate state variables and the species mass fractions variables, and J_m and J_f the J parts containing the measured mass fractions and flowrates, respectively. Since the upper minimization level is non-linear, one can apply the substitution/PNL method to decrease the number of search variables. This method is depicted in Figure 2.10.

To illustrate the method, let us again consider the introductory example of Section 2.1.1. Flowrates F_1 and F_2 can be selected as independent flowrates of the upper level, since F_3 can be deduced from

$$F_3 = F_1 - F_2. \tag{2.98}$$

Giving values to these three variables leads to a linear structure for the copper and zinc mass conservation equations:

$$\begin{aligned} F_1c_1 - F_2c_2 - F_3c_3 &= 0; \\ F_1z_1 - F_2z_2 - F_3z_3 &= 0. \end{aligned} \tag{2.99}$$

The calculation of the reconciled metal mass fractions is then a LQ problem given the values of the flowrates. The solution to this problem is obtained by forming the Lagrange function:

$$\mathscr{L} = J_c + 2\lambda_c(F_1c_1 - F_2c_2 - F_3c_3) + 2\lambda_z(F_1z_1 - F_2z_2 - F_3z_3), \tag{2.100}$$

where J_c is the part of the criterion which contains the metal assays:

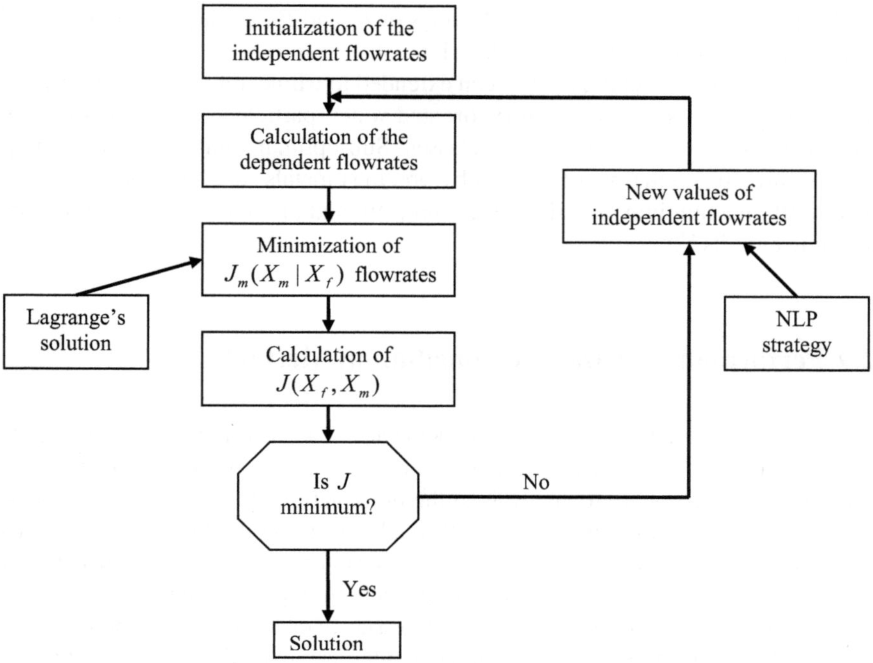

Figure 2.10 BILMAT™ reconciliation algorithm

$$J_c = \sum_{i=1}^{3} \frac{(c_i - c_i^m)^2}{\sigma_{c_i}^2} + \sum_{i=1}^{3} \frac{(z_i - z_i^m)^2}{\sigma_{z_i}^2}. \tag{2.101}$$

Then the Lagrange function is derived with respect to the six metal mass fractions and the two Lagrange multipliers, and the derivative equations set to 0. The solution of this system of eight unknown variables into eight equations leads to

$$\begin{aligned} \hat{c}_1 &= c_1^m - \lambda_c F_1 \sigma_{c_1}^2 \\ \hat{c}_2 &= c_2^m - \lambda_c F_2 \sigma_{c_2}^2 \\ \hat{c}_3 &= c_3^m - \lambda_c F_3 \sigma_{c_3}^2 \end{aligned} \tag{2.102}$$

with

$$\lambda_c = \frac{F_1 c_1^m - F_2 c_2^m - F_3 c_3^m}{F_1^2 \sigma_{c_1}^2 + F_2^2 \sigma_{c_2}^2 + F_3^2 \sigma_{c_3}^2} \tag{2.103}$$

for copper, and similar expressions for zinc. The reconciled values obtained from Equations (2.102) are optimum only for the selected particular values of the flowrates. To minimize the overall criterion

$$J = J_c + \sum_{j=1}^{3} \frac{F_j - F_j^m}{\sigma_{F_j}^2}, \tag{2.104}$$

a NLP is required to find the optimal values of F_1 and F_2, as shown in Figure 2.10. The optimal values are given in Table 2.1.

This hierarchical technique has been extended to trilinear and quadrilinear problems, *i.e.*, problems involving slurry, ore and water, particle size distributions, and chemical assays within particle size classes. Sub-optimal solutions have been applied to investigate the path followed by precious metals in a base metal sulphide plant [59], in a uranium ore grinding circuit [59], and in gold ore comminution and leaching circuits [28].

2.9 Performance of Data Reconciliation Methods

When the reconciliation calculation is achieved, it is worthwhile to estimate the reliability of the results in comparison with those of the raw measurements. This is performed by a sensitivity analysis of the propagation of the measurement errors and rate of accumulation uncertainties through the reconciliation process [34] and [44, 88]. In other words, the variance–covariance matrix of the reconciled states has to be calculated for an assessment of the reconciliation procedure benefits.

There are two approaches to calculate the variance matrix of the reconciled states, $V_{\hat{X}}$, *i.e.*, the variance of the estimation error. One can explicitly formulate the reconciled states as functions of Y and then calculate the variance $V_{\hat{X}}$ as a function of V and V_{ε}. When the reconciliation method leads to a linear estimator, *i.e.*, when the reconciled states are linear functions of Y, then the usual linear algebra of variance calculation can be applied. Otherwise, one can linearize the expression around the process steady-state values and apply linear variance algebra. The second method consists in randomly generating synthetic values of Y according to a normal distribution $N(\hat{Y}, V)$ and repeating the reconciliation procedure. If this is done a sufficiently large number of times, the statistical properties of $\hat{X}$ can be subsequently estimated. This method is known as a Monte-Carlo sensitivity analysis.

When the X estimator is linear, one can calculate the variance of $\hat{X}$ using the relationship between $\hat{X}$ and Y. For instance, in the linear steady-state case, Equation 2.65 gives

$$\hat{X} = \Gamma Y, \tag{2.105}$$

where Γ is given by

$$\Gamma = \Pi^{-1}[I - M(M\Pi^{-1}M^T)^{-1}M\Pi^{-1}]C^T V^{-1}. \tag{2.106}$$

Therefore $\hat{X}$ variance is

$$V_{\hat{X}} = \Gamma V \Gamma^T. \tag{2.107}$$

Since $\hat{X}$, in the assumed Gaussian context, is a maximum likelihood estimate, it is obvious that this estimate is such that necessarily the diagonal terms of $V_{\hat{X}}$ are lower than the diagonal terms of the variance of reconciled values obtained by any other estimator. It has been shown [44] that the variance of the reconciled measured variables is

$$V_{\hat{Y}} = V - V_{in}, \tag{2.108}$$

where V_{in} is the variance of the innovations in, which are residuals of the observer, *i.e.*, the differences between the measured and reconciled values:

$$in = Y - \hat{Y} = Y - C\hat{X}. \tag{2.109}$$

Since V_{in} has, by definition, positive diagonal terms, one obtains

$$V_{\hat{Y}}(i,i) \leq V(i,i) \text{ for } i = 1 \text{ to } m, \tag{2.110}$$

an inequality showing the variance reduction due to data reconciliation. A global variance reduction factor ρ_v can be defined and calculated for the case where a subset of states is measured ($Z = X_m$):

$$\rho_v = \frac{1}{m} \sum_{i=1}^{m} \frac{V(i,i) - V_{\hat{Y}}(i,i)}{V(i,i)} = \frac{1}{m}[q + m - n_X]. \tag{2.111}$$

This equation shows that the reduction factor is directly related to the degree of redundancy ρ_r of the information on the system (see Equation 2.34). The expression (2.111) is valid also for bilinear SSR problems, providing that the mass conservation constraints are linearized around the reconciled states [44].

In addition to this difference between the variances before and after reconciliation, there is a strong difference that appears in the covariance terms. It is here assumed that the measurement errors are not correlated. However, the estimation errors are now correlated because of the correlation induced by the conservation equations. Thus, matrices $V_{\hat{X}}$ and $V_{\hat{Y}}$ exhibit covariance terms that correspond to the state consistency induced by the reconciliation procedure. This last property has a strong impact on the reliability of the process performance indicators or process model parameters subsequently calculated using reconciled states instead of raw measurements [89, 90]. When the stationary reconciliation method is used rather than the steady-state method, the covariance terms are usually smaller since the conservation equations are given more flexibility by relaxing these constraints through the tuning of V_ε.

Example 1. The impact of the variance reduction and the role of the covariance terms have been illustrated for the calculation of separation plant recoveries in [89]. Figure 2.11 shows a separation node whose separation efficiency of component i can be calculated by one of the two possible formulae:

$$R_i = \frac{F_2 x_{2i}}{F_1 x_{1i}}, \tag{2.112}$$

$$R_i = \frac{x_{3i} - x_{1i}}{x_{3i} - x_{2i}} \frac{x_{2i}}{x_{1i}}. \tag{2.113}$$

This last formula is known as the two-product formula. When using raw data, one obtains quite different results depending on the expression used to calculate the efficiency (for complex plants and data sets there is a very large number of paths to calculate separation efficiencies or recoveries, Equations 2.112 and 2.113 being just two simple examples). A major advantage of using reconciled data is that there is a unique efficiency value whatever the formula used for calculation. Moreover, its estimate is more reliable.

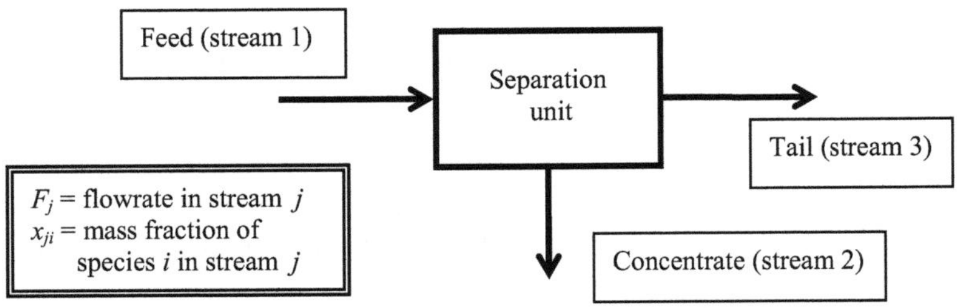

Figure 2.11 A separation unit

Example 2. Let us consider again the flotation circuit of Figure 2.7. Table 2.4 compares the standard deviations of the experimental and reconciled values of the mineral flowrates. The results show that reconciled data is more accurate than raw data. In addition to this positive effect and to the estimation of unmeasured variables, a drastic improvement of the process performance evaluation is usually brought on by the reconciled data. In this example, the relative standard deviation of the metal recovery (ratio of metal in the concentrate to metal in the feed) goes from 15% down to 6%, an improvement that would allow much more efficient decision for plant optimization.

Table 2.4 Reliability of measured and reconciled values (error standard deviations) for the plant of Figure 2.7

Stream	s.d of measured value	s.d. of reconciled value
1	0.6	0.44
2	0.8	0.45
3	∞	0.47
4	0.3	0.27
5	0.2	0.19
6	0.3	0.23
7	0.5	0.46
8	0.2	0.2

Example 3. Table 2.5 shows the results obtained for part of a copper–lead–zinc flotation plant as depicted in Figure 2.12 [90]. Clearly, when using raw data, one finds different values of the recoveries depending upon the formula used. Furthermore some calculations are very unreliable (standard deviations of 31 and 38%). The role

of the $V_{\hat{X}}$ variance and covariance terms is considerable in the drastic increase of the precision when using reconciled data.

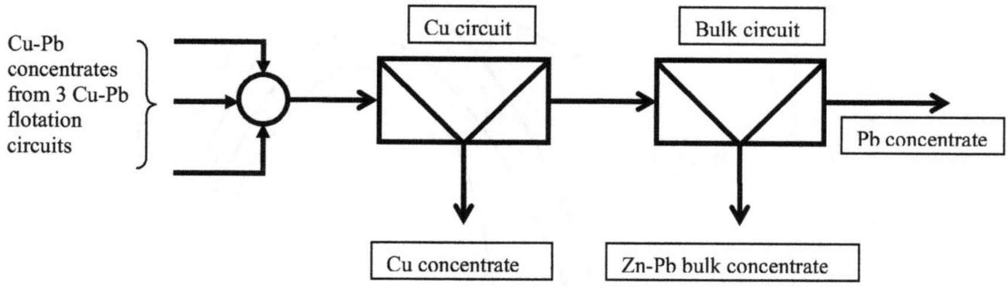

Figure 2.12 Copper–lead–zinc flotation plant

Table 2.5 Recoveries R and their reliabilities (standard deviations σ_R) as calculated with reconciled and raw data

Method	Steady-state reconcilation		Raw data and (2.112)		Raw data and (2.113)	
Recovery of	$R(\%)$	σ_R (%)	$R(\%)$	σ_R (%)	$R(\%)$	σ_R (%)
Copper in copper concentrate of copper circuit	78.0	4.0	59.4	31	79.8	7.2
Lead in lead concentrate of bulk circuit	95.3	3.0	89.3	13	79.0	38
Zinc in bulk concentrate of bulk circuit	56.0	7.0	29.6	13	77.8	10

Example 4. Another use of reconciled data is the estimation of model parameters [91] such as grinding, leaching, and flotation rate constants [58], and hydrocyclone selectivity. For example, the confidence intervals of the estimated rates of breakage in an industrial ball mill of a closed circuit are presented in Figure 2.13 as functions of particle size [90]. Using reconciled values instead of raw data drastically improves the model reliability, particularly in the fine size range where the sensitivity of the rate of breakage is extremely high, because of the typical structure of the breakage phenomena, which involves chain reactions.

2.10 An Overview of Dynamic Reconciliation Methods

Steady-state data reconciliation is frequently used to enhance process information quality. However steady-state operating conditions do not exist in practice and corresponding methods cannot fulfill the need for reliable dynamic state estimation,

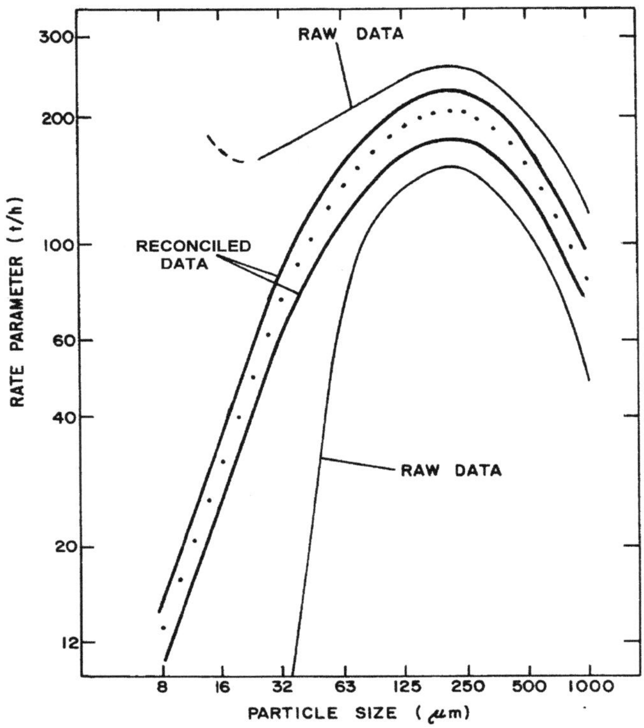

Figure 2.13 Standard deviation intervals for grinding rate parameters calculated from either raw or reconciled data

particularly when the state variable amplitude variations due to process dynamics are larger than the inherent measurement errors.

Various options are available to tackle the dynamic data reconciliation problem [40]. On-line application of SSR is the first option [54, 92], but it usually generates estimation errors larger than the measurement errors [76]. Averaging techniques coupled to steady-state data reconciliation can be used to compensate dynamic variations. However, this usually induces phase lags that deteriorate the process control performances [93], and tuning problems of the time window [65, 94]. Dynamic reconciliation is obviously required for batch or semi-batch processes [31, 95].

Mass conservation constraints incorporating material accumulation terms can be used [36, 37], but usually metallurgical inventories are not measurable or extremely inaccurate. Furthermore, the observability of these inventory variables, when unmeasured, is quite problematic. Treatment of accumulation rates as stochastic elements, as in the stationary method presented in Section 2.7.2, is a powerful alternative to the explicit use of inventories. However, the tuning of the variance of the uncertainties related to the neglected dynamics is a difficult task [96] for on-line application of stationary reconciliation methods [97].

Full model observer (FMO) based on phenomenological or empirical [121] process models can also be constructed, but the parameters are frequently so inaccurate and time varying that the observer performances are poor, therefore justifying the use of sub-models, more reliable than full models [98]. A number of intermediate methods between the SSR procedure and the full model-based observer have been experimented with and deserve to be clarified, since they are, in practice, usually more efficient than SSR or FMO approaches.

The objective of this section is to make a short survey of the possible dynamic data reconciliation methods, classified according to the models and reconciliation criteria used.

Two main classes of data reconciliation models will be considered: full (or causal) models, *i.e.*, models that can simulate process states and outputs from input variables and initial states, and sub-models that express mass or energy conservation constraints to be obeyed by the process variables. In both types of models, uncertainties and disturbances are considered since the observers to be designed must have the ability to process the uncertainties about the process behavior. The full model can be phenomenological or empirical, or a mixture of both approaches. The minimal sub-model necessarily contains phenomenological mass conservation constraints, but it can be augmented by other modeling considerations that may be phenomenological or empirical.

2.10.1 Phenomenological Causal Models

Although phenomenological models are obviously specific to a given metallurgical process, they nevertheless contain ingredients that are common to most chemical reactors (mixing and transport properties, reaction kinetics and equilibrium, material or energy balance equations). The flowing material is divided into $n+1$ components i ($i=0$ to $n+1$), and the process variables are the component flowrate vectors f_i, in the p plant streams, stacked into the vector f, and the n_n unit process inventory vectors m_i. The mass conservation constraints are then (see Equation 2.6 where the structural uncertainty terms have been omitted):

$$\frac{dm_i}{dt} = M_i f_i + P_i \quad i \in \{0 \ldots n\}, \tag{2.114}$$

where M_i is the incidence matrix for species i, and P_i the vector of production rates. The flowrates and production rates are usually related to the equipment inventories through kinetic parameters (such as breakage, flotation, leaching, reducing rate constants), and through the mixing and material transport properties:

$$\psi f = \varphi(m), \tag{2.115}$$

where ψ is a matrix that extracts the node output steams, and φ expresses the relationship between the flowrate and the equipment load. Finally, the disturbances or model uncertainties must be modeled. The plant feed stream variations are modeled by the following linear stochastic equations:

$$z(t+1) = Az(t) + \xi(t), \tag{2.116}$$

$$\Omega f(t) = Bz(t) + \Omega \bar{f}, \tag{2.117}$$

where A is a state matrix, $\xi(t)$ a white noise of variance V_f, Ω the matrix that extracts the component feed rates from the plant stream flowrates stacked into f, $\bar{f}$ the mean value of f, and B the matrix defining the feed rates from the state variable z. Parameter and model disturbances in Equations 2.114 and 2.115 are represented by additive uncertainties $\delta(t)$ that can be modeled by equations similar to Equations 2.116 and 2.117 driven by a white noise $\xi_\delta(t)$ of variance V_δ.

The model described by Equations 2.114 to 2.117 is a causal model since it can be used to simulate the process from the known inputs $\Omega\bar{f}$, $\xi_\delta(t)$, and $\xi(t)$. However, it contains a very large number of parameters (rate constants and feed coefficients, parameter and model disturbances), and is non-linear. An alternative to this approach is to construct empirical causal models.

2.10.2 Empirical Causal Model

To illustrate an example of an empirical model, let us consider any mineral processing plant flowsheet, linearized around nominal stream flowrates $\bar{f}$. It can be modeled by a network of connected basic units including (see Figures 2.14 and 2.15):

- stream junctions, *i.e.*, zero dynamics elements that combine flow streams;
- stream separators, *i.e.*, zero dynamics elements characterized by separation coefficients s_{ki} representing the split of component i between the two products of node k;
- unit gain transfer functions (usually pure delays or low order transfer functions) representing the dynamic relationships that exist between the node feed and its output streams;
- feed disturbance generators driven by white noise signals ξ_i, as considered in Equation 2.116;
- internal disturbance generators, driven by white noise $\xi_s(t)$ used to model the separation efficiency disturbances. They are designed to preserve material conservation by adding to one stream the same amount of material that is removed from the other one.

The various connected elements can be gathered into a generic linear state space model:

$$z(t+1) = A'(t) + \xi'(t), \tag{2.118}$$

$$f(t) = B'z(t) + \bar{f}. \tag{2.119}$$

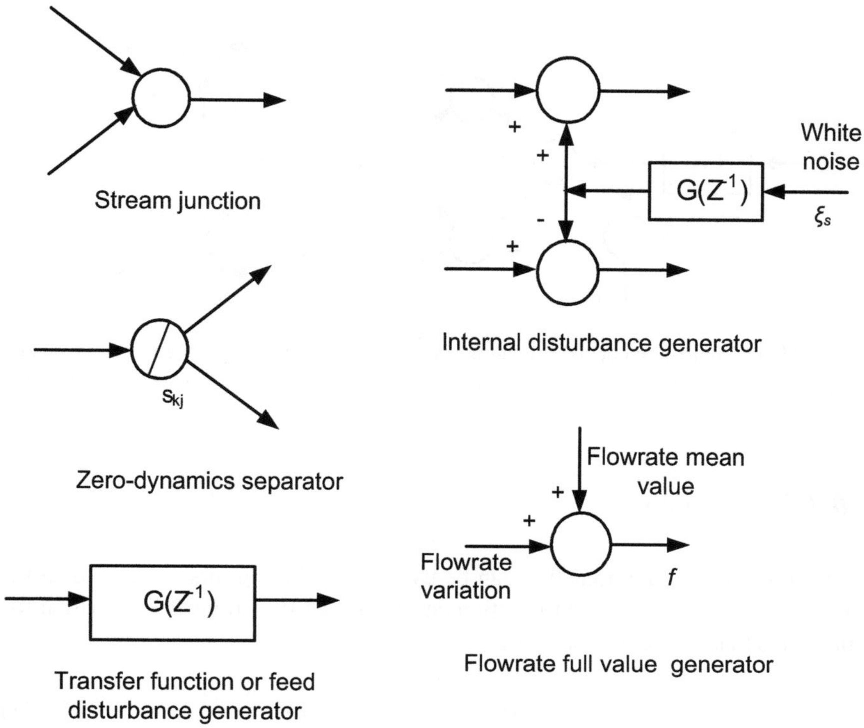

Figure 2.14 Basic elements of an empirical model of a mineral separation plant

This empirical full model implicitly contains the mass conservation Equations 2.8, in the absence of component transformation as assumed in the operating conditions defined by the previous selected basic units. Mass conversation for component i is written

$$\frac{dm_i}{dt} = M_i f_i = M_i(f_i - \bar{f_i}) = M_i B_i' z(t) \quad \text{for } i = 0 \text{ to } n+1 \tag{2.120}$$

since the constraint $M_i \bar{f_i} = 0$ is valid for the nominal state values. A representation of the model for a single flotation cell is given in Figure 2.15. The same structure can be used for all the stream components and for all the nodes of a complex plant.

The model requires, in addition to the steady-state values $\bar{f}$, the knowledge of the separation coefficients s_{kj}, the transfer function coefficients, and the variance of the driving white noise vector ξ' including ξ_i and ξ_s. Both the empirical and the phenomenological models contain numerous parameters to be calibrated. This is why it is interesting to look at sub-models for dynamic data reconciliation.

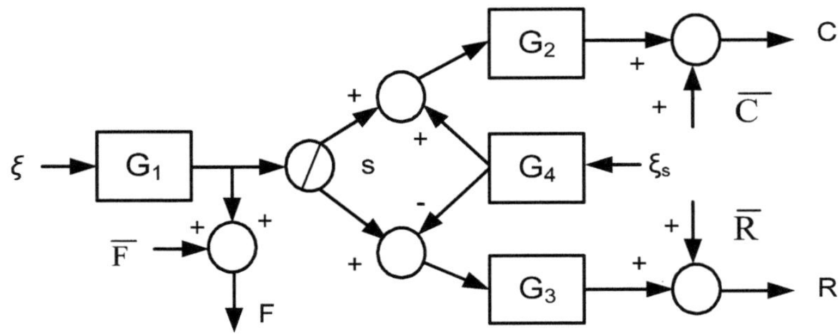

Figure 2.15 An empirical model of a flotation single unit (F, R and C stand for feed, concentrate and reject streams, and G for transfer functions, s is species separation coefficient, and ξs are white noise)

2.10.3 Sub-models

The minimal set of equations that can be used for designing model-based data filtering methods is the mass conservation constraints of Equations 2.114 written for components that are not transformed:

$$\frac{dm_i}{dt} = M_i f_i \quad i \in \{0 \ldots n+1\} \tag{2.121}$$

or

$$\frac{dm}{dt} = Mf \tag{2.122}$$

using the stacked vectors f and m, and the block-diagonal matrix of the M_is. This equation does not contain any parameters to be calibrated or uncertainties to be estimated. Obviously these would be ideal conditions for data reconciliation. Unfortunately the number of equations q $(= n_n(n+1))$ is low compared with the number n_X $(= (p+n_n)(n+1))$ of variables to be estimated. Hence – even if all the process variables are measured – this creates a low redundancy level, which in turn leads to limited improvements of the process variable reliability. In addition, the inventories m_j are usually not available for measurement.

To cope with the problem of inventory unavailability, it is possible to consider the stationary model described in Section 2.7.2 for instantaneous data reconciliation, while adding time correlation through the statistical behavior of the accumulation rates (node imbalances):

$$Mf = \varepsilon \quad \varepsilon \sim N(0, V_\varepsilon(\tau)), \tag{2.123}$$

where τ is the time lag used for calculating $V(\tau)$ which is the autocovariance of ε. This stationary dynamic model degenerates into the stationary model of Section 2.7.1 when $V_\varepsilon(\tau) = 0$ for $\tau > 0$, or into the steady-state mass balance when $V_\varepsilon = 0$.

To add time links between successive values of stream variables included in Equations 2.122, it has been proposed to add empirical evolution models for each flowrate [36]. Random walk, for instance, can be used:

$$f(t+1) = f(t) + \omega(t) \quad \omega \sim N(0, V_\omega), \tag{2.124}$$

where V_ω is considered to be a diagonal matrix. This model is rather a rough and arbitrary approximation since it does not take account of the physical links that must exist between the input and output flows of a process. Stationary autoregressive models should be preferred for describing the time evolution of the variables.

Instead of adding empirical information to the mass conservation constraint of Equation 2.122, one may add parts of the phenomenological model. For instance, the assumption of perfect mixing creates a link between the outlet streams and the inventories, therefore limiting the problems generated by the difficulties measuring inventories.

The various models and sub-models presented above must now be coupled with measurement values to subsequently perform data reconciliation. The next section describes some of the options for on-line and dynamic data reconciliation.

2.10.4 Reconciliation Methods

The state variables are usually not directly measurable. As in the steady-state and stationary methods, two main options can be used: either reconstructing pseudo-measurements of state variables by combining primary measured variables, or using the actually measured variables and expressing them as functions of the state variables. The second approach has the drawback of generating non-linear observation equations but the merit of using directly the raw information. As usual the measurement equation is

$$Y(t) = g[X(t)] + e \quad e \sim N(0, V), \tag{2.125}$$

where e is the measurement error and X the state variables, either f_i and m_i in the linear state equation case, or f_0 and c_i, total mass flowrates and component concentrations, in the bilinear case.

A variety of reconciliation methods are possible depending on the model and criterion used in the filter. The process equations can be those of a full model or a sub-model, empirical or phenomenological, steady-state, stationary, or fully dynamic. The reconciliation criterion usually contains weighted quadratic terms derived from the maximum likelihood estimation of the state variables X from the measurement Y, in a Gaussian context for model uncertainties, measurement errors, and driving white noise. The criterion can be instantaneous, or expressed in a finite time window, or launched at time zero. Depending on the consistency between the modeling assumptions and the criterion formulation and resolution, the resulting filter can be optimal or sub-optimal.

Let's now define the possible terms of a reconciliation criterion. They are written below at time t only, but can be extended in a finite size window, or a window beginning at time zero, thus usually leading to optimal observation. First, there are terms representing measurement errors, *i.e.* squared residuals according to

$$J_m(t) = [Y(t) - g[X(t)]]^T V^{-1} [Y(t) - g[X(t)]]. \tag{2.126}$$

Squared node imbalances can be put into the criterion when appropriate. They are

$$J_\varepsilon(t) = f(t)^T M^T V_\varepsilon^{-1}(0) M f(t). \tag{2.127}$$

When time correlation between the node imbalances ($V_\varepsilon(\tau) \neq 0$ for $\tau > 0$) is to be taken into account, the vector f is replaced by the stacked values of f at previous times, and the weighting matrix accordingly augmented with the autocovariance terms.

The model uncertainties δ and the driving white noise signals, either ξ, ξ' or ω, which are functions of the nature of the state variables to be estimated, should also be part of the reconciliation criteria and lead to the following quadratic terms:

$$J_\varsigma(t) = \varsigma(t)^T V_\varsigma^{-1} \varsigma(t), \tag{2.128}$$

where ς is either δ, ξ, ξ', ω, depending upon the model used. Moreover, if a smoothing horizon is considered in the criterion formulation, past terms can also be used.

Table 2.6 summarizes some of the possibilities of coupling state equations, observation equations and quadratic reconciliation criteria. It gives the models and measurements that are used as well as the quadratic criteria. The methods are optimal or sub-optimal depending on the consistency between the modeling assumptions and the reconciliation criterion used. For instance, when the time correlation is quantified in the model while the criterion contains only present information (thus freezing past estimates) the method is sub-optimal. On the contrary, when the smoothing horizon is extended to all past information, the methods are optimal.

The criterion minimization may be a LQ optimization problem, but most frequently it is a non-linear optimization problem that requires using non-linear programming methods (see for instance [38]) or sequentially linearizing the observation equations. In the case where the estimation horizon starts at time zero and a full model is used, one could apply either the Kalman filter in the LQ case [99], or the extended Kalman filter (EKF) in the non-linear case. When a sub-model is used the generalized version of the Kalman filter (GKF) can be used [37, 96, 100].

It is also possible to use sub-models for the steady-state part of the model (thus ignoring the process gains) and a full model for its dynamic part [93]. This allows some kind of synchronization of the data, while avoiding constructing a complete model of the process. This technique has been used to deal with processes exhibiting large pure delays, as those occurring in pipe-lines [101].

There are also methods that keep the bilinear structure of the mass conservation equations (products of flowrates and concentrations), instead of using species flowrates as above, thus leading to linear observation equations. Estimation algo-

Table 2.6 Some of the possible dynamic reconciliation methods (h = smoothing horizon)

Method	Model	State equations	Criterion	Time window	References
On-line steady-state mass balance constraints	Sub-model	(2.123), $V_\varepsilon = 0$	(2.126)	$(t-h,t)$ or t	Makni and Hodouin, 1994 [92]
On-line stationary mass balance constraints (also called node imbalance method)	Sub-model	(2.123), $V_\varepsilon(0) \neq 0$	(2.126), (2.127)	$(t-h,t)$ or t	Bazin *et al.* 1995 [102], Hodouin and Makni 1998[24], Makni *et al.* 1995 [104], Lachance *et al.* 2006, 2007 [43, 76]
Idem with correlated node imbalances	Sub-model	(2.123) $V_\varepsilon(k) \neq 0$	(2.126), (2.127)	$(t-h,t)$ or t	Hodouin (unpublished)
Dynamic mass balance constraints	Sub-model	(2.122)	(2.126)	(0, t)	Almasy 1990 [36], Darouach and Zasadzinski 1991 [37], Lachance *et al.* 2006 [100]
Dynamic mass balance constraints plus empirical dynamics	Sub-model	(2.122), (2.124)	(2.126), (2.127), (2.128)	$(0,t)$	Almasy, 1990 [36], Lachance *et al.* 2006 [100]
Full phenomenological model	Full-model	(2.114) to (2.117)	(2.126), (2.127)	$(0,t)$	Liebman *et al.* 1992 [38]
Full empirical model	Full-model	(2.118) (2.119)	(2.126), (2.128)	$(0,t)$	Bai *et al.* 2006 [39], Hodouin and Makni 1996 [93], Makni *et al.* 1995 [97]

rithms derived from the BILMAT™ steady-state method [4, 92] can be used, which hierarchically decompose the optimization problem. The state variables are estimated in two steps. At the upper level the flowrates are estimated by minimizing the node imbalances calculated with the measured values of the concentrations. At the lower level, flowrate estimates are frozen, thus leading to a LQ problem, and the concentrations are reconciled. The procedure can be iterated between the two levels to improve the minimum localization. These techniques have been illustrated by numerical tests based on simulated or industrial data [46] and their drawbacks and advantages discussed [42].

The stationary method (node imbalance method) is recommended for industrial applications since it does not require any model parameter calibration, does not rely on inventory measurement, and allows reasonable adjustments of the dynamic (V_ε) and measurement error (V) variances (see, for instance, [43, 78], and [96]), which gives sufficient filtering performances to the method with respect to practical industrial needs.

2.10.5 An Example of Dynamic Reconciliation for a Simulated Flotation Circuit

A Pb–Cu–Zn flotation plant is depicted in Figure 2.16. It produces four different concentrates. On-stream analyzes (OSA) for lead, copper and zinc are available for the main streams of the plant (all the streams of Figure 2.16), but only the plant ore feedrate is measured. The plant is simulated using a phenomenological kinetic model [40] and random variations of the process variables are imposed as well as measurement errors. For data reconciliation a sub-model is considered that involves an empirical transfer function to describe the feed-to-concentrate and feed-to-reject dynamics for the ore and the three metals (only the ore flow model is given in Figure 2.17 but the flow models for the three metals have similar structures). However, the separation coefficients for the ore and the three strategic metals are not modeled. Consequently the model is incomplete.

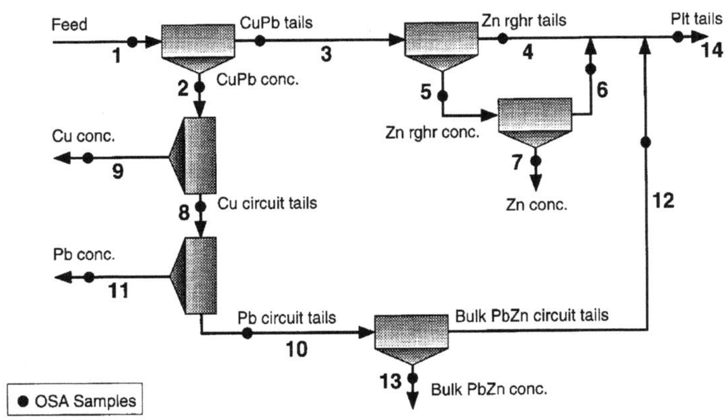

Figure 2.16 Flowsheet of a lead–zinc–copper flotation plant

The reconciliation procedure has a hierarchical structure (see Figure 2.18). At the upper level, metal assays are used to estimate flowrates in order to minimize model equation residuals at time t (this is a one-step method using criterion J of Equation 2.127). Assuming that ore flowrates are set at their estimated values, metal assays are reconciled at the lower level by minimizing squared residuals of the measurement equations and squared residuals of metal flowrate dynamic models. This is again a one-step method. There is also a possibility of refining flowrate estimation by iterating between the two levels, but this is usually unnecessary since the process variables do not vary much between two subsequent sampling times. Some results are given in Figures 2.19 to 2.22. The first one shows the estimated values of the unmeasured flowrates compared with the simulated values ("true values"). The other figures show the reconciled values of the metal assays for selected streams.

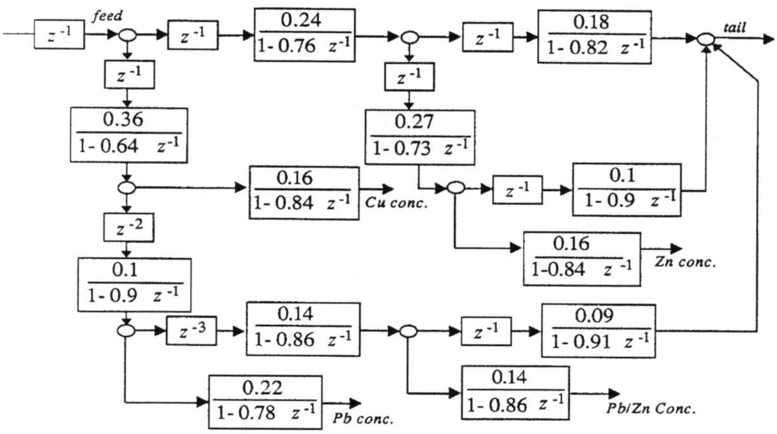

Figure 2.17 Empirical sub-model of the ore flow dynamics in the plant (z^{-1} in transfer functions is the backshift operator)

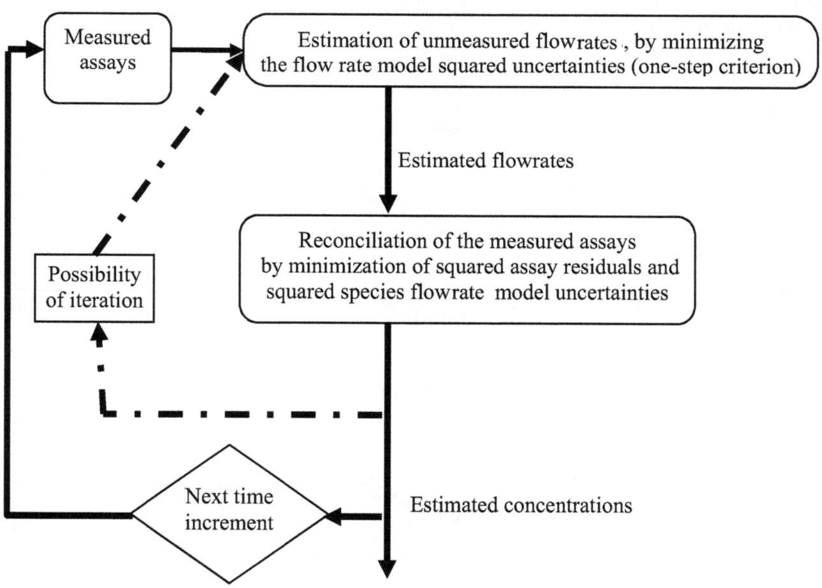

Figure 2.18 Computation scheme for reconciliation of Figure 2.17 plant data

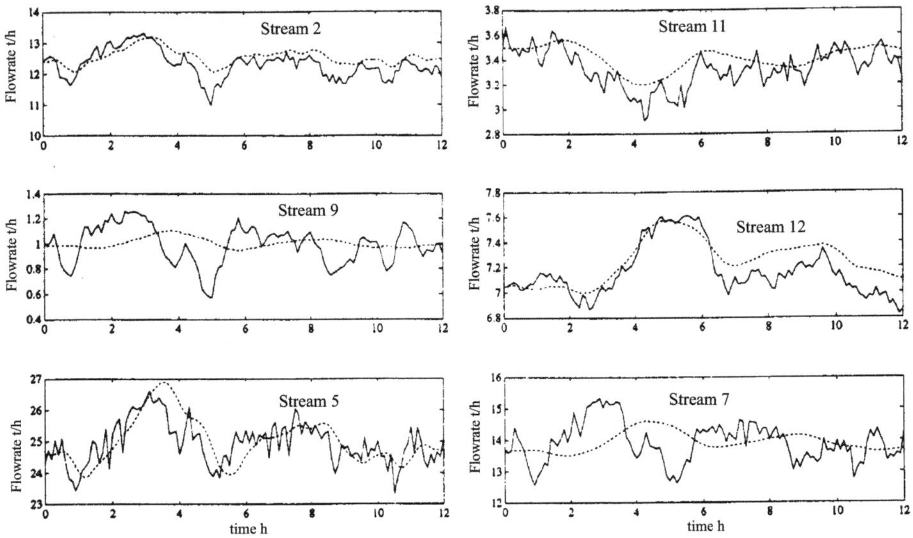

Figure 2.19 Comparison of estimated (dotted line) and simulated (continuous line) ore flowrates in six different streams

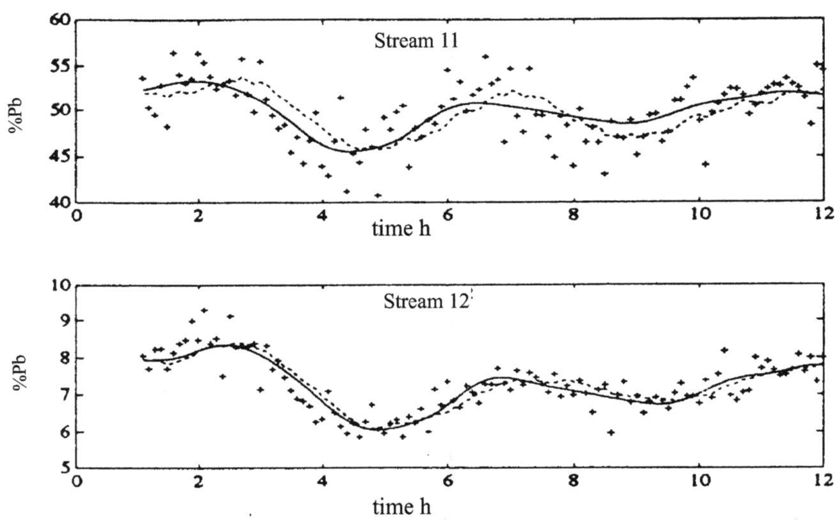

Figure 2.20 Comparison of measured (crosses), estimated (dotted line) and simulated (continuous line) lead grades in two selected streams

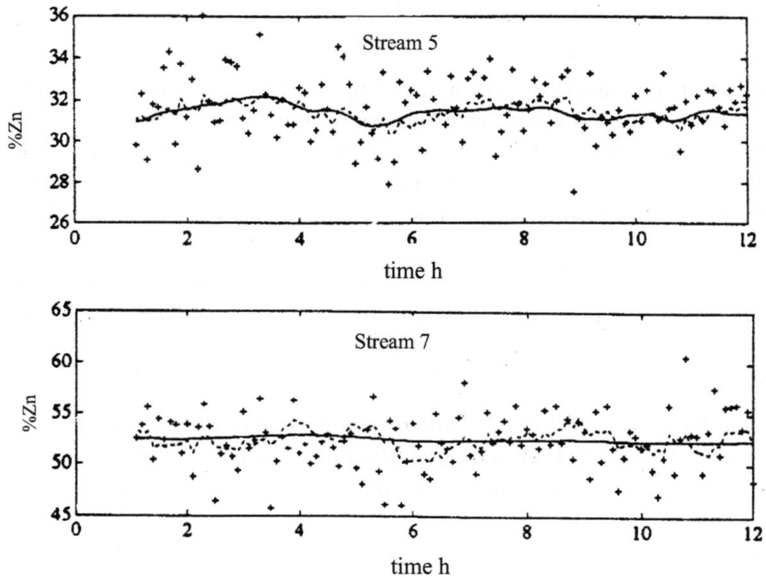

Figure 2.21 Comparison of measured (crosses), estimated (dotted line) and simulated (continuous line) zinc grades in two selected streams

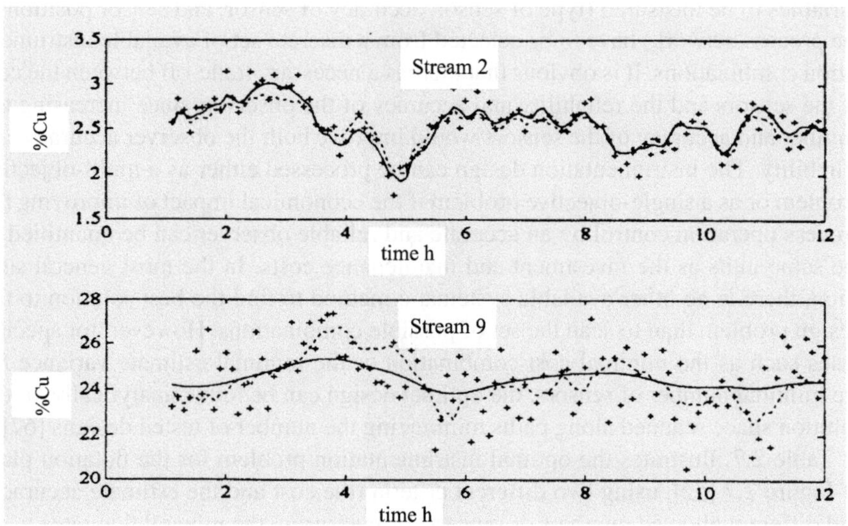

Figure 2.22 Comparison of measured (crosses), estimated (dotted line) and simulated (continuous line) copper grades in two selected streams

2.11 Instrumentation Strategy Design

If raw data delivered by sensors is to be updated by a reconciliation procedure, which at the same time generates estimates of unmeasured variables, the sensors, or more generally the variables to be measured, must be selected in a way to maximize the reconciliation procedure performance [51, 105–107]. The instrumentation placement or measurement strategy design problem is defined as [64, 108–110]:

> *Knowing the state equation $f(X) = \varepsilon$, or $CX = \varepsilon$, and V_ε (usually $= 0$ at the instrumentation design stage), find $g(.)$, or C, and V_e in such a way that the following properties of the data reconciliation procedure are obtained, subject to the constraints that states X are observable:*
>
> - o *maximum accuracy of the estimated states and of the process performance indices that are subsequently calculated with these estimates;*
> - o *Minimum instrumentation cost (capital investment, labor and maintenance costs);*
> - o *maximum reliability of the process observer (maximum operating life without loss of observability, in the presence of possible total failures of sensors).*

This is a multi-criterion discrete optimization problem since the properties of the variables to be measured (type of sensor, accuracy of sensor, and sensor position in the process network) have to be selected from a discrete set of available instrumentation combinations. It is obvious that there is a necessary trade-off between the cost of the sensors and the reliability and accuracy of the observer, since increasing the number and accuracy of the sensors would improve both the observer accuracy and reliability. The instrumentation design can be processed either as a multi-objective problem or as a single-objective problem if the economical impact of improving the process operation control by an accurate and reliable observer can be quantified in the same units as the investment and maintenance costs. In the most general situation, there is no other available systematic method to find the best solution to the design problem than to scan the set of possible combinations. However, for specific cases such as the minimal cost combination or the minimal estimate variance for the minimal number of sensors, the optimal design can be found analytically, or the solution space scanned along paths minimizing the number of tested designs [67].

Table 2.7 illustrates the optimal instrumentation problem for the flotation plant of Figure 2.7 [52], using two different criteria (the cost and the estimate accuracy) and different allowed numbers of sensors for measuring the mineral flowrates.

Table 2.7 Optimal sensor placement for the plant of Figure 2.7 (numbers of the streams to be instrumented)

Number of available sensors	4	5	6
Minimal cost configuration	1, 6, 7, 8	1, 4, 6, 7, 8	1, 2, 4, 6. 7, 8
Minimal estimate variance configuration	5, 6, 7, 8	4, 5, 6, 7, 8	1, 4, 5, 6, 7, 8

2.12 Fault Diagnosis

The above data reconciliation procedures are unbiased insofar as the assumption that random components e and ε of the process model have zero means is valid. In the absence of such a property, the X estimates become biased and the data reconciliation, instead of improving the data and the subsequent control actions on the process, might distort quite significantly the process state observation and consequently deteriorate the decision making process involved in the manual or automatic control procedures [111].

Non-centered rates of accumulation may arise from omitted streams, corresponding either to infiltration of matter into the nodes, or, on the contrary, to material leakages, or to secondary inputs or intermittently active outputs, or to persistent deviations from the stationary operating conditions. Gross errors or accidental errors are also improbable events that may distort the reconciliation results [49, 112–116]. Either these faults should be first detected and corrected, or the reconciliation procedure robustified to attenuate their impacts on the reconciled values. Providing for such faults, in the linear data reconciliation case, Equation 2.21 can be more exactly rewritten as [70, 117]

$$MX = \varepsilon + L_F F_X, \tag{2.129}$$

where F_X is the vector of potential sources of non-centered accumulation rates, and L_F the matrix that distributes the impact of these faults on the various nodes. When L_F is the identity matrix, it is assumed that each node is potentially faulty and that all the potential faults around a node k are lumped into a single variable $F_X(k)$.

Non-centered measurements may arise from sensor biases due either to calibration deficiencies, matter sampling systematic errors, or to sensor misplacement in the streams to be measured. Providing for such faults, Equation 2.29 in the linear case can be more exactly rewritten as

$$Y = CX + e + K_F F_Y, \tag{2.130}$$

where K_F and F_Y have the same meanings for the measurement biases as L_F and F_X for the node balance faults. The fault detection problem can be defined as follows:

Knowing M, C, Y, V, V_ε, L_F, and K_F, decide between the two following hypotheses:

$$H0: \; F_X = F_Y = 0, \text{ i.e., there are no faults in the process;}$$
$$H1: \text{ either } F_X \text{ or } F_Y \text{ or both are } \neq 0\text{, i.e., there is at least one fault in the process.}$$

Most of the time, detection of gross errors is performed after the reconciliation procedure by testing innovations, *i.e.*, corrections brought to the measured process variable values. When the statistical tests are sequentially applied to single innovations, these methods are incorrect since they ignore the fundamental correlation brought to the measured variable estimates by the reconciliation procedure. A gross error in any process variable usually contaminates the whole set of state estimates, and therefore might lead to wrong diagnosis. An alternative to the innovation residuals obtained through data reconciliation is to directly use the residuals of the redundancy equations. Various fault detection tests applied to the residuals of the redundancy equations (see Figure 2.5 and Equation 2.38) are available [7, 8]. Only the parity space approach is presented here. Each residual (element of the parity vector, [114]) can be tested against a normal distribution, or they can be tested together. The latter approach, called the global detection test, is applied to the following quadratic term:

$$J_r = r^T V_r^{-1} r, \tag{2.131}$$

where r represents the residuals of the redundancy equations (Figure 2.5 and Equation 2.38), hence a normal centred vector in the absence of faults since it depends only on e and ε. V_r, the variance matrix of the residuals, is directly calculable, in the linear case, since r consists of linear functions of e and ε (Equation 2.45). This quadratic term is a fault signature in the most usual case where the set of potential faults is not specified (no L_F and K_F matrices structuring the fault distribution). Since the observer uncertainties have been assumed to be normal, J_r follows a χ^2 statistical distribution with $m + q - n_X$ degrees of freedom and a non-central parameter only depending on F_X and F_Y, which are zero when there is no fault in the process. J_r is tested against a given level of false alarms.

When $H1$ is the conclusion of the fault detection step, the fault isolation step becomes [118]

Deciding between the two following hypotheses:

$$H0: \; \mathscr{F}_a = 0, \mathscr{F}_b \neq 0,$$
$$H1: \; \mathscr{F}_a \neq 0, \mathscr{F}_b = 0,$$

where $\mathscr{F}_a$, $\mathscr{F}_b$ are two exhaustive subsets of the $\mathscr{F}$ set of faults ($F_X \bigcup F_Y$).

By repetitive application of this diagnosis test to various fault subsets, one can isolate the most probable active faults. The statistical isolation tests are performed on residuals derived from the redundant equations (Equation 2.45). The generalized likelihood ratio test [118] is applied to the residuals of the assumed fault subsets, and the subset exhibiting the maximum score is assumed to be responsible for the detected faults [70, 117].

Example. Figure 2.23 gives an illustration of the method of fault isolation for the plant of Figure 2.7. A bias that has an amplitude of three times the measurement error standard deviation has been simulated in the sensor of stream 4 (in the data reconciliation example of Section 2.7.2, the stream 3 metal flowrate was assumed to be unmeasured; here all the states are measured, thus increasing the redundancy degree). The set of potential faults consists of the biases $F_Y(1)$ to $F_Y(8)$ for the eight streams and the node imbalances faults $F_X(1)$ to $F_X(4)$ for the four nodes. The diagnosis test is performed by sequentially splitting the fault set into one particular fault and the remaining potential faults. Knowing the statistical properties of e and ε as well as the fault amplitude, it is possible to calculate the probability of isolation of the fault. This is what has been done in Figure 2.23, assuming that a 5% level of false alarms is tolerated. It shows the probability of fault detection (D), while the other probability bars represent the probability of diagnosing the twelve potential faults as being the actual fault. It is clear that the assumption that the fault is in sensor 4 has the largest probability. However, wrong isolation decisions are also possible. The isolation test performance would obviously increase if the bias were of larger amplitude, or the data acquisition process repeated in a larger time window.

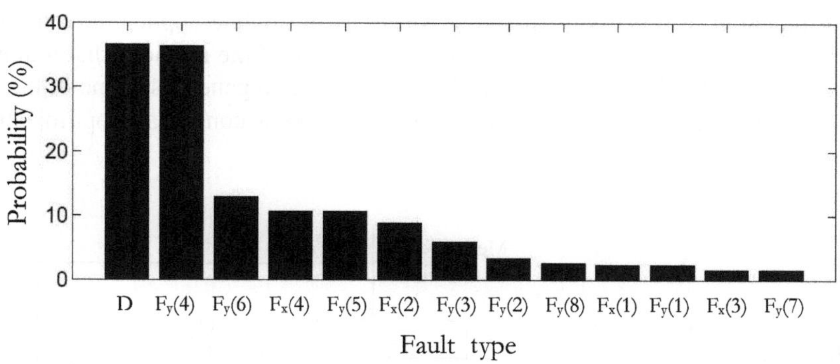

Figure 2.23 Probabilities of detection (D) and isolation of a bias in sensor 4, and of other potential faults F_X and F_Y

Figure 2.24 shows the same bar diagram for a simulated leakage at node number 2 of Figure 2.7. The amplitude of the leakage is three times the standard deviation of the node imbalance. Again the right diagnosis is the most probable, but wrong fault isolations are possible.

The same method can be applied to dynamic data reconciliation [116], [119, 120]) increasing the size of the vector Y by adding past state values, or by pre-filtering the measurements on a given past time horizon such that the node input and output measurements are synchronized [50].

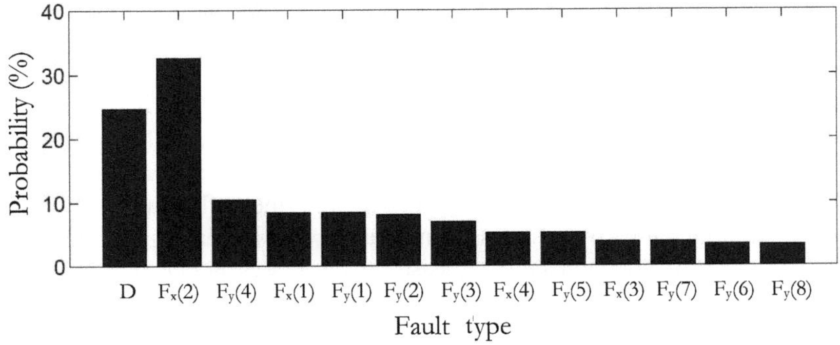

Figure 2.24 Probability of detection and isolation of a leakage at node 2

2.13 Coupling Data Reconciliation with Process Control and Optimization

As discussed in Chapter 2 introduction, data reconciliation has off-line applications such as process audit, metallurgical inventories, process modeling, plant design or reconfiguration, and tuning of operating conditions. Applications of on-line data reconciliation to automatic control and real-time optimization have also an interesting potential in process engineering, since, as in off-line applications, the use of better information would improve the performance of the methods that requires experimental data [47, 48, 121, 122]. Figure 2.25 gives a general scheme of the integration of a data reconciliation procedure into process control and optimization loops.

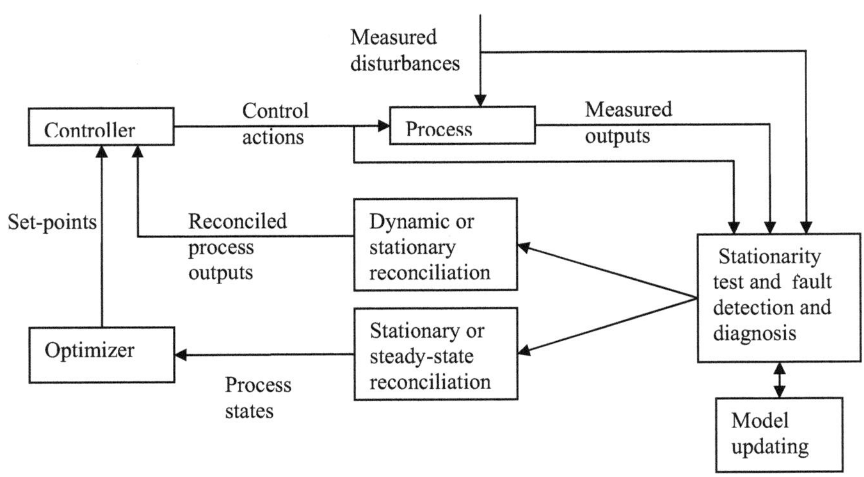

Figure 2.25 General scheme for the integration of data reconciliation techniques into automatic control and real-time optimization loops

Data reconciliation techniques are involved at two different levels when automatic control and real-time optimization are simultaneously implemented. The two observers must be different since the dynamics of the two loops are quite different, optimization being performed with slower dynamics than automatic control. This implies that the variances of the measurement errors and of the accumulation rates (node imbalances) must be tuned differently. Also, the models as well as the state variable of the two observers may be slightly different. Even if the same stationary reconciliation structure is used for both observers, the variances in the reconciliation criterion must be different. As the measured values in the optimization observer are obtained through an averaging technique involving a moving time window with several samples, their variances are usually divided by the number of samples in the window. Also, the accumulation rate variances are necessarily lower because the averaging process in the time window decreases the magnitude of the node imbalance variations, by attenuating the signal dynamics. Ultimately, the optimization observer could be steady-state.

In addition to the two observers, Figure 2.25 shows peripheral tools for data preprocessing. Sensor failures or abnormal process behavior must be detected before feeding the reconciled values to the optimizer or the controller. As the optimization observer is assumed working in stationary regime, it is important to test that the process variable means are statistically constant before reconciliation. Also when persistent mean changes are detected, it might be helpful to adapt model parameters, when, for instance, permanent changes occur to operating conditions such as ore grindability and grade, tonnage, chemical reagent type. If an adaptation procedure is integrated into the loops, it should be activated only when permanent changes due to persistent disturbance means or set-point changes are detected.

Flotation plant example: to illustrate the concept of control– optimization–reconciliation coupling, a simple example for a flotation plant is depicted in Figure 2.26. Data reconciliation is performed only at the optimization level and the control loop limited to a single-input-single-output system, where the collector addition is the manipulated variable and the concentrate grade the controlled variable. The grade set-point is supervised through the maximization of an economic index. Although there is no documented study of the performance of such a real-time optimization strategy, the concept has certainly a potential that should be investigated.

2.14 Conclusion

The objective of this chapter was to point out a problem that faces most metallurgical engineers and mineral processors who are involved with metal production and willing to understand and optimize the processes they are dealing with: the available measurements are uncertain, incomplete, and inconsistent with process behavior prior knowledge. The emphasis is put here on data reconciliation with mass conservation constraints, however this topic belongs to the universal problem of matching raw data and prior theoretical knowledge. The subject is superficially

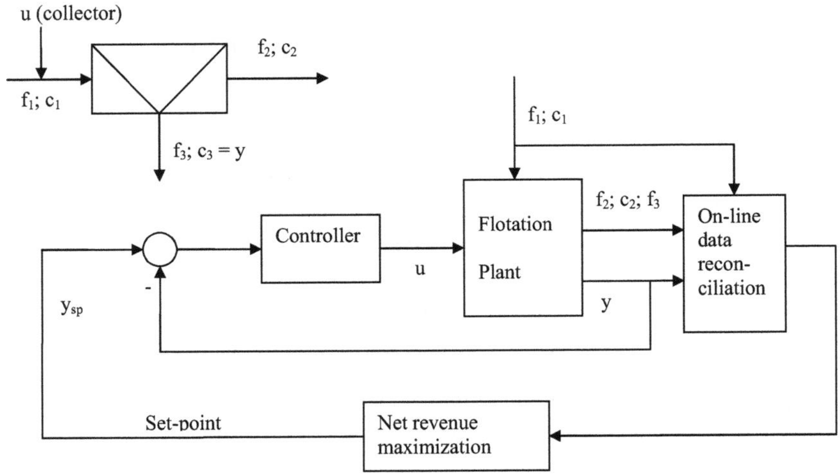

Figure 2.26 Real-time optimization of a flotation plant (f_j = ore flowrate; c_i = ore grade)

covered but the chapter explores techniques that could be helpful to practitioners as well as to researchers willing to deepen the concepts. The following issues arise from this limited presentation:

1. Measurement error sources are many and their variances are additives. There is a complex intricacy between the true measurement errors and the true dynamic variations of the process variables. It is most important to clarify these concepts at the beginning of a reconciliation procedure, to properly define the pursued reconciliation objectives.
2. The reconciliation objectives vary from the estimation of the underlying process steady-state operating regime to the fast tracking of its real instantaneous dynamic state. The method to be used for reconciliation must be adapted to the subsequent utilization of the reconciled data, which are mainly monitoring, modeling, control, and optimization;.
3. Assumptions that process variables variations and measurement errors are Gaussian and unbiased are frequently made. This is obviously not exactly true. This is why it is so important to make a prior detection of abnormal data or process behavior and a posterior analysis of the reconciliation residuals.
4. The reconciliation criterion may contain different types of residuals and weighting factors. Its formulation is essential to adequately match the reconciliation objectives, as well as the structure of data statistical properties. It must be rigorously designed, avoiding as much as possible users subjectivity or empirical tunings.
5. The reconciliation feasibility must be carefully investigated, by looking at redundancy of data and constraint information content.

6. As the success of data reconciliation methods strongly depends on the database structure, it can be used to design measurement strategies that promote estimate accuracy and reliability.

Nomenclature

The notation that is specific to illustrative examples in the text is not defined below.

Roman letters

A, B, A', B': matrices of coefficients in a state space stochastic model
b: measurement bias
$B(.)$: matrix function of the separation coefficients for steady-state modeling of a separation plant
c: composition factor
c_i and $\hat{c}_i$: concentration of component i and reconciled value
C: measurement coefficient matrix in the linear case
C_f: measurement coefficient matrix for the total mass flowrates
C_i: measurement coefficient matrix for the component i concentrations
C_i^d: diagonal matrix constructed with the vector c_i
d: sieve opening retaining 25% of the particles
d_{lib}: particle liberation size
D_i, E_i, F_i: matrices of coefficients associated to component i
e_I: integration error
e_r: measurement error of the redundant measured variables
e_f: measurement error of the total mass flowrates
$E(.)$: mathematical expectation
$f(X)$: constraints of mass or energy conservation
f: vector of all the component flowrates (or shape factor in Section 2.4)
f_i: vector of component i flowrates in the plant network
$f_i(.)$: i^{th} reconciliation constraint
$\underline{f_0}$: total mass flowrate vector
$\underline{f_0}$: f_0 mean value
f: f mean value
F_0^d: diagonal matrix of the vector f_0
F_X, F_Y : Faults in the conservation and measurements equations
$\mathscr{F}, \mathscr{F}_a, \mathscr{F}_b$: Union of the faults sets F_X and F_Y and two complementary subsets of it.
g: size distribution factor
$g(X)$: expression defining the measured variables
$G(z^{-1})$: discrete transfer function
h: time window width
$h(.)$: function of the independent variables
h_i: component i mass fraction in node hold-ups
i: index of the steam components i
in: innovations of the reconciliation procedure
I: identity matrix
$J(X)$: reconciliation criterion
J_m: part of the reconciliation criterion that contains measurements

J_ε: part of the reconciliation criterion that contains node imbalances
J_ζ: part of the reconciliation criterion that contains driving white noise
K_F: projection matrix of the faults in the measurements equations
k: node index
L: matrix for expressing state variables for independent state variables
L_F: projection matrix of the faults in the conservation equations
$\mathscr{L}$: Lagrangian function
m: number of measured process variables
m_i: vector of component i hold-up in the various nodes of the plant network
m_0: mass hold-ups of total material
M_{dep} and M_{ind}: matrices associated to dependent and independent state variables
M_i: network incidence matrix for component i
M: incidence matrix of the plant network for all the stream components
M_m and M_{um}: incidence matrices for measured and unmeasured state variables
M_0: incidence matrix for total mass
M_s: sample mass
n: number of components in plant streams (total mass not included)
n_X: number of state variables
n_n : number of plant network nodes
p: number of streams in the plant network
P_i: production rate of the component i
q: number of conservation equations
Q, Q', Q'': functions or coefficient matrices in the deductible part of the constraint equations
r: residuals of the redundancy equations (vector in parity space)
R or $R(Y)$: coefficient matrix or function in the redundancy equations
s: vector of node separation coefficient of a mineral separation plant
s_{ki}: component i separation coefficient at node k
$S(e)$: contributions of the measurement errors to the parity vector
T or $T(\varepsilon)$: contributions of the constraint uncertainties to the parity vector
$Var(.)$: statistical estimate of the variance of a process variable
$V_{\hat{X}}$: variance of the reconciled states
$V_x(k)$, V_x: autocovariance of any process variable x and variance (*i.e.*, $V_x(0)$)
V: variance of the measurement errors e
$V_\varepsilon, V_{\varepsilon i}$: variances of ε and ε_i
V_f: variance of f
V_i: variance of c_i
V_{in}: variance of the innovations
V_r: variance of Y_r
$V_{\hat{Y}}$: variance of the reconciled measured variables
V_δ, V_ω: variance of white noise $\delta(t)$ or $\omega(t)$
X: state variable
X_m: measured state variables
X_{mr}, $\hat{X}_{mr}$: redundant measured state variables and reconciled value
X_o, X_{or}, X_{onr}: observable state variables (redundant and non-redundant)

$\hat{X}_o$: reconciled value of X_o
X_{no}: non-observable state variables
X_{um}: unmeasured state variables
x_i: state variable associated to component i
x: any generic process variable (scalar or vector)
x_s: sample mineral content
x^*: true value of the mineral content of a sampled batch
$\hat{X}$: reconciled state variable
X_{dep}: dependent state variables
X_{ind}: independent state variables
$\hat{X}_{ind}$: reconciled independent state variables
$\hat{X}_{dep}$: reconciled dependent state variables
Y: measurement values
Y_f: measurement values of f
Y_i: measurement values of component i mass fractions
Y_{nr}: measured values of the non-redundant measured variables
Y_r: measured values of the redundant measured variables
z: state space variables of a stochastic system
z^{-1}: backshift operator
Z: measured variables
Z_{nr}: non-redundant measured variables
Z_r: redundant measured variables

Greek letters
ε: overall conservation constraint uncertainties
ε_0: uncertainty of the total mass conservation equation
ε_i: uncertainties in the conservation constraints of component i in the plant network
γ_i: auxiliary matrix in the node imbalance solution
$\delta(t)$: uncertainties in the process dynamic model
λ: Lagrange's multipliers
$\xi_i(t)$, $\xi(t)$, $\xi'(t)$: white noise
$\xi_d(t)$: white noise driving uncertainties in the process dynamic model
μ_x: mean value of x
σ_e: standard deviation of the measurement error e
σ_F: standard deviation of the fundamental sampling error
$\sigma_{\hat{X}}$: standard deviation of the reconciled states
Γ: auxiliary matrix in the linear reconciliation case
ρ_v: degree of variance reduction
$\rho(k)$: autocorrelation matrix
ρ_{gan}: gangue density
ρ_{min}: mineral density
ρ_r: degree of redundancy
Ψ and Φ: coefficient matrices in the linear case reconciliation information system
ψ: matrix that extracts the node output streams from the vector f_i
$\varphi(.)$: function defining the node hold-ups

Π: auxiliary matrix in the expression of $\hat{X}$
τ: time lag for autocovariance definition
$\omega(t)$: white noise
Ω: matrix for the extraction of the feed stream flowrates from the plant stream flowrates
$\zeta(t)$: generic name for driving white noise

Acknowledgements This chapter is based on studies which have been undertaken by students and on ideas discussed with colleagues. The list is long but their names deserve to be given: Nicole Alliot, Gilles Barbery, Claude Bazin, Steve Bellec, Antoine Berton, Yves Bérubé, Charles-Eudore Boudreault, Cameron Crowe, Luiz Rogerio Pinho de Andrade Lima, André Desbiens, Yan Guang Du, Marc-Denis Everell, Frédéric Flament, Thiérry Gelpe, Luc Lachance, Chefi Ketata, Daniel Laguitton, Sami Makni, El Hassan Mazzour, Jacques McMullen, Azar Mirabedini, Éric Poulin, Éric Plamondon, Simon Rochon-Tremblay, Jules Thibault, and Solange Vaz Coehlo.

References

[1] Hodouin D, Jämsä-Jounela SL, Carvalho MT, Bergh L (2001) State of the art and challenges in mineral processing control. Control Engineering Practice 9(9):995–1005
[2] Wiegel RL (1972) Advances in mineral processing material balances. Canadian Metallurical Quarterly 11(2):413–424
[3] Smith HW, Ichiyen N (1973) Computer adjustment of metallurgical balances. CIM Bulletin 66:97–100
[4] Hodouin D, Everell MD (1980) A hierarchical procedure for adjustment and material balancing of mineral process data. International Journal of Mineral Processing 7(2):91–116
[5] Kuehn DR, Davidson H (1961) Computer control II: Mathematics of control. Chemical Engineering Progress 57(6):44–47
[6] Crowe CM (1996) Data reconciliation-progress and challenges, Journal of Process Control 6:89–98
[7] Narasimhan S, Jordache C (1999), Data reconciliation & gross error detection: an intelligent use of process data. Gulf Pub. Co., Houston.
[8] Romagnoli JA, Sanchez MC (2000) Data processing and reconciliation for chemical process operations. Academic Press
[9] Algosys Inc. (2004) Bilmat Technical Guide Version 9.0: An advance computer program for data reconciliation through mass balancing, Quebec City, Canada
[10] Caspeo (2008) www.caspeo.net
[11] JKTech: www.jktech.com
[12] Banisi: www.banisi.ir
[13] OSIsoft: www.osisoft.com
[14] IPS: www.ips.invensys.com
[15] AspenTech: www.aspentech.com
[16] Belsim: www.belsim.com
[17] Du YG, Hodouin D, Thibault J (1997a) Use of a novel autoassociative neural network for nonlinear steady-state data reconciliation. AIChE Journal 43(7):1785–1796
[18] Du YG, Thibault J, Hodouin D (1997b) Data reconciliation for simulated flotation process. Artificial Intelligence in Engineering 11(4):357–364
[19] Aldrich C, Van Deventer J (1994) Identification of gross errors in material balance measurements by means of neural nets. Chemical Engineering Science 49(9):1357–1368

[20] Meert K (1998) A real-time recurrent learning network structure for data reconciliation. Artificial Intelligence in Engineering 12(3):213–218
[21] Mandel D, Abdollahzadeh A, Maquin D, Ragot J (1998) Data reconciliation by inequality balance equilibration: a LMI approach. International Journal of Mineral Processing 53(3):157–169
[22] Hodouin D, Kasongo T, Kouamé É, Everell MD (1981) BILMAT: an algorithm for material balancing mineral processing circuits: applications to comminution, desliming and flotation units. CIM Bulletin 74:123–131
[23] Bazin C, Hodouin D (1996) Processing assays of size fractions from sieve and cyclosizer analyses. Minerals Engineering 9(7):753–763
[24] Hodouin D, Makni S (1998) Data reconciliation for stationary multi-phase multi-stream metallurgical processes. Proceedings of the IFAC Automation in Mining, Mineral and Metal Processing, Köln, Deutschland, 49–54
[25] Laplante AR (1984) Plant sampling and mass balancing for gold ores. Proceedings of the 1st Int. Symp. on Precious Metals Recovery, 1–25
[26] De Andrade LRP (2006), Nonlinear data reconciliation in gold processing plant. Minerals Engineering 19:938–951
[27] Cimon D, Barbery G, Flament F, Hodouin D (1987) Materials balance in gold processing plants. Proceedings Gold Metal. Int. Symp., 9–21
[28] Bellec S, Hodouin D Bazin C, Duchesne C (2007) Multi-level data reconciliation - application to a gold ore processing plant. Proceedings of IFAC MMM07 Symposium, 39–44
[29] Bazin C, El-Ouassiti K, Hodouin D, Zouadi M (2005a) Data reconciliation in hydrometallurgy: applications to leaching of clay and copper solvent extraction. Proceedings Conference of Metallurgists, Calgary, ed. METSOC CIM
[30] Bazin C, Hodouin D, Zouadi M (2005b) Data reconciliation and equilibrium constant estimation: Application to copper solvent extraction. Hydrometallurgy 80(1):43–53
[31] Eksteen JJ, Frank SJ, Reuter MA (2002) Dynamic structures in variance based data reconciliation adjustments for a chromite smelting furnace. Minerals Engineering 15:931–943
[32] Bazin C, Hodouin D, Duchesne C, Thibault J, Trusiak AR (1998) Reconciliation of mass and energy data measurements: application to a rotary dryer. Canadian Metallurgical Quarterly 37(3-4):333–342
[33] Bazin C, Rochon-Tremblay S, Gosselin C (2003) Estimation of gas flow rates and pellets temperature in an iron oxide induration furnace. Canadian Metallurgical Quarterly 42:301–312
[34] Hodouin D, Gelpe T, Everell MD (1982) Sensitivity analysis of material balance calculations - an application to a cement clinker grinding circuit. Powder Technology 3:139–153
[35] Maquin D, Adrot O, Ragot J (2000) Data reconciliation with uncertain models. ISA Transactions 39:35–45
[36] Almasy GA (1990) Principles of dynamic balancing. AIChE Journal 36:1321–1330
[37] Darouach M, Zasadzinski M (1991) Data reconciliation in generalized linear dynamic systems. AIChE Journal 37(2):193–201
[38] Liebman MJ, Edgar TF, Lasdon LS (1992) Efficient data reconciliation and estimation for dynamic processes using nonlinear programming techniques. Computers and Chemical Engineering 16:963–986
[39] Bai S, Thibault J, McLean DD (2006) Dynamic data reconciliation: alternative to Kalman filter. Journal of Process Control 16:485–498
[40] Hodouin D, Lachance L, Desbiens A (2007) Dynamic Data Reconciliation: From the full model observer to the stationary on-line mass conservation filter - the flotation case. Proceedings of IFAC Symposium on Automation in Mining, Mineral and Metal Processing, Québec, Canada, 87–92
[41] Darouach M, Ragot J, Zasadzinski M, Krzakala G (1989) Maximum likelihood estimator of measurement error variances in data reconciliation. IFAC AIPAC Symposium, Nancy, France, 109–112.

[42] Hodouin D, Bazin C, Makni S (1993) On-line reconciliation of mineral processing data. Proceedings of the AIME/SME Symposium Emerging Computer Techniques for the Mineral Industry. Reno, Nevada, 101–110
[43] Lachance L, Poulin É, Hodouin D, Desbiens A (2007) Tuning stationary observers: Application to a flotation unit simulator. Proc. of IFAC MMM Automation, Qubec City, Canada.
[44] Hodouin D, Flament, Bazin C (1989) Reliability of material balance calculations - a sensitivity approach. Minerals Engineering 2(2):157–170
[45] Bazin C, Franklin M (1996) Real-time material balance for flotation plants using a least-squares recursive algorithm. International Journal of Mineral Processing 46(3-4):231–244
[46] Hodouin D, Bazin C, Makni S (1997) Dynamic material-balance algorithms: application to industrial flotation circuits. Minerals and Metallurgical Processing 14(2):21–28
[47] Abu-el-zeet ZH, Roberts PD, Becerra VM (2002). Enhancing model predictive control using dynamic data reconciliation. AIChE Journal 48:324–332
[48] Ramamurthi Y, Sistu PB, Bequette BW (1993) Control-relevant dynamic data reconciliation and parameter estimation. Computers and Chemical Engineering 17(1):41–59
[49] Tong H, Crowe CM (1997) Detecting persistent gross errors by sequential analysis of principal components. AIChE Journal 43(5):1242–1249
[50] Berton A, Hodouin D (2007) Synchronized node imbalances for fault detection and isolation in plant networks involving material recirculation. Computers and Chemical Engineering 31:815–832
[51] Bagajewicz M (2001) Process plant instrumentation design and upgrade. Technomic Publishing Company
[52] Mazzour EH, Hodouin D, Makni S (2002) Optimal sensor implementation in metallurgical plants - an application to a generic mineral separation plant. International Journal of Mineral Processing 69:185–203
[53] Hodouin D, Vaz Coehlo S (1987) Mass balance calculations around mineral processing units using composition analyses within particle-size classes. International Journal of Mineral Processing 21:65–82
[54] Bellec S, Jiang T, Kerr B, Diamond M, Stuart P (2007) On-line processing and steady-state data reconciliation of pulp and paper mill process data. Pulp & Paper Canada 108:36–40
[55] Bazin C, Hodouin D (2001) Importance of covariance in mass balancing of particle size distribution data. Minerals Engineering 14(8):851–860
[56] Gy P (1979) Sampling of particulate materials. Elsevier, Amsterdam.
[57] Pitard (1992) Pierre Gy's sampling theory and sampling practice. CRC Press, Baton Rouge, LA, Vols 1 and 2
[58] Bazin C, Hodouin D, Grant R (1996) Semi-batch flotation tests: reliability of performance indices. Canadian Metallurgical Quarterly 35(4):321–327
[59] Hodouin D, Garon M, Rémillard M, Thérien M (1988) Assessment of precious metal distribution in Lac Mattagami flotation plant by computer mass balance calculation. CIM Bulletin 81:62–69
[60] Alhaj-Dibo M, Maquin D, Ragot J (2008) Data reconciliation: a robust approach using a contaminated distribution. Control Engineering Practice 16(2):159–170
[61] Karjala TW, Himmelblau DM (1994) Dynamic data rectification by recurrent neural networks vs. traditional methods. AIChE Journal 40(11):1865–1875
[62] Kretsovalis A, Mah RSH (1988) Observability and redundancy classification in generalized process networks. Computer and Chemical Engineering 16:689-703
[63] Ragot J, Luong M, Maquin D (1996) Observability of systems involving flow circulation. International Journal of Mineral Processing 47:125–140
[64] Hodouin D, Mazzour EH (2006) Robustness of metallurgical process observers based on mass and energy conservation constraints. Proceedings of a Workshop on Automation in Mining, Mineral and Metal Processing, Krakow, Poland, 167–172
[65] Vaclavek V (1974) How long to take the balance periods at balance calculations of quasi-stationary continuous processes. Chemical Engineering Science 29:2307–2313

[66] Vaclavek V, Loucka M (1976) Selection of measurements necessary to achieve multicomponent mass balance in chemical plants. Chemical Engineering Science 31:1199–1205
[67] Mazzour EH, Hodouin D, Makni S (2003) Une aide algorithmique á l'optimisation du placement des capteurs dans un procéd. Journal Européén des systemes automatisés 37:1251–1276
[68] Crowe CM, Garcia Campos GY, Hrymak A (1983) Reconciliation of process flow rates by matrix projection. AIChE Journal 29: 881–888
[69] Kelly JD (1998) On finding the matrix projection in the data reconciliation solution. Computers and Chemical Engineering 22(11):1553-1557
[70] Berton A, Hodouin D (2002) Linear and bilinear fault detection and diagnosis based on mass and energy balance equations. Control and Engineering Practice 11:103–113
[71] Sanchez M, Romagnoli JA (1996) Use of orthogonal transformations in data classification-reconciliation. Computers and Chemical Engineering 20(5):483–493
[72] Hodouin D, Alliot N, Flament F (1991) Redundancy analysis of complex sets of mineral processing data for mass balance computation. International Journal of Mineral Processing 32(3-4):213–231
[73] Almasy GA, Mah RSH (1984) Estimation of measurement error variances from process data. Industrial Engineering Chemistry Process Research and Development. 23:779–784
[74] Keller JY, Zasadzinski M, Darouach M (1992). Analytical estimator of measurement error variances in data reconciliation. Computers and Chemical Engineering 16:185–188
[75] Chen J, Bandoni A, Romagnoli JA (1997) Robust estimation of measurement error variance/covariance from process sampling data. Computers and Chemical Engineering 21:593–600
[76] Lachance L, Poulin É, Hodouin D, Desbiens A (2006b) Performance of steady-state and stationary data reconciliation as a function of process disturbances dynamics. Mineral Process Modeling, Simulation and Control Conference, Laurentian University, Sudbury, Canada, 393–408
[77] Poulin É, Hodouin D, Lachance L (2009) Impact of plant dynamics on the performance of steady-state data reconciliation. submitted to Computers and Chemical Engineering
[78] Poulin É, Hodouin D, Lachance L (2009) Estimation of measurement error variances in data reconciliation using a flow distribution model. Proc. IFAC SYSID'2009, St Malo, France
[79] Hodouin D, Ketata C (1993) Variance of average stream compositions obtained by automatic incremental sampling. International Journal of Mineral Processing 40(3-4):199–223
[80] Mirabedini A, Hodouin D (1998) Calculation of variance and covariance of sampling errors in complex mineral processing systems, using state-space dynamic models. International Journal of Mineral Processing 55(1):1–20
[81] Maquin D, Ragot J, Darouach M, Fayolle J (1988) Validation des mesures par q́uilibrage hiérarchis é de bilans-matière. International Journal of Mineral Processing 23:241–252
[82] Schraa OJ, Crowe CM (1996) The numerical solution of bilinear problems using unconstrained optimization methods. Event of the European Federation of Chemical Engineers 20:S727–S730.
[83] Stanley GM, Mah RSH (1977) Estimation of flows and temperatures in process networks. AIChE Journal 23(5):642–650
[84] Kelly JD (2004) Techniques for solving industrial nonlinear data reconciliation problems. Computers and Chemical Engineering 28:2837–2843
[85] Arora N, Biegler LT, Heyen G (2002) Data reconciliation framework. Software architecture and tools for computer aided process engineering. Computer Aided Chemical Engineering 11:193–212
[86] Schladt M, Hu B (2007) Soft sensors based on nonlinear steady-state data reconciliation in the process industry. Chemical Engineering and Processing 46(11):1107–1115
[87] Bazin C (1999) Personal communication.
[88] Flament F, Hodouin D, Bazin C (1986) Propagation of measurement errors in mass balance calculation of mineral processing data. Proceedings of APCOM'86, Pennsylvania State University Editor, SME-AIME.

[89] Hodouin D, Flament F (1991) Influence of data collection and conditioning strategies on the significancy of performance indices in mineral processing plants. Proceedings Int. Symp. Evaluation and Optimization of Metallurgical Performance, SME/AIME, 195–208
[90] Hodouin D, Bazin C, Trusiak A (1984) Reliability of calculation of mineral process efficiencies and rate parameters from balanced data. Proceedings of Control 84, JA Herbst Editor, SME-AIME, Chapter 16, 133–144
[91] Reimers C, Werther J, Gruhn G (2008) Flowsheet simulation of solids processes. data reconciliation and adjustment of model Parameters. Chemical Engineering and Processing 47:138–158
[92] Makni S, Hodouin D (1994) Recursive BILMAT algorithm: an on-line extension of data reconciliation techniques for steady-state bilinear material balance. Mineral Engineering 7(9):1179–1191
[93] Hodouin D, Makni S (1996) Real-time reconciliation of mineral processing plant data using bilinear material balance equations coupled to empirical dynamic models. International Journal of Mineral Processing 48:245–264
[94] Hlavacek V (1977) Analysis of a complex plant steady state and transient behavior. Computers and Chemical Engineering 1:75–100
[95] Fillon M, Meyer M, Pingaud H, Enjalbert M (1996) Efficient formulation for batch reactor data reconciliation. Industrial and Engineering Chemistry Research 35(7):2288–2298
[96] Lachance L, (2007) Observation de procé ds basée sur des sous-modèles - Applications au traitement et au transport de la matière. PhD. thesis, Université Laval, Québec City, Canada
[97] Makni S, Hodouin D, Bazin C (1995) On-line data reconciliation by minimization of a weighted sum of squared residuals and node imbalances. Proceedings of the XIX International Minereral Processing Conference, San Francisco, USA
[98] Dochain D (2003) State and parameter estimation in chemical and biochemical processes: a tutorial. Journal of Process Control 13: 801–818
[99] Makni S (1990) Le filtrage Kalman des ateliers de flottation. Master thesis, Université Laval, Quebec, Canada
[100] Lachance L, Hodouin D , Desbiens A (2006a) Using sub-models for dynamic data reconciliation. Proceedings of IFAC ADCHEM Int. Symp., Gramado, Brazil, 711–716
[101] Lachance L, Hodouin D, Desbiens A, Poulin É (2008) Data reconciliation through mass balancing for flowsheets involving plug-flow units. International Mineral Processing Congress, Beijing, China
[102] Bazin C, Trusiak AR, Hodouin D (1995) Application of a dynamic material balance program to an industrial flotation circuit. CIM Bullentin. 88:47–54
[103] Hodouin D, Mirabedini A, Makni S, Bazin C (1998) Reconciliation of mineral processing data containing correlated measurement errors. International Journal of Mineral Processing 32:201–215
[104] Makni S, Hodouin D, Bazin C (1995b) A recursive node imbalance method incorporating a model of flowrate dynamics for on-line material balance of complex flowsheets. Minerals Engineering 8(7):753–766
[105] Ali Y, Narasimhan S (1995) Redundant sensor network design for linear processes. AIChE Journal 41(10):2237–2249
[106] Bagajewicz M, Sanchez M (2000a) Reallocation and upgrade of instrumentation in process plants. Computers and Chemical Engineering 24(8):1945–1959
[107] Bagajewicz M, Sanchez M (2000b) Cost-optimal design of reliable sensor networks. Computers and Chemical Engineering 23(11-12):1757–1762
[108] Hodouin D, Mazzour EH (2004) Impact of sensor implementation strategies on the efficiency of data reconciliation. Proceedings of the IFAC Symposium on Mineral, Metal and Materials Processing Automation, Nancy, France
[109] Hodouin D, Berton A, Mazzour EH (2004) Sensor placement, fault detection and data reconciliation using heat and mass conservation constraints. Proceedings of Control 2004, PAPTAC, Québec City, 303–306

[110] Hodouin D, Mazzour EH (2008) Measurement accuracy selection for designing observers of metallurgical plant performances. Proceedings of MCA-IEEE MED'08 Symposium, Ajaccio, France
[111] Özyurt DB, Pike RW (2004) Theory and practice of simultaneous data reconciliation and gross error detection for chemical processes. Computers and Chemical Engineering 28:381–402
[112] Romagnoli JA, Stephanopoulos G (1981) Rectification of process measurement data in the presence of gross errors. Chemical Engineering Science 36:1849-1863
[113] Tamhane AC, Mah RSH (1985) Data reconciliation and gross error detection in chemical process networks. Technometrics 27:409–422
[114] Ragot J, Aitouche A, Kratz F, Maquin D (1991) Detection and location of gross errors in instruments using parity space technique. International Journal of Mineral Processing 31:281–299
[115] Bagajewicz M, Jiang Q (1998) Gross error modeling and detection in plant linear dynamic reconciliation. Computers and Chemical Engineering 22(12):1789–1809
[116] Chen J, Romagnoli JA (1998) A strategy for simultaneous dynamic data reconciliation and outlier detection, Computers and Chemical Engineering 22(4-5):559–562
[117] Berton A, Hodouin D, (2004) A FDI procedure based on linearized bilinear mass and heat conservation models. Proceedings of the IFAC Symposium on Mineral, Metal and Materials Processing Automation, Nancy, France
[118] Basseville M (1997) Information criteria for residual generation and fault detection and isolation. Automatica 33:783–803
[119] Albuquerque JS, Biegler LT (1996) Data reconciliation and gross-error detection for dynamic systems. AIChE Journal 42(10):2841–2856
[120] Bagajewicz M, Jiang Q (2000) Comparison of Steady State and Integral Dynamic Data Reconciliation. Computers and Chemical Engineering 24:2367–2383
[121] Bai S, McLean DD, Thibault J (2005) Enhancing controller performance via dynamic data reconciliation. Canadian Journal of Chemical Engineering 83:515–526
[122] Bai S, McLean DD, Thibault J (2007) Impact of model structure on the performance of dynamic data reconciliation. Computers and Chemical Engineering 31:127–135

Chapter 3
Multivariate Image Analysis in Mineral Processing

Carl Duchesne

Abstract In several process industries including mineral processing, where the materials are solids or slurries, some important measurements cannot be obtained using standard instrumentation (*e.g.*, flow, temperature, pressure, pH, power draw, *etc.*), but can be visually appraised, and could be automatically quantified using machine vision techniques. In general, the information to extract from process images is not well defined and is stochastic in nature compared with those typically encountered in the manufacturing industry for automatic inspection, which have well defined deterministic features. Multivariate image analysis (MIA) as well as Multiresolution analysis (MRA) have been shown to be very efficient methods for spectral/textural analysis of process images. The objective of this chapter is to illustrate these methods using three mineral processing problems: (1) on-line estimation of froth grade in a flotation process; (2) flotation froth health monitoring for appropriate reagent dosage; and (3) on-line estimation of run-of-mine ore lithotype composition on conveyor belts. In all cases, the extracted image information could be used for developing new vision sensors for advanced control of mineral processing plants.

3.1 Introduction

Multivariate statistical methods such as principal component analysis (PCA) and projection to latent structures (PLS), also known as multivariate projection methods or latent variable methods, have been widely used in the process industry since the early 1990s for analyzing and extracting meaningful information from large, noisy and highly collinear datasets. A few good reviews covering both theoretical and practical aspects of latent variable methods for a variety of data structures, problems and applications can be found in [1–4]. In the mineral processing industry in

Carl Duchesne
Department of Chemical Engineering, Université Laval, Québec, Canada, G1V 0A6, e-mail: carl.duchesne@gch.ulaval.ca

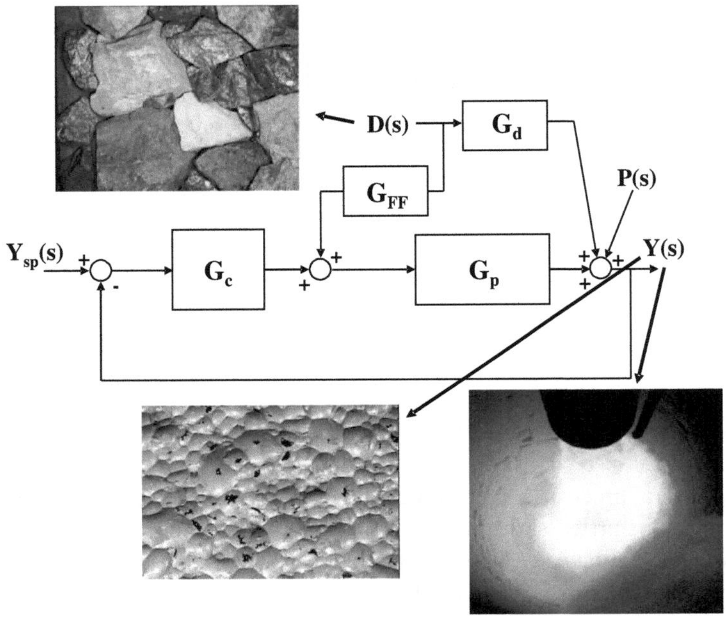

Figure 3.1 Standard feedback/feedforward control loop with controlled and disturbance variables measured using machine vision

particular, these methods have been used for analyzing spectral data for ore characterization and sorting [5] and for monitoring of crushing and grinding mills [6, 7], for post-analysis and monitoring of grinding mills and flotation plants [8–11], and for soft-sensor development [12]. In these applications the relevant process states could be extracted from the large data sets obtained from standard process sensors (*e.g.*, flow, pressure, pH, power draw, *etc.*) and/or from analytical instruments (*e.g.*, vibrations, assays, size distributions, *etc.*). However, in some situations, and particularly in mineral processing where the materials are solids or slurries, the desired process information cannot be easily obtained from standard instrumentation but can be *visually* interpreted and evaluated using machine vision techniques.

For example, consider Figure 3.1 showing a standard feedback/feedforward control loop. Variables **Y**, $\mathbf{Y}_{sp}$, **D** and **P** correspond to the controlled variable, its set-point, a measured, and an unmeasured disturbance, respectively. The **G**s are the appropriate transfer functions. Tracking changes in the composition of run-of-mine ore (*i.e.*, disturbance **D**), monitoring variations in the quality or "health" of a froth flotation process, or in the amount of heat transferred to solids by combustion in a rotary kiln (both **Y**s) are examples of difficult to measure variables using conventional instrumentation that can be visually appraised by experienced operators and engineers. Once automated using machine vision techniques, the information extracted from these images can be used in process control loops. Indeed, a very common example in mineral processing is on-line particle size analysis using commercial products, such as VisioRock™[1] (Metso Minerals Oy, Helsinki, Finland) and WipFrag™[2] (WipWare Inc., Ontario, Canada). Size measurements obtained from these imaging systems are typically used to manipulate crusher jaw opening or SAG mill throughput to meet product size specifications.

Machine vision has been successfully used for several years in the manufacturing industry for automatic inspection and assembly, but had limited success in the process industry since the intrinsic nature of the information to be extracted from the images collected in each field are fundamentally different [13]. In the former, the relevant information typically consists of inspecting the shape, position, orientation, structure, *etc.*, of an object of interest within an imaged scene (*e.g.*, a micro-chip) and to compare against a pre-defined assumption of what the desired object should look like; the desired informations are *deterministic* in nature [13]. Traditional imaging techniques involving gray level images (*i.e.*, monochrome or univariate images) and algorithms for image enhancement/restoration and pattern recognition techniques using morphological operations and segmentation algorithms are efficient methods for extracting these deterministic features from the image itself (*i.e.*, in the image space) for which specific assumptions can be made about the scene [14].

In contrast, the desired information to obtain from process images such as those shown in Figure 3.1 are often ill-defined and essentially *stochastic* [13]. Variations in froth texture and color are known to be related to separation performance (*i.e.*, grade, recovery) in flotation circuits, but it is difficult to make pre-defined assumptions on what the perfect froth should be. Rock fragments from the same lithotype often greatly differ in color and surface texture, even a single fragment shows variations across different faces. Classifying run-of-mine ore (ROM) into mineral types (*i.e.*, estimating ROM composition) cannot rely on a well defined visual appearance of each rock type. Thus, the features to be extracted from process images are more complex and require images carrying more information, that is multivariate images (*i.e.*, images containing more than one spectral channel for each pixel, such as RGB). Moreover, a combination of a few methods are often necessary for extracting the desired information for these multivariate images (*i.e.*, color, textural features, *etc.*) and for correlating them to key process variables. This chapter presents a machine vision framework for efficiently extracting color and textural features from multivariate images for monitoring and control of mineral processing plants. This framework heavily relies on latent variable methods.

This chapter is organized as follows. A brief overview of the two most frequently used latent variable methods (*i.e.*, PCA and PLS) is presented in Section 3.2. Section 3.3 discusses the nature of multivariate images followed by a description of the machine vision framework and the multivariate methods used for color and textural feature extraction, reduction, and analysis (Section 3.4). To illustrate the concepts, three mineral processing case studies are then presented in Section 3.5, two froth

---

[1] VisioRock is copyright of Metso Minerals Cisa BP 6009, 45060 Orlans Cedex 2, France, www.svedala-cisa.com

[2] WipFrag is a registered trademark of WipWare Inc., North Bay, Ontario, Canada, www.wipware.com

flotation examples in which froth grade is predicted and froth health is monitored, and one application to ROM in which it is attempted to estimate the proportions of the various mineral types on a conveyor belt. Some conclusions are drawn in Section 3.6.

3.2 Background on Latent Variable Methods

3.2.1 Principal Component Analysis

Principal component analysis is a classical multivariate data analysis approach and a tutorial with some chemical examples can be found in [15]. Figure 3.2 (a) shows a data table, **X**, that consists of *I* measurements taken on *J* different variables. PCA takes advantage of the correlation structure among the *J* variables to summarize **X** into a few principal components or latent variables. The number of principal components, *A*, is generally smaller than the number variables *J* and is often viewed as an estimate of the effective rank of **X**.

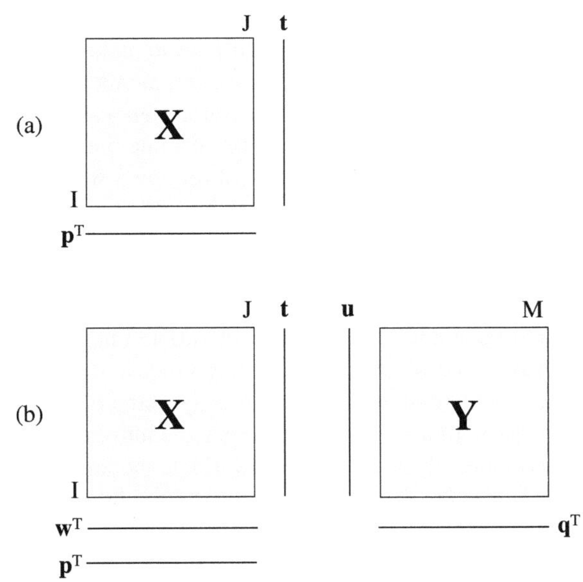

Figure 3.2 Data vectors and matrices involved in projection methods. (a) PCA. (b) PLS

From a rigorous mathematical point of view, the PCA model building procedure starts with finding the direction in **X**, or alternatively, finding a linear combination of x-variables $\mathbf{p_1}$ that explains the most variance in **X**:

$$\max_{\mathbf{p}_1} \left\{ \mathbf{p}_1^\top \mathbf{X}^\top \mathbf{X} \mathbf{p}_1 \right\} \qquad \text{subject to} \qquad \mathbf{p}_1^\top \mathbf{p}_1 = 1.0 \tag{3.1}$$

A first summary variable or score, $\mathbf{t_1}$, is obtained simply by projecting $\mathbf{X}$ in the direction of $\mathbf{p_1}$, $\mathbf{t_1} = \mathbf{X}\,\mathbf{p_1}$. This high variance direction is then removed from $\mathbf{X}$, leaving a residual matrix $\mathbf{E_1} = \mathbf{X} - \mathbf{t_1}\,\mathbf{p_1}^\top$, containing the variance of $\mathbf{X}$ that is not explained by the first component. The construction of the PCA model can continue with the computation of a second linear combination $\mathbf{p_2}$, explaining the second highest amount of variance in $\mathbf{X}$. The objective in this case is the same as shown in Equation 3.1, but replacing $\mathbf{p_1}$ by $\mathbf{p_2}$ and $\mathbf{X}$ by $\mathbf{E_1}$ and imposing the additional constraint that the second component be orthogonal to the first one (*e.g.*, $\mathbf{p}_1^\top\,\mathbf{p}_2 = 0$). This procedure is repeated until the desired number of components is computed. The final structure of the model is $\mathbf{X} = \mathbf{T}\,\mathbf{P}^\top + \mathbf{E}$, which can be seen as an eigenvector or singular value decomposition (SVD) of $\mathbf{X}^\top\mathbf{X}$. In fact, the $\mathbf{p}$ vectors are just the eigenvectors of $\mathbf{X}^\top\mathbf{X}$ and the $\mathbf{t}$ vectors are the eigenvectors of $\mathbf{X}\,\mathbf{X}^\top$. When as many components are computed as there is variables (*e.g.*, A=J), the decomposition of $\mathbf{X}$ is perfect and $\mathbf{E} = 0$.

An alternative approach for computing the $\mathbf{p}$ and $\mathbf{t}$ vectors sequentially is to use the Nonlinear iterative partial least squares (NIPALS) algorithm. The starting point of this algorithm typically (but not always required) consists of mean-centering and scaling of matrix $\mathbf{X}$ (discussed later in this section). The following steps are outlined below:

1. set $\mathbf{t}$ to be one column of $\mathbf{X}$;
2. $\mathbf{p} = \mathbf{X}^\top\,\mathbf{t}/\mathbf{t}^\top\,\mathbf{t}$;
3. $\mathbf{p} = \mathbf{p}/(\mathbf{p}^\top\,\mathbf{p})$;
4. $\mathbf{t} = \mathbf{X}\,\mathbf{p}/(\mathbf{p}^\top\,\mathbf{p})$;
5. continue iterating between 2. and 4. until convergence on $\mathbf{t}$ or $\mathbf{p}$;
6. residual matrix: $\mathbf{E} = \mathbf{X} - \mathbf{t}\,\mathbf{p}^\top$;
7. store $\mathbf{p}$ and $\mathbf{t}$ in $\mathbf{P}$ and $\mathbf{T}$, respectively;
8. calculate next dimensions by returning to 1, using $\mathbf{E}$ as the new $\mathbf{X}$.

After computing each latent variable, one needs to decide whether another dimension should be added to the PCA model. Cross-validation [16] is often a typically used criterion for selecting the number of components to keep in the model.

3.2.2 Projection to Latent Structures (PLS)

Projection to latent structures, or alternatively, partial least squares is a truly multivariate latent variable regression method. PLS is used to model relationships both within and between two blocks of data, $\mathbf{X}$ and $\mathbf{Y}$. A tutorial on PLS is found in [17] and a review of PLS history is available in [18]. Some mathematical and statistical properties of PLS were also addressed in [19, 20].

In PLS, the covariance structures of $\mathbf{X}$ and $\mathbf{Y}$ are modeled via a set of A latent variables, $\mathbf{t}$ and $\mathbf{u}$ respectively, as shown in Figure 3.2(b). However, these latent variables are computed in such a way that the covariance between the two blocks **X** and **Y** is modeled as well. Mathematically, this is achieved by selecting a set of linear combinations of **X** variables, $\mathbf{w}_i, i = 1, \ldots, A$, that maximizes the covariance between the matrix of descriptors **X** and the matrix of responses **Y**, under the following constraints:

$$\max_{\mathbf{w}_i} \left\{ \mathbf{w}_i \, \mathbf{X}^\top \, \mathbf{Y}\mathbf{Y}^\top \, \mathbf{X} \, \mathbf{w}_i^\top \right\} \quad \begin{array}{ll} \text{subject to} & \mathbf{w}_i^\top \, \mathbf{w}_i = 1 \\ \text{subject to} & \mathbf{w}_i^\top \, \mathbf{w}_j = 0 \quad \text{for} \quad i \neq j. \end{array} \tag{3.2}$$

The structure of the PLS model is shown below:

$$\begin{aligned} \mathbf{X} &= \mathbf{T} \, \mathbf{P}^\top + \mathbf{E} \\ \mathbf{Y} &= \mathbf{T} \, \mathbf{Q}^\top + \mathbf{F} \\ \mathbf{T} &= \mathbf{X} \, \mathbf{W} \, (\mathbf{P}^\top \mathbf{W})^{-1}, \end{aligned} \tag{3.3}$$

where the A columns of **P** and **Q** define the linear combinations of the **X** and **Y** variables modeling their covariance structure. **E** and **F** are just model residuals. It has been shown by [19] that the vectors **w**, **q**, **t** and **u** are eigenvectors of $\mathbf{X}^\top\mathbf{Y}\mathbf{Y}^\top\mathbf{X}$, $\mathbf{Y}^\top\mathbf{X}\mathbf{X}^\top\mathbf{Y}$, $\mathbf{X}\mathbf{X}^\top\mathbf{Y}\mathbf{Y}^\top$ and $\mathbf{Y}\mathbf{Y}^\top\mathbf{X}\mathbf{X}^\top$, respectively.

Instead of computing all the latent variables at once, a version of the NIPALS algorithm was adapted for sequentially computing the PLS latent variables, one at a time. This algorithm is outlined below, with the starting point being mean-centering and scaling both **X** and **Y** matrices:

1. set **u** to be one column of **Y**;
2. $\mathbf{w} = \mathbf{X}^\top \, \mathbf{u} / \mathbf{u}^\top \, \mathbf{u}$;
3. $\mathbf{w} = \mathbf{w} / (\mathbf{w}^\top \, \mathbf{w})$;
4. $\mathbf{t} = \mathbf{X} \, \mathbf{w} / (\mathbf{w}^\top \, \mathbf{w})$;
5. $\mathbf{q} = \mathbf{Y}^\top \, \mathbf{t} / \mathbf{t}^\top \, \mathbf{t}$;
6. $\mathbf{u} = \mathbf{Y} \, \mathbf{q} / (\mathbf{q}^\top \, \mathbf{q})$;
7. continue iterating between 2. and 6. until convergence on **t** or **u**;
8. residual matrix: $\mathbf{E} = \mathbf{X} - \mathbf{t} \, \mathbf{p}^\top, \mathbf{F} = \mathbf{Y} - \mathbf{t} \, \mathbf{q}^\top$;
9. store **w**, **p**, **t** and **u** in **W**, **P**, **T** and **U**, respectively;
10. calculate next dimensions by returning to 1, using **E** and **F**.
    as the new **X** and **Y**

3.2.3 Statistics and Diagnostic Tools Used With Latent Variable Models

This section presents some statistics and diagnostic tools used in this chapter to assess the performance of the latent variable models and for interpreting them. In particular, the scaling of the data matrices, the selection of the number of components by cross-validation, and the distance to the model diagnostic tool are discussed in turn. More details on performance assessment and LV model diagnostic tools can be found in [4, 21].

3.2.3.1 Scaling of Data Matrices

All projection methods described previously are scale dependent and therefore, appropriate scaling needs to be performed on all measurements prior to analyzing them. When no prior process knowledge is available, one common practice is to scale all variables to unit variance, as this gives them equal importance in the model, with respect to one another. However, if prior knowledge exists, then scaling should be modified accordingly. For example, if it is common knowledge that a particular set of variables is roughly twice as important as another set, the most important set could be scaled to twice the variance of the less important set of variables. In this chapter, mean-centering and scaling to unit variance was applied prior to building the latent variable models unless otherwise stated.

3.2.3.2 Number of Components

Another important issue in building empirical models with projection methods is to select the number of components to keep in the model in a meaningful way. The most widely used method for selecting the number of components in projection methods is cross-validation [16]. This method suggests to keep adding latent variables to the model as long as they significantly improve the predictions of the model (PCA or PLS). Model predictive ability can be evaluated using the predictive multiple correlation coefficient $Q^2(a) = 1 - PRESS(a)/SS_r(a-1)$. $PRESS(a)$ is the total prediction error sum of squares obtained by cross-validating a model with a latent variables. This is performed by dividing the I observations included in the database (**X** and/or **Y**, see Figure 3.2(b) into g groups of size q ($I = gq$). Then, each group is deleted one at a time and a PCA or PLS model with a latent variables is built on the remaining $g-1$ groups. The prediction error sum of squares is then computed for the group not used to build the model. $PRESS(a)$ is the total of the prediction error sum of squares for all groups. $SS_r(a-1)$ is just the residual sum of squares of a model with $a-1$ latent variables. As long as Q^2 is greater than zero, the ath dimension is improving the predictive power of the model. Therefore, one should keep adding latent variables until Q^2 is consistently lower than zero. Statistical hypothesis tests are sometimes used to verify if the ath dimension has led to a sufficient increase in Q^2 to be added to the model. Several other criteria have been developed for selecting an appropriate number of latent variables. Overviews of criteria available in the signal processing and chemometrics literature are available in [22, 23].

Another statistic used for assessing the fit of the latent variable models is the fit multiple correlation coefficient $R^2(a)$, or alternatively, the explained variance for a model with a latent variables, $R^2(a) = 1 - SS_r(a)/SS_{tot}$. In this statistic, $SS_r(a)$ is the residual sum of squares of a model (PCA or PLS) with a latent variables, while

SS_{tot} is the total sum of squares of the original data (before the model is built). As R^2 reaches values close to one, a very good fit of the data is obtained. The fit is poor when R^2 values approach zero.

3.2.3.3 Distance to the Model Statistic

The distance to the model is an essential tool in the interpretation of latent variable models since it allow one to verify how well each observation in the dataset projects onto the reduced space of the model (plane or hyperplane). This is useful to identify outliers and the presence of a different correlation structure in specific observations.

The distance of an observation (measurement on each variables) in the **X** space, $\mathbf{x}_i$, from the PCA or PLS model is given by the square prediction error of this observation, defined as $SPE_i = (\mathbf{x}_i - \hat{\mathbf{x}}_i)^\top(\mathbf{x}_i - \hat{\mathbf{x}}_i)$, where $\hat{\mathbf{x}}_i$ is the projection of $\mathbf{x}_i$ onto the reduced space of the model (plane): $\hat{\mathbf{x}}_i = \mathbf{t}_i\mathbf{P}^\top$. SPE_i is therefore a measure of the perpendicular distance of the observation $\mathbf{x}_i$ from the plane defined by the PCA model with A latent variables. Note that we have chosen the distance in **X** space, but the equations are the same for the distance in **Y** space, just by replacing $\mathbf{x}_i$ by $\mathbf{y}_i$ and $\hat{\mathbf{x}}_i = \mathbf{t}_i\mathbf{P}^\top$ by $\hat{\mathbf{y}}_i = \mathbf{t}_i\mathbf{Q}^\top$. When the distance is large, this indicates that observation $\mathbf{x}_i$ has a different correlation structure from that normally seen in the historical database used to build the PCA model (*i.e.*, not well captured by the model) and should be investigated closely.

Another very similar measure of perpendicular distance to the model is the $DMOD$ statistic [21]. This is just a normalized SPE. For the ith observation in the **X** space, the distance is computed as $DMODX_i = \sqrt{SPE_i/(J-A)}$, where J is the number of variables in **X** and A is the number of latent variables. Note that $DMODY_i$ can be computed similarly. This measure of perpendicular distance to the model is shown here since it is used in some commercial software packages (*e.g.*, SIMCA-P, Umetrics Inc.).

Statistical upper limits for the SPE and $DMOD$ statistics can also be used to establish a threshold for discriminating normal/abnormal data. These involve reference or theoretical (*i.e.*, F) distributions [21, 24].

3.3 Nature of Multivariate Digital Images

A digital multivariate image is a stack of congruent univariate images taken at various wavelengths as shown in Figure 3.3. It consists of a three-way array of data $\underline{\mathbf{X}}$ $(x \times y \times \lambda)$ having two spatial directions (*i.e.*, x and y), together defining the spatial resolution of the image (*i.e.* number of pixels), and a third direction λ, referred to as the spectral direction, corresponding to light intensities captured by the camera CCD (charge-coupled device) at different wavelengths (or spectral channels). For example, the RGB color image of froth shown in Figure 3.3 has a spatial resolution of 516×346 pixels, which yields an array of data of dimensions $516 \times 346 \times 3$.

Here, $\lambda = 3$ since the froth image was captured using a three-color RGB camera collecting light intensities in the red (R), the green (G), and the blue (B) colors.

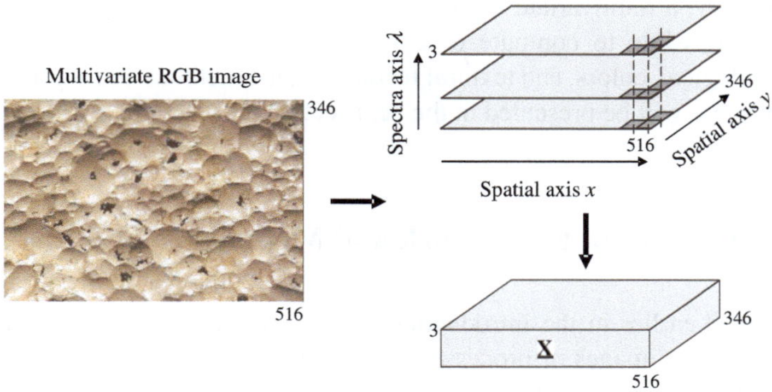

Figure 3.3 A multivariate image is a stack of congruent images collected at various wavelengths. An example of a three-channel RGB image of a froth flotation process is shown. Adapted from Bharati and MacGregor [25]

The camera CCD is responsible for this three-way array structure. It basically consists of a matrix of photo-sensitive detectors each giving rise to a pixel within the image. The number of detectors in both spatial directions define the spatial resolution of the image. Each detector is sensitive to a certain spectral range within the electromagnetic spectrum. If all detectors of the CCD have a similar spectral sensitivity, then a monochrome or gray-level image of dimensions $(x \times y \times 1)$ is obtained. When the CCD consists of a mosaic of detectors having different spectral sensitivities or when a spectroscope is mounted in front of a monochrome camera (in this case λ can be in the order of a few hundred), then a full three-way array structure $(x \times y \times \lambda)$ is generated. The images used in this chapter were all collected using 1-CCD RGB cameras, that is a single CCD made of a mosaic of three types of detectors, some sensitive to the red, the green, and the blue colors, and spatially distributed across the CCD matrix according to the the well known Bayer pattern. To fill the empty pixels within the three-way array $\underline{\mathbf{X}}$ (*e.g.*, a red-sensitive detector is less sensitive to green and blue colors) interpolation using appropriate neighboring detectors is usually performed (within the camera electronics, transparent to the user). Cameras having three different CCDs, each specialized in one spectral range (*i.e.*, each pixel has a true reading in each of the RGB channel), are also available on the market but are much more expansive.

The readings provided by each detector of the CCD is proportional to the number of photons accumulated by the detector during the user-defined exposure time, which are then converted into numbers (*i.e.*, integers) using an analog–digital converter. The resolution of this AD converter defines the number of gray levels with which light intensities will be quantified. For example, a typical 8-bit camera will result in light intensities described by integers varying from 0–255 gray levels, 0 corresponding to black and 255 to white. In Figure 3.3, each univariate image or slice of $\underline{\mathbf{X}}$ along λ is filled with such numbers. Linear combinations of these numbers for the red, green, and blue channels will give rise to any colors of the visible spectra.

In summary, a multivariate digital image is a three-way array of data $\underline{\mathbf{X}}$ from which it is possible to compute or extract various features, such as geometrical/morphological, colors, and textural features. Some methods and algorithms used for this purpose will be presented in the next section.

3.4 Machine Vision Framework and Methods

As mentioned earlier in the introduction section, the information to be extracted from multivariate images of processes and products are essentially stochastic, are generally more complex, and often require combining different methods for extracting relevant image features (*i.e.* color, textural features, *etc.*) and for correlating them to key process variables. A general machine vision framework proposed by Liu [13] and Tessier *et al.* [26] is presented in Figure 3.4. This three-step framework will be used to discuss the role of each multivariate method and how they interconnect in order to obtain valuable information from process images.

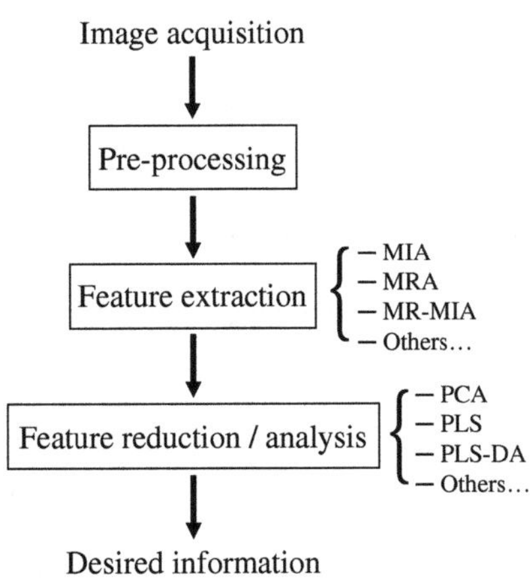

Figure 3.4 A general machine vision framework used in multivariate imaging. Typical multivariate methods used in each step are also listed. Adapted from Liu [13]

The steps to be performed following image acquisition are described below:

- **Step 1: Image pre-processing.** When necessary, image quality is enhanced to improve feature extraction. Examples include contrast and brightness adjustments, correction for non-uniform illumination, noise removal, edge enhancement and segmentation. These operations are performed using well known traditional image processing techniques discussed in [27], for example, and implemented in software such as MATLAB® (The MathWorks Inc.).

- **Step 2: Feature extraction.** This step is the core of the framework where image characteristics are computed and stored for further analysis, such as for assessing the state of the process or for predicting key process variables. The desired visual appearance information to extract from images may consist of color variations (*i.e.* spectral signatures), textural changes or both. Geometrical features are best extracted using traditional image processing techniques and are well covered elsewhere [27]. In this chapter, the MIA technique will be used for extracting color (*i.e.* spectral) signatures. Textural information will be quantified using MRA) and multiresolution MIA (MR-MIA) will be presented as a method for combining both color and textural information.

- **Step 3: Feature reduction and analysis.** Even though modern computer capabilities allow for rapid computation of a large number of features from several hundreds to thousands of multivariate images of a set, their interpretation with respect to the ultimate machine vision objective is not straightforward. Image features are often highly collinear and efficient methods for exploring their clustering patterns, for classifying and monitoring them (non-supervised or supervised methods), or for establishing relationships between these features and key process variables (*i.e.*, regression on images) are necessary. Multivariate latent variable methods, such as PCA, PLS and their extensions, are very efficient methods for making sense of such large image feature database by projecting the information down to a lower dimensional feature space where it is easier to visualize and interpret the image information and process variations.

Some of the multivariate methods for extracting color and textural image features will be presented in the remainder of this section. These methods will be later used in various combinations in a few case studies related to mineral processing (Section 3.5). This section does not provide an extensive account of all multivariate techniques that can be used in image analysis, but rather provides an introduction to the field. Additional references for further readings will be provided when appropriate.

3.4.1 Feature Extraction Methods

3.4.1.1 Multivariate Image Analysis

The MIA technique allows one to explore and quantify the spectral characteristics of an image based on unsupervised classification (*i.e.*, PCA) of each individual pixel of a multivariate image according to their spectral signature (*i.e.*, colors for a RGB image). It was originally proposed by Esbensen and Geladi almost 20 years ago [28], and was later described in greater details by Geladi and Grahn [29]. It was introduced in the process industries in the late 1990s [25] and since then, it has been used in various areas such as in the forest products and pulp and paper fields [30, 31], in the snack food industry [32, 33], in steam plants [34], and in the mineral processing industry, in both ore concentration [26, 35, 36] and pyrometallurgical areas [38, 39].

Multivariate image analysis consists of two steps: (1) a decomposition of the image information using multi-way PCA (MPCA) which yields a lower-dimensional feature space (or latent variable space); and (2) feature exploration by iterative segmentation of regions-of-interest (ROI) within the feature space followed by mapping of these selected ROIs back in the original image space for visual interpretation of the identified spectral features. The MPCA decomposition of a multivariate image is illustrated in Figure 3.5 using the froth image shown in Figure 3.3. First, it involves reorganizing the image data cube (*i.e.*, three-way array) $\underline{\mathbf{X}}$ $(x \times y \times \lambda)$ into a matrix $\mathbf{X}$ of dimensions $((x \times y) \times \lambda)$ by collecting the light intensities at the various wavelengths row-wise for all pixels of the image. The image data $\mathbf{X}$ is then decomposed using PCA into a set of A orthogonal score vectors $\mathbf{t}_a$ and loading vectors $\mathbf{p}_a$ together defining a lower dimensional subspace where most of the image information lies:

$$\mathbf{X} = \sum_{a=1}^{A} \mathbf{t}_a \, \mathbf{p}_a^\top + \mathbf{E}. \tag{3.4}$$

When the number of principal components (PCs) $A < \lambda$, $\mathbf{E}$ contains the projection residuals of each row of $\mathbf{X}$ (*i.e.*, distance of each observation in the original $\Re^\lambda$ space and their projection into the lower dimensional space). In image analysis, $\mathbf{X}$ is generally not centered and scaled prior to applying PCA since the average color intensities explained by the first PC is often meaningful and the light intensities captured in each channel are measured in the same units and span the same range (0–255 for an 8-bit camera). As mentioned in Section 3.2.1, the loading and score vectors can be computed either by the iterative NIPALS algorithm or using SVD. Since the number of rows of $\mathbf{X}$ is typically very large (*i.e.*, equal to the number of pixels of the image), it is more computationally efficient to compute the loading vectors $\mathbf{p}_a$ by applying SVD on the kernel matrix $(\mathbf{X}^\top\mathbf{X})$ [29], which is only a 3×3 matrix for a RGB image. The score vectors are then computed as $\mathbf{t}_a = \mathbf{X}\,\mathbf{p}_a$, $a = 1, 2, \ldots, A$. In the analysis of RGB images, $A = 2$ is typically used since the explained variance of $\mathbf{X}$ with two components is often in excess of 99%.

A shown in Figure 3.5, the results of the MPCA decomposition of $\underline{\mathbf{X}}$ can also be expressed in multi-way notation by reorganizing the score vectors $\mathbf{t}_a$ $((x \times y) \times 1)$ into a matrix $\mathbf{T}_a$ $(x \times y \times 1)$ according to the original location of each pixel of the digital image:

$$\underline{\mathbf{X}} = \sum_{a=1}^{A} \mathbf{T}_a \otimes \mathbf{p}_a + \underline{\mathbf{E}}, \tag{3.5}$$

where symbol $\otimes$ is the Kronecker product. The advantage of this representation is that the score matrix $\mathbf{T}_a$ can be displayed as univariate (gray-scale) images, which enables direct visual interpretation of the information captured by each principal component.

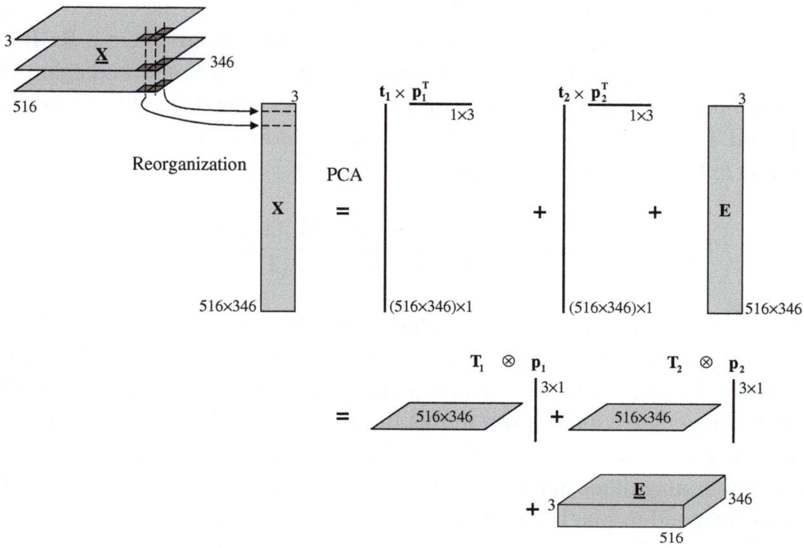

Figure 3.5 Application of multi-way PCA (MPCA) to decompose the spectral information contained within a multivariate image. Adapted from Bharati and MacGregor [25]

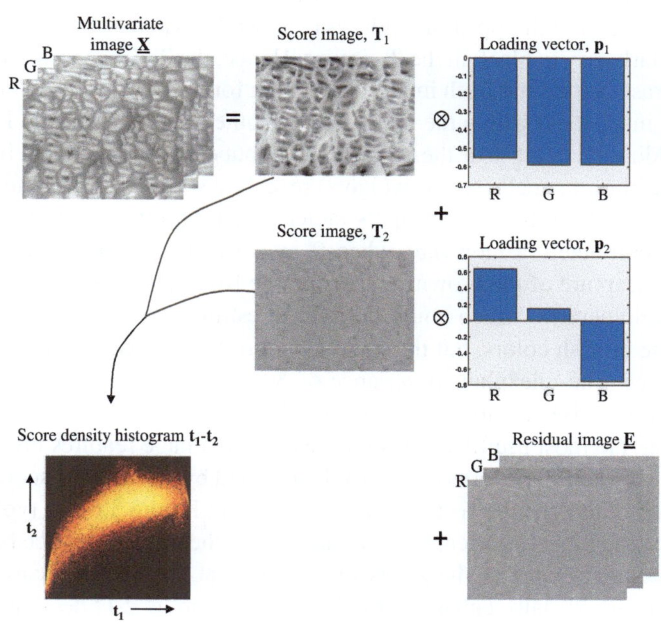

Figure 3.6 Visual interpretation of score and loading vectors obtained from MPCA when applied for image analysis. Adapted from Liu *et al.* [37]

To illustrate how to interpret the score and loading vectors in the context of digital image analysis, consider the example presented in Figure 3.6 where a RGB froth image taken from a zinc (*i.e.* sphalerite) flotation plant is decomposed using PCA. Only the first two components are considered in this example. This figure shows the $\mathbf{T}_1$ and $\mathbf{T}_2$ score images obtained after scaling score values between 0 and 255 according to the following equation:

$$\mathbf{T}_a(x,y) = Round\left(\frac{\mathbf{T}_a(x,y) - min(\mathbf{T}_a(x,y))}{max(\mathbf{T}_a(x,y)) - min(\mathbf{T}_a(x,y))} \times 255\right), \tag{3.6}$$

where x and y correspond to the pixel location in both spatial directions. The corresponding loading vectors ($\mathbf{p}_1$ and $\mathbf{p}_2$) containing the weights of the linear combinations of the three color channels explaining most of the color variations across the RGB image $\underline{\mathbf{X}}$ are also shown in this figure. The first PC clearly capture variations in gray colors across the froth (*i.e.*, the dominant color) since all $\mathbf{p}_1$ weights have a similar magnitude and the same sign. The darker regions in the original froth image correspond to pixels having low RGB intensities. Taking a linear combination of these low RGB intensities according to the $\mathbf{p}_1$ weights will result in small negative $\mathbf{t}_1$ values (*i.e.*, close to zero). On the other hand, light gray regions correspond to high RGB values and, therefore, to strong negative $\mathbf{t}_1$ values. Once scaled between 0 (min($\mathbf{t}_1$)) and 255 (max($\mathbf{t}_1$)) and displayed as a $\mathbf{T}_1$ image, darker regions in the froth image will appear as light gray to white, whereas clearer regions will appear as dark gray to black in the $\mathbf{T}_1$ image. Hence, the $\mathbf{T}_1$ image captures the gray color contrasts within the froth image accounting for 99.88% of the color variations (*i.e.*, RGB intensities) within the image. On the other hand, the second PC explains only an additional 0.085% of the variance of $\underline{\mathbf{X}}$, but captures a contrast between the red and the blue channels (see $\mathbf{p}_2$ weights) revealing some interesting features. The opposite signs of the R and B weights indicate that these two intensities vary in a negatively correlated fashion (*i.e.*, when R increases B decreases and *vice versa*). This allows capture of the brownish areas on the froth, corresponding to sphalerite particle agglomerates, which could be used to estimate froth grade. This PC also captures the blueish colors, but these are most probably caused by light reflections from bubbles. The unexplained variance of $\underline{\mathbf{X}}$ after two PCs (*i.e.*, $\mathbf{T}_3$) essentially corresponds to noise and are left as residuals $\mathbf{E}$.

However, the most useful approach for exploring image features involves investigating the clustering patterns of the pixels obtained by plotting the score vectors $\mathbf{t}_a$ against each other (see the $\mathbf{t}_1$–$\mathbf{t}_2$ score scatter plot in Figure 3.6), as typically done when analyzing large historical process databases. The only difference between traditional data and image analysis lies in the fact that, in the latter case, the score scatter plots are visually enhanced by displaying them as 2-D density histograms due to the very large number of observations projected into this plot. Indeed, there are as many observations (*i.e.*, points) in the score plots as the number of pixels forming the image, that is $516 \times 346 = 178536$ observations in the $\mathbf{t}_1$–$\mathbf{t}_2$ plot for the image shown in Figure 3.5. It would therefore be very difficult to see any cluster in such a plot, especially considering that 516×346 pixels is a very low resolution using today's imaging technology, which currently provides spatial resolutions as high as 14 megapixels. Constructing a 2-D density histograms from a $\mathbf{t}_i$–$\mathbf{t}_j$ scatter plot ($i \neq j$) involves: (1) discretizing the score plot into a certain number of bins by applying a grid of a user-defined mesh size; (2) counting the number of occurrences (*i.e.* number of pixels) falling into each bin; and (3) applying a color map showing pixel density variations. A grid of 256×256 typically provides sufficient resolution for exploring the information contained within a single image [25, 29]. A lower resolution is often used after the initial exploration stage, when the features of interest are known and are used to develop classification of regression models, for example.

The feature exploration stage of MIA is illustrated in Figure 3.7 using another froth image, again collected from a zinc flotation plant. After decomposing the froth image using PCA, a 256×256 $\mathbf{t}_1$–$\mathbf{t}_2$ score density histogram is computed and displayed using a hot color map. The regions in black indicate that no pixels have that particular combination of $\mathbf{t}_1$ and $\mathbf{t}_2$ values, whereas colors of increasing brightness, from dark red to yellow and, eventually, to white indicate an increasing density of pixels falling into specific areas of that plot. Then, the feature exploration proceeds iteratively with (1) drawing of a mask (*i.e.*, ROIs) on the score density histogram, followed by (2) identification of the pixels falling under that mask (*i.e.*, pixels having $\mathbf{t}_1$ and $\mathbf{t}_2$ values falling within the region segmented by the mask) and, (3) displaying these selected pixels on the original image using an easily distinguishable color. Pixels falling within a certain neighborhood have similar features or spectral signatures whereas those pixels falling in distinct regions of the density histogram will capture different features.

This procedure is illustrated in Figure 3.7 with the blue and green masks. These masks are initially of arbitrary size and shape, but are refined in later stages using either automatic masking techniques [40] or using regression [32, 38]. After identifying the pixels falling underneath each mask and overlaying them onto the original image, one can visually interpret the color features extracted by each mask; the blue and green masks, respectively, capture the sphalerite agglomerates (*i.e.* brown areas) and the clear (black) windows on the bubbles associated with poor bubble loading. Since MIA is performed on a pixel-by-pixel basis, these features are easily extracted regardless of their size, shape, and location within the image. This is an important advantage of the method since these features are stochastically distributed from one image to the other.

Once the initial feature exploration stage of individual images is completed, the subsequent steps usually involve collecting images under different operating conditions according to some designed experiments (or not) for the purpose of on-line monitoring or classification of features of interest, or for predicting some key process variables. Comparing features of a set of J images requires building a common PCA model for all the images within the set. This is accomplished by applying SVD to the global kernel matrix $\mathbf{Z} = \sum_{j=1}^{J} \mathbf{X}_j^\top \mathbf{X}_j$ [29], which will result in a common set of loading vectors $\mathbf{p}_a$ for all J images. The score vectors of the jth image $\mathbf{t}_a^j$ are then computed as $\mathbf{t}_a^j = \mathbf{X}_j\ \mathbf{p}_a$. A common scaling range is also required for the scores prior to displaying them as density histograms. This scaling range corresponds to

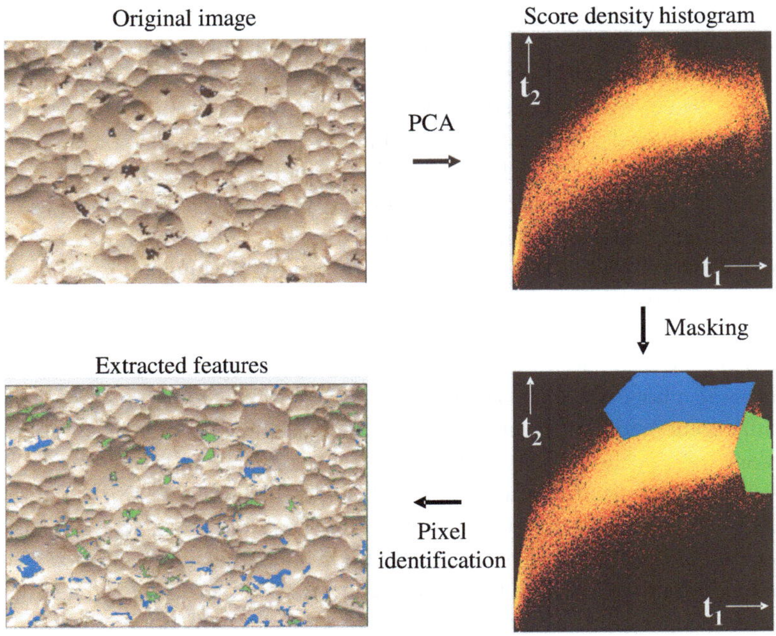

Figure 3.7 Example of MIA applied to a froth flotation image. The blue and green masks identify pixels belonging to brown areas (*i.e.*, sphalerite) and clear windows (appear as black), respectively

the minimum and maximum values of all $\mathbf{t}_1^j$ and $\mathbf{t}_2^j$ score vectors of the set of images (*i.e.*, training set). Such a scaling range may be exceeded when the model is used on-line and new operating conditions, falling outside the operating range included in the training set, are experienced. This situation can, however, easily be monitored and, when detected, new conditions may be incorporated in the model through some updating strategy.

The spectral features extracted from the set of images using MIA can be formulated in various ways depending on the degree of complexity of the features of interest. This was investigated by Yu and MacGregor [32] who classified the features in two categories: overall and distribution features. If one is only interested in tracking overall color variations from one image to the other, then the weights of the loading vectors $\mathbf{p}_a^j$ obtained from each individual image $\mathbf{X}_j$ (not from $\mathbf{Z}$) can be used as representative color features of the jth image for further analysis. If, on the other hand, the most interesting features consists of one or few, more subtle, color characteristics distributed within the image, then using the score density histograms based on $\mathbf{Z}$ (*i.e.*, a common set of loading vectors), normalized as mentioned previously, will be the most appropriate way to extract the desired information. These features could consist of a vector of pixel densities corresponding to each combination of $\mathbf{t}_i$–$\mathbf{t}_j$ values ($i \neq j$) obtained directly from the score histogram with a certain mesh size, or counts of pixels falling underneath one or a few masks (see Figure 3.7). Which type of feature to use is application dependent and often requires some trial and error unless prior knowledge is available on the features of interest. Note that only those feature extraction methods that will be used in Section 3.5 are presented here. For more details or for a more extensive account of the various possibilities, the reader is referred to Yu and MacGregor [32].

A final but important remark needs to be made about MIA. Applying PCA to an image requires unfolding (*i.e.* reorganizing) the original three-way array image $\underline{\mathbf{X}}$ into a matrix $\mathbf{X}$ (see Figure 3.5) by collecting the light intensities of each pixel row-wise in $\mathbf{X}$. Since interchanging rows in $\mathbf{X}$ does not affect the results of the PCA decomposition, this unfolding operation destroys the information related to the spatial organization of the pixel intensities (*i.e.*, the image texture). Therefore, MIA is a very useful technique for extracting spectral characteristics from images, but cannot be used to extract textural features. The latter is performed using textural methods, such as wavelet texture analysis (WTA).

3.4.1.2 Multiresolution Analysis for Texture Analysis

Images collected in various fields often show objects or structures of different sizes [41]. For example, consider the brick wall shown in Figure 3.8. When looking at the wall from a certain distance (Figure 3.8 (a)), one can clearly recognize the bricks and mortar joints as well as the visual pattern defined by those. One could also detect cracks in mortar joints at this level (*i.e.*, distance from the wall). However, when zooming on a single brick (see Figure 3.8 (b)), visual information at a much finer level of details appear: the brick surface texture with pits and subtle gray color variations. That is the brick wall image contains very different textural information depending on the distance from the wall. If one was to use this image to detect joint cracks (*i.e.*, coarser texture) or to grade the quality of the brick surfaces, textural information at a very different level will need to be extracted. This could be accomplished by taking several pictures of the wall at various distances to obtain the desired resolution. However, when the optimal resolution is initially unknown, or when several resolutions are required to obtain the desired information, then collecting several images can be cumbersome. A multiresolution framework was jointly developed by Stéphane Mallat [41–44] and Yves Meyer [45] in the late 1980s [46, 47] for extraction of coarser to finer textural information based on a single version of the image (*i.e.*, image captured only once at a fixed distance), allowing scale-invariant interpretation of the images [41].

In multiresolution analysis (MRA) of a 1-Dl signal, the latter is sequentially decomposed at different levels, also called scales or frequencies. At each level of decomposition $j = 1, 2, \ldots, J$, the signal is decomposed into two orthogonal signals: an approximation signal containing coarser information, and a detail signal capturing finer characteristics. This decomposition is performed such that the approximation signal at level j includes all detail signals at higher levels of decomposition $j+1, j+2, \ldots, J$ (*i.e.*, the detail signals at higher levels than j are embedded into the approximation signals at level j). Such a decomposition allows for reconstruction of the original signal by combining the approximation at the highest level J and

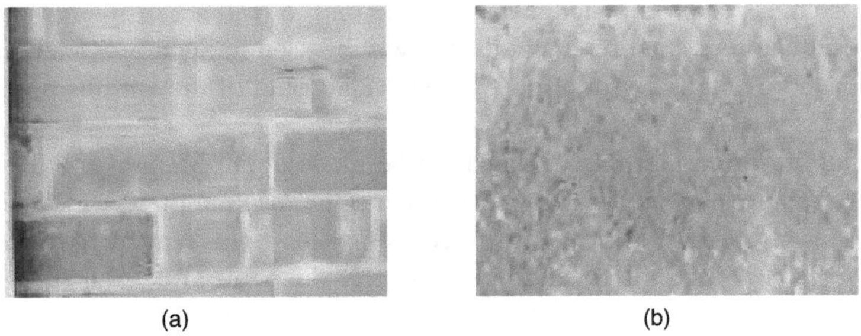

Figure 3.8 Brick wall image. (a) coarse pattern defined by the bricks and mortar joints. (b) fine surface texture of a brick. From Tessier *et al.* [26]

all detail signals from levels 1 to J. Wavelet decomposition is one method satisfying the MRA conditions and is frequently used for signal compression, denoising, but also for texture analysis of 2-D signals (*i.e.* images).

In Fourier analysis, a continuous 1-D signal or function $f(x)$ (*e.g.*, a time series) is decomposed into a series of periodic functions or waves of infinite duration represented by sines and cosines at different frequencies. The wavelet transform, on the other hand, uses waveforms of *finite length*. This allows space/time-frequency decomposition of $f(x)$ instead of frequency decomposition only as obtained by the Fourier transform. Many waveforms are available for this purpose (*e.g.* Haar, Daubechies, Coiflet, Symlet, Mexican hat, *etc.*), and each is mathematically represented by a function called the *mother wavelet* $\psi(x)$. Translation and dilation of the mother wavelet $\psi(x)$ generates a family of wavelets $\psi_{a,b}(x)$, where a and b are integers, respectively scaling and translating the mother wavelet as described below. Since the scaling coefficient a compresses or stretches wavelet $\psi_{a,b}(x)$, it effectively changes the frequency of the wavelet function.

$$\psi_{a,b}(x) = \frac{1}{\sqrt{|a|}}\psi\left(\frac{x-b}{a}\right). \tag{3.7}$$

The continuous wavelet transform (CWT) is the convolution of signal $f(x)$ with the scaled and dilated wavelet function $\psi_{a,b}(x)$ resulting in a scalar quantity called the wavelet coefficient $d_{a,b}$. The convolution integral can also be expressed as a scalar product between the two functions represented by vectors:

$$d_{a,b} = \int_{\Re} f(x)\psi_{a,b}(x)dx = \langle \psi_{a,b}(x), f(x) \rangle. \tag{3.8}$$

Hence, the wavelet coefficient $d_{a,b}$ can be viewed as a measure of similarity between the signal $f(x)$ and the wavelet function $\psi_{a,b}(x)$ [13], at a particular location in space or time, determined by the value of b, and at a specific frequency, given by the value of a. Applying this transform to several values of a and b yields a set of wavelet coefficients characterizing the space/time-frequency decomposition of signal $f(x)$. Note that in the CWT , the values of b are selected to cover the entire signal (*i.e.*, space or length of time covered by $f(x)$), but a can take any values.

To reduce computing time and the amount of data generated by this analysis, the discrete wavelet transform (DWT) was proposed, which is performed only at a smaller number of discrete locations and frequencies. It was shown that selecting shifting and dilations (*i.e.*, discrete frequencies) as a power of 2, called dyadic, would yield great computational economy with similar accuracy as the CWT. The DWT using dyadic locations and frequencies is obtained by setting $1/a = 2^{-m}$ or $a = 2^m$ and $-b/a = -n$ or $b = n2^m$, where both m and n are integers.

$$\psi_{m,n}(x) = 2^{-m/2}\psi(2^{-m}x - n). \tag{3.9}$$

The wavelet coefficients are computed as for the CWT but only at specific m and n values:

$$d_{m,n} = \int_{\Re} f(x)\psi_{m,n}(x)dx = \langle \psi_{m,n}(x), f(x)\rangle. \tag{3.10}$$

Practical implementation of DWT algorithms are, however, performed using quadrature mirror filters. Indeed, a very important contribution of Mallat was to link orthogonal wavelets to the filters traditionally used in the signal processing field [47]. Through this work, a new function called *scaling function* $\phi(x)$, orthogonal to the wavelet function $\psi(x)$, was introduced to capture the low frequency information (*i.e.* coarse) whereas the high frequency detail is extracted by the wavelet function. These two functions are expressed in terms of orthogonal filters, a low-pass filter h_0 related to the scaling function $\phi(x)$, and a high-pass filter h_1, related to the wavelet function $\psi(x)$, as shown below:

$$\psi_{j,l}[k] = 2^{j/2}h_1[k - 2^j l] \quad \text{and} \quad \phi_{j,l}[k] = 2^{j/2}h_0[k - 2^j l], \tag{3.11}$$

where j is the decomposition level, l is the shifting parameter, and k is the discrete location, either in time or space, on the signal to be decomposed (all integers).

From these two functions, the wavelet coefficients can be computed as the inner product (or scalar product) of the signal $f[k]$ and the wavelet and the scaling functions. This results in two sets of coefficients, those capturing the low frequency information (*i.e.*, associated with the low-pass filter), called the approximation coefficients $a_{(j)}[k]$, and the second set representing the details or high-frequency information called the detail coefficients $d_{(j)}[k]$:

$$a_{(j)}[k] = \langle f[k], \phi_{j,l}[k]\rangle \quad \text{and} \quad d_{(j)}[k] = \langle f[k], \psi_{j,l}[k]\rangle. \tag{3.12}$$

These coefficients are calculated sequentially, for each level of decomposition j, using filter banks [46]. How to implement wavelet series expansion using filter banks was shown by Mallat [41], while Daubechies [48, 49] showed how to construct wavelets from filter banks [46].

Applying DWT to a multidimensional signal, such as a 2-D image, can be readily accomplished by using 1-D DWT along each dimension, one at a time [46]. This is referred to as a *separable solution*, which uses separable filters (*i.e.*, h_0 and h_1 within a multidimensional filter bank, as shown in Figure 3.9). This approach is one of the most frequently used in image processing [46].

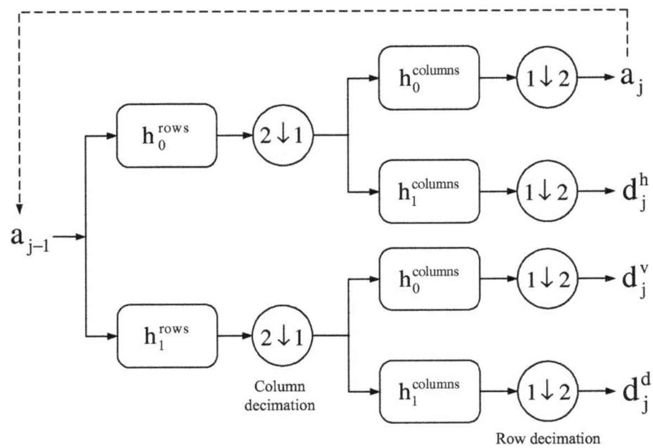

Figure 3.9 A 2-D separable solution for applying DWT to an image. Adapted from Liu *et al.* [37]

The algorithm begins with the original image, a_j, $j = 0$. Note that in texture analysis, only gray scale or univariate images are used. Therefore, a_0 could be an image captured using a monochrome camera or an RGB image converted into gray levels. Each row of this image is filtered using the low-pass (h_0) and the high-pass (h_1) filters, followed by column decimation keeping only one column out of two for the next step. The resulting low and high frequency information along the horizontal direction is then filtered vertically (*i.e.*, along each column) using again both filters. A final row decimation step removes one row out of two. The output of this algorithm consists of four new images containing the wavelet coefficients. The low-pass information in both directions yields an approximation image, a_{j+1} or a_1, capturing low frequency information. The horizontal low-pass and vertical high-pass extracts vertical edges in the image, thus capturing the horizontal details (*i.e.* high frequency information) into a detail image d_1^h. On the other hand, horizontal edges (or vertical details) are characterized by the high pass horizontal and low pass vertical information and are observed in detail image d_1^v. Finally, the high frequency information in both directions corresponds to diagonal details, d_1^d. The approximation and detail images corresponding to the first level of decomposition now have a four times smaller resolution due to the column and row decimation (*i.e.*, the number of horizontal and vertical pixels are divided by 2). Such image resolution will be further divided by 4 at each level of decomposition. This is how the dyadic shifting and scaling of the discrete wavelet is implemented within the filter bank solution. Further decomposition of the textural information at higher levels (or towards lower frequencies) can be performed using approximation image a_{j+1} as the input of the filter bank (see dashed line in Figure 3.9).

The link between the decomposition level and the resolution of the textural information is as follows, assuming that each image has its own distribution of textural features at different resolutions. In the first level of decomposition $j = 1$, the input of the filter bank is the original image a_0 containing textural features at any resolution (*i.e.* both coarse and fine textures). At this level, the detail coefficients d_1^h, d_1^v, and d_1^d will capture high frequency information, or the finest textural details of the image. The lower frequency information not used at this level correspond to the approximation coefficients a_1, which will be further decomposed at level $j+1$ or 2 in this case. As the analysis proceeds towards higher level of decomposition, the detail coefficients d_j^h, d_j^v, and d_1^d will progressively extract coarser textural features due to the column and row decimation. In practice, this analysis proceeds until the approximation image a_{j-1} has a spatial resolution greater than 10×10 pixels [50, 51].

To illustrate this procedure, Figure 3.10 shows an application of 2-D DWT to a gray level image of a manufactured stone walkway using three decomposition levels ($j = 3$). The A_j images represent the approximation images at each level whereas the D_j^q images show the textural details extracted at each level j and direction $q = h, v, d$. These images were obtained after an appropriate scaling was applied to the wavelet coefficient matrices for display as gray level images. One can clearly see which details are extracted at each level, their directionality, and also the decimation (*i.e.*, 2×2 binning of pixels) occurring at each level allowing extraction of finer to coarser textural features.

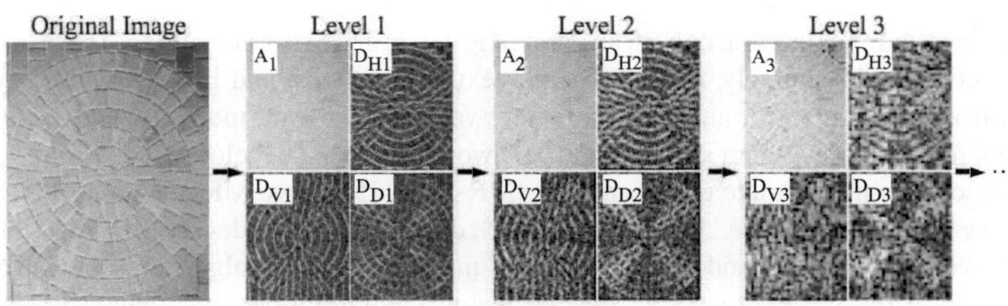

Figure 3.10 2-D DWT applied to an image of a manufactured stone walkway. Three levels of decomposition are shown ($j = 3$). From Gosselin *et al.* [52]

Wavelet texture analysis (WTA) using J decomposition levels, in all three directions, yields $3J + 1$ images: $3J$ images of details and the final approximation image. The size of these images vary according to the dyadic scaling discussed previously. In order to analyze and compare the textural information contained in each image of a set, one needs to reduce the detail images D_j^q $j = 1, 2, \ldots, J$, and the approximation image A_J (if relevant), to a vector of textural features representing each image of the set. One of the most frequently used textural feature is the energy, which consists of the Frobenius norm of the detail images:

$$E_{jq} = ||D_j^q||_F^2. \tag{3.13}$$

Several other features have also been proposed and used for representing multiscale image texture, such as wavelet histogram signatures or simply *wavelet signatures* (*i.e.*, first order statistics of the sub-image histograms) [53], and higher-order statistics of the co-occurrence matrix of the detail sub-images called wavelet co-occurrence signatures (*e.g.*, contrast, correlation, entropy, angular second moment, *etc.*) [54]. Some guidelines as to which features to use can be found in [13, 53, 54]. However, the main objective is to use those which provide the desired degree of texture discrimination between images.

Finally, other textural methods are also available for extracting the spatial organization of pixels intensities from images, such as statistical methods (*e.g.*, gray-level co-occurrence matrices or GLCM [54], the angle measurement technique [55]), structural based methods, model based methods and filter-based methods (*e.g.*, Fourier and wavelet transforms) [13]. A comparison between some of these image texture extraction methods was performed by [51] using images of steel sheets. Wavelet texture analysis was shown to be one of the most powerful technique for image texture analysis.

3.4.1.3 Multiresolution Multivariate Image Analysis (MR-MIA)

The complexity of the information to be obtained from multivariate process images often requires extraction of both spectral (*e.g.* color) and textural features. These can be computed separately, using MIA and textural methods, and later combined by concatenation of color and textural feature vectors. However, more efficient methods for combining them in a unified framework have been developed, such as multiresolution multivariate image analysis (MR-MIA) [56]. Two MR-MIA algorithms have been proposed (*i.e.*, MR-MIA I and II), depending on the degree of interaction between the spectral and textural (spatial) information. Both algorithms use MIA (*i.e.* PCA on images) to extract spectral features and multiresolution analysis (*i.e.*, wavelets) for texture, but in a different order.

MR-MIA I. When spatial and spectral features are interacting, for example when some textures are observable only at specific wavelengths within a multivariate image, and not in other regions of the spectra, then MR-MIA I should be used [56]. It consists of applying 2-D DWT to a given decomposition level $j = 1, 2, \ldots, J$ (*i.e.*, scale) and orientations $q = h, v, d$ separately, on each spectral channel l of a multivariate image, $l = 1, 2, \ldots, \lambda$. Since the detail images obtained for each wavelength at the same scale j and orientation q, are congruent, these can be combined into a new multivariate image. Considering all scales and orientations, applying 2-D DWT to each spectral channel results in $3J$ new, congruent multivariate detail images $\underline{\mathbf{D}}_j^q$, $j = 1, 2, \ldots, J$ and $q = h, v, d$, plus a multivariate approximation image $\underline{\mathbf{A}}_J$. These images are then analyzed using MIA to extract textural–spectral (or spatial–spectral) features.

This algorithm is especially useful for hyperspectral images, where spectral resolution is often very high and features of interest (*e.g.*, some chemical groups in the NIR region) absorb light in a narrow wavelength band. For RGB images, however, the spectral sensitivity of the red, green and blue detectors of the camera CCD matrix (*i.e.*, detectors in the Bayer pattern) are strongly overlapping. Therefore, textures often appear similarly in each channel (*i.e.*, there is little interaction between spatial and spectral features). In this case, MR-MIA I and II algorithms will yield similar results, and since MR-MIA II is less computationally intensive, it will be the most efficient algorithm for RGB images [56].

MR-MIA II. In this algorithm, MIA is performed first and the score images (*e.g.*, $\mathbf{T}_1$ and $\mathbf{T}_2$) are obtained. The first score image is very close to a gray-scale version of the image, whereas subsequent components extract color informations that are independent of the intensity component [56]. These score images capture the dominant color features of the RGB image. Texture methods (*e.g.*, 2-D DWT) are then applied to the score images. If A score images are extracted using MIA, and texture analysis is performed to the Jth level of decomposition, and along all three orientations ($q = h, v, d$), then $3AJ$ univariate detail images $\mathbf{D}_j^{q,a}$ are obtained plus A approximation images $\mathbf{A}_J^a$. Therefore, textural features are obtained for each color contrast captured by each principal component. One then computes wavelet signatures (*e.g.*, energy, Equation 3.13) on each of the $3AJ$ detail images, and combines them into a single spectral–spatial feature vector. This approach will be illustrated in one of the case studies (Section 3.5.2). For a more detailed analysis of the MR-MIA algorithms, the reader is referred to the work of Liu and Liu and MacGregor [13, 56].

3.4.2 Feature Reduction and Analysis

Most applications of machine vision in the process industries can be formulated as either a *classification* problem or a *regression* problem, aimed at classifying the state of the process or at predicting some key process variables based on a set of images and, perhaps, on some data obtained from process instrumentation. Images are typically collected under various process conditions selected so as to capture the effect of disturbances or manipulated variables, using designed experiments or simply during normal or abnormal production situations. After extracting spectral and/or textural features from each image of the set, and synchronizing these with process data obtained from data historians, one obtains the dataset shown in Figure 3.11, which is composed of three blocks or data matrices: $\mathbf{X_F}$, $\mathbf{Y}$, and $\mathbf{T_F}$. The nature of the data included within each of these blocks is discussed below.

- $\mathbf{X_F}$. This block contains a collection of feature vectors of size $1 \times K$ computed for each of the N images of the set. These image features could represent: (1) color (or spectral) feature only, such as the loading vectors of PCA applied on each image separately, or counts of pixels falling under one or a few masks segmenting a particular region of the score space (density histogram obtained by

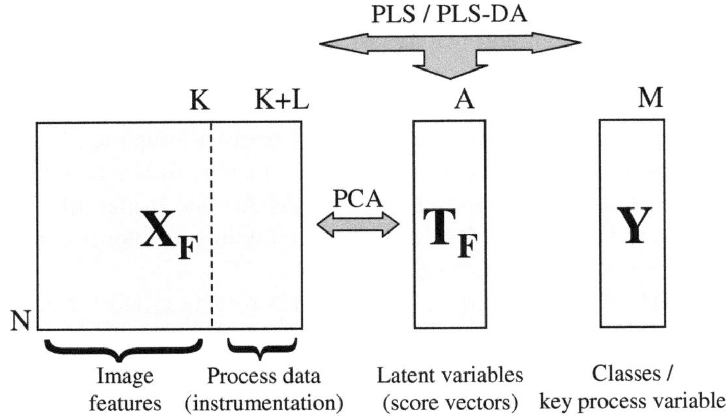

Figure 3.11 Typical applications of machine vision in the process industries can be formulated into either a classification or a regression problem

MIA, see Figure 3.7), or pixel densities obtained from a rectangular grid applied to the score histogram (*i.e.*, discretized 2-D density histogram, see Section 3.4.1.1); (2) textural features only, such as the wavelet signatures (*e.g.*, energy) and/or the co-occurrence signatures obtained from detail images; and (3) both spectral and textural features, either obtained separately and subsequently concatenated, or computed jointly using one of the MR-MIA algorithms. When useful for classification or prediction purposes, $\mathbf{X_F}$ may be augmented with a set of L measurements obtained from current process instrumentation, after appropriate synchronization with the images.

- **Y**. This second block is a matrix of responses that may, or may not be available depending on the machine vision objective and the application *per se*. Typical responses may be key process variables which we would like to predict using the images, such as froth grade in flotation, or categorical response (*i.e.*, dummy or binary variables) used to associate each image with some pre-defined classes, such as process states, type of raw materials, *etc*. This information may be useful in supervised classification problems.

- $\mathbf{T_F}$. The third block consists of the lower dimensional latent variable space which will be used for classifying the images, establishing multivariate statistical process control schemes for process monitoring and fault detection, or for predicting response variables of interest. These score vectors are obtained using the latent variable method, such as PCA or PLS, which is the most appropriate for the problem at hand. Note that to simplify the discussion, Figure 3.11 focuses on the score vectors since they represent the projection of the multivariate features onto the latent variable space. However, one should also consider the square prediction errors (or DMODX, see Section 3.2) to diagnose observations that are at a significant distance above or below the hyperplane defined by the latent variable space (*i.e.* observations showing a different correlation structure, inconsistent with that of the latent variable space).

The latent variable method used to compute $\mathbf{T_F}$ (and other related vectors) depends upon the existence of a response matrix **Y**, as well as the nature of the response variables (*i.e.*, categorical or not). The following discusses which method to use for three generic types of machine vision problems.

- **Unsupervised classification.** This situation arises when no response matrix **Y** is available, or when one simply wants to assess the natural clustering pattern of the image based on their features. This is accomplished by applying PCA on the feature matrix $\mathbf{X_F}$. An example of such a situation will be presented in Section 3.5.2.

- **Supervised classification.** When the machine vision objective is to classify new, incoming images in pre-defined categories, and when the N images of the training set can be assigned to a known class, then a supervised classification problem may be formulated. In this case, the response matrix **Y** is filled with binary numbers assigning each image to a particular class. It has as many columns as the number of categories. If the nth image belongs to the mth class, then a value of 1 will appear in position (n,m) in matrix **Y** and zeros elsewhere. When such a categorical response matrix is available, PLS may be used to extract the information in the feature matrix $\mathbf{X_F}$, which is the most highly correlated with the classes in **Y**, that is, to find the combinations of features that maximize class discrimination. Applying PLS regression to a categorical response matrix **Y** is called PLS-DA, for PLS-discriminant analysis. A supervised classification problem using PLS-DA is illustrated in Section 3.5.3.

- **Regression**. Image regression can be used for rapid and non-intrusive prediction of key process variables (*i.e.*, **Y**), when these are difficult to measure on-line or when they are obtained after long dead-times, such as assays requiring the use of analytical instruments. In this case, PLS regression is used to build prediction models for **Y** based on image features $\mathbf{X_F}$. An example image regression problem is presented in Section 3.5.1. Note that one may need to account for the dynamics between $\mathbf{X_F}$ and **Y**. This can be accomplished by augmenting $\mathbf{X_F}$ with past lags (*i.e.*, time-shifts) of image features and process data before using PLS (*i.e.*, finite impulse response model). There are other ways of introducing dynamics in the model, but this subject is outside the scope of this chapter (see system identification literature as well as [57–59]).

3.5 Case Studies

Three case studies are presented in this section to illustrate the various multivariate imaging concepts presented in this chapter. The first two cases investigate froth grade prediction and froth health monitoring based on RGB color images collected from the zinc flotation circuit (*i.e.*, sphalerite flotation) at the Agnico-Eagle/Laronde plant (Abitibi, Québec, Canada). The third example consists of estimating the proportions of three different rock types within a ROM mixture incoming Raglan's mine concentrator (Xstrata Nickel Canada).

3.5.1 Prediction of Flotation Froth Grade Using MIA

Flotation is one of the most widely used mineral separation processes. Mechanical cells and/or columns are organized in flotation circuits in order to achieve a desired grade and recovery of valuable minerals, the two key process variables related to metallurgical and economical performance. Recovery is measured when the flotation circuit operates in steady-state, and requires measuring the flow and grade of the head, tailings, and concentrate streams of the circuit (*i.e.*, using a mass balance). Grade is traditionally measured using analytical techniques in the laboratory. However, in the last 15–20 years, on-line stream analyzers (OSA) were developed and introduced in concentrators to measure the concentration of various minerals in a stream. Most of them are based on X-ray fluorescence, such as the Courier™[3] elemental analyzer (Outotec Oyj). These analyzers are expensive and used for assaying many streams (10–12), which result in sampling times varying from 10–25 min for a given stream, depending on the number of streams in the OSA cycle. For process control purposes, there was an incentive to reduce sampling time to better capture the dynamics of the process and the impact of fast disturbances. Furthermore, ways of increasing the number of monitored streams or flotation cells/columns at a low cost was also considered. Machine vision sensors were therefore investigated intensively during the 1990–2005 period, essentially, as one solution to infer concentrate grade using froth images.

Indeed, several authors have shown over the years that concentrate grade is correlated with color and textural features of the froth, see [35, 36, 60–65] to cite a few. An example is given in Figure 3.12 showing four different froth images corresponding to zinc grades of 44, 49, 52, and 54%. Sphalerite minerals on the froth surface have a brownish color. Note that as zinc grade increases, more brown areas appear on the froth surface. A calibration model between the color features of each image (*i.e.*, number of pixels showing a brownish color) and concentrate grade can be developed for on-line grade estimation.

An experimental plan was designed for collecting a set of froth images and grade data for training such a model, using Agnico-Eagle/Laronde's zinc flotation circuit as a test site. A schematic process diagram of the plant is presented in Figure 3.13. The head stream (*i.e.*, feed) is first conditioned using three mechanically agitated tanks. Lime and a copper sulfate ($CuSO_4$) solution are fed into the first tank to control pH and to activate sphalerite mineral surfaces, respectively. Sphalerite does

[3] Courier is a registered trademark of Outotec Riihitontuntie 7 D, PO Box 86, FI-02200 Espoo, Finland, www.outotech.com

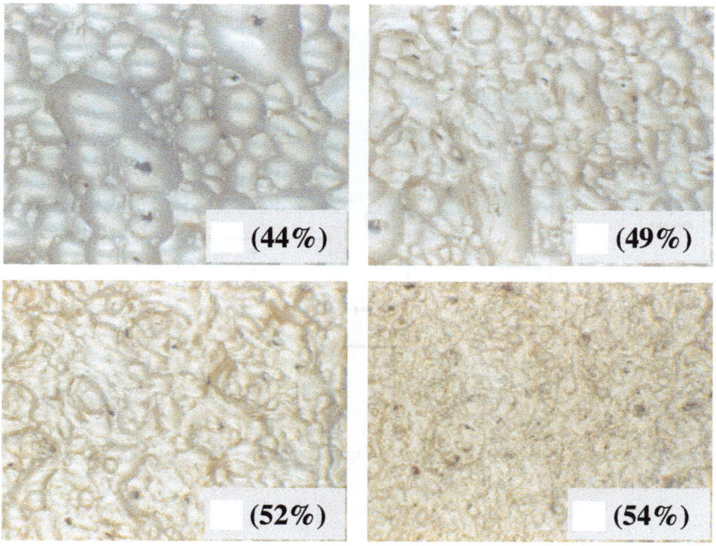

Figure 3.12 Froth images of increasing sphalerite grade (elemental Zn %) collected at the zinc flotation plant of Agnico-Eagle/Laronde (Abitibi, Québec, Canada)

not float well without pre-conditioning with an activator agent. The activator agent is therefore crucial for maintaining a high collector selectivity. The collector agent (potassium xanthate or KAX) is then added into the second tank. The last tank is used for mixing only (*i.e.* additional residence time for conditioning). Finally, an air stream is mixed with the conditioned feed stream before entering the first rougher cell, called a contact cell (*i.e.*, a hybrid between conventional flotation cells and columns). This rougher cell was selected for the test since it was known to be difficult to stabilize. A digital RGB camera as well as a lighting system were installed above the cell as shown in Figure 3.14, close to the junction of the center and peripherical downcomers. Each image had a resolution of 640×480 pixels, and captured a froth surface approximately 10×15 cm in size.

To span a wide range of operating conditions, froth color and texture, and concentrate grade, three tests of about 2–4 h each were performed. Step changes were implemented on the flow rates of collector (KAX), air, and activator agent ($CuSO_4$). Froth images were collected every 30 s (except 15 s for the air flow test). A total of 2050 froth images were captured: 643, 872, and 535 during the KAX, air, and activator tests, respectively. Since the contact cell was not equipped with an OSA, a froth sampling strategy was used to make representative composite samples for concentrate grade assays. The 12-points sampling strategy, also shown in Figure 3.14, was designed to obtain a representative sample of the concentrate overflowing the downcomers lips all around the cell and in the center, at least the regions that were physically accessible. This also minimized the dead-time between the concentrate

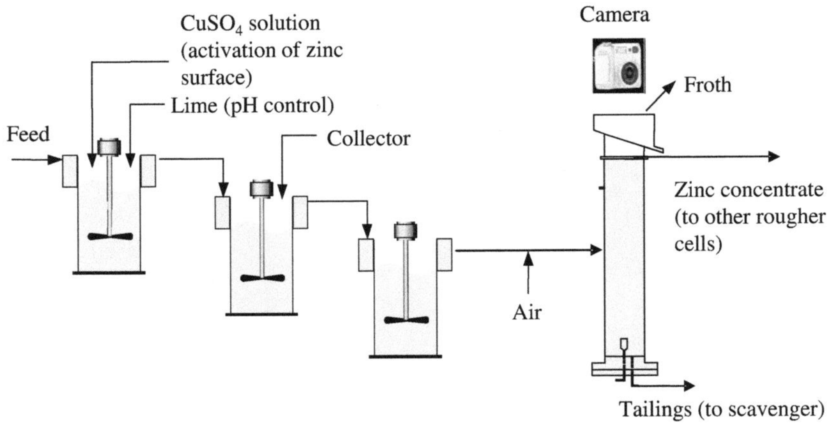

Figure 3.13 A schematic diagram of Agnico-Eagle/Laronde's zinc circuit and image acquisition system. From Liu *et al.* [37]

sample and the froth images, which would have been much larger if the concentrate samples had been collected after several meters of piping. The disadvantage of this sampling approach, however, is that the calibration model establishes a relationship between a local image of froth and an average grade measurement. If grade distribution on the froth surface is not homogeneous (as was later observed by Bartolacci *et al.* [36]), a lack-of-fit should be expected from the model. Nevertheless, this approach was selected to verify the robustness of a model based on a local froth image. In total, 50 composite samples were assayed for their content of zinc (Zn–sphalerite), lead (Pb–galena), iron (Fe–mainly pyrite), and silver (Ag–several minerals). Most of these samples were gathered during the collector and activator tests as air flow rate was not expected to have a significant effect on grade.

The PLS calibration model uses the color features of the froth images (regressor variables, $\mathbf{X_F}$) corresponding to the 50 grade measurements (response variable, $\mathbf{Y}$) collected during the tests. Synchronization of the images and grade measurement was performed before the analysis. The color features were extracted by applying a 2-D 30×30 rectangular grid to the $\mathbf{t}_1$–$\mathbf{t}_2$ score density histogram of each image (see Figure 3.7), computed based on a global MIA model (*i.e.*, using the entire set of froth images). The columns of the feature matrix $\mathbf{X_F}$ correspond to the number of pixels falling in each bin of the grid (*i.e.*, particular combination of $\mathbf{t}_1$–$\mathbf{t}_2$ values), and hence, to a particular color. The 2-D score histograms of all images are congruent since they were obtained based on a common set of MIA loading weights and a common scaling range for score values. This means that variations in the number of pixels falling within a given bin across a set of images reflect a variation in the amount of the corresponding color within those images. The resulting matrices had the following dimensions: $\mathbf{X_F}$ $(50\times(30\times30))$, $\mathbf{Y}$ (50×1).

A four latent variable PLS model was built on the $\mathbf{X_F}$–$\mathbf{Y}$ dataset. Leave-one-out cross-validation was used to select the number of components. The statistics

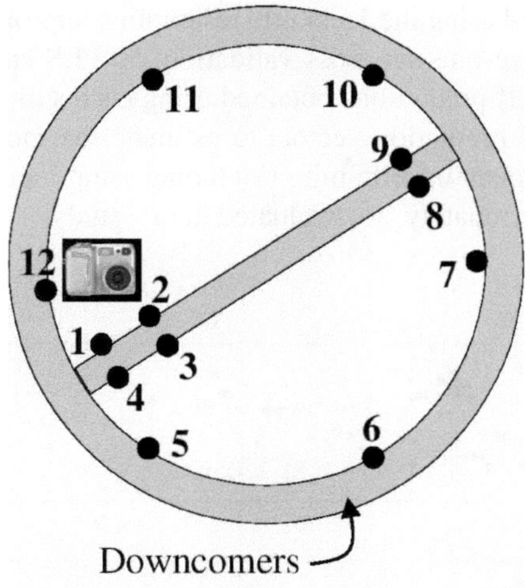

Figure 3.14 Froth sampling strategy used for the contact cell at Laronde's plant

describing the model predictive ability are summarized in Table 3.1. The model explained 67% of the variations in **Y** using 44% of the information in $\mathbf{X_F}$ (see cumulative R_Y^2 and R_X^2). About 80% of the variations in lead and silver contents are explained by the color features of the froth (see R_Y^2 values). Cross-validation results (Q_Y^2) are also very similar. Sphalerite content (Zn) had a lower explained variance, however, model predictions follow the trends quite well, as is shown in Figure 3.15. Pyrite was the most difficult to predict using froth color. This reminds us of the basis and limitations of the approach: a mineral cannot be detected by imaging if changes in its concentration do not introduce a distinguishable color change on the froth surface. One cannot expect that such a simple low cost solution will replace more expensive X-ray (*i.e.*, OSA) or laboratory analyses. However, prediction results obtained for some elements are acceptable for detecting disturbances occurring between OSA measurements.

Table 3.1 Statistics describing the predictive ability of the PLS model with four latent variables

Statistics	Zn (%)	Pb (%)	Fe (%)	Ag (g/t)	Cumulative
$R_Y^2(\%)$	0.61	0.87	0.41	0.81	0.67
$Q_Y^2(\%)$	0.51	0.84	0.27	0.79	0.60
$R_X^2(\%)$	–	–	–	–	0.44
$SE_{Jackknife}$	1.63	0.34	0.79	30.90	–

The last row in Table 3.1 provides an estimate of standard prediction error for each element, obtained using the Jackknife resampling approach [66–68]. Basically, when performing leave-one-out cross validation, 50 PLS sub-models are fitted to subsets of $\mathbf{X_F}$ and $\mathbf{Y}$. If predictions obtained using each sub-model are stored, then one can use these 50 prediction vectors to estimate the model uncertainties (*i.e.*, $SE_{Jackknife}$). Measurement uncertainties (including sampling, handling, and analytical errors) were, unfortunately, not evaluated in this study.

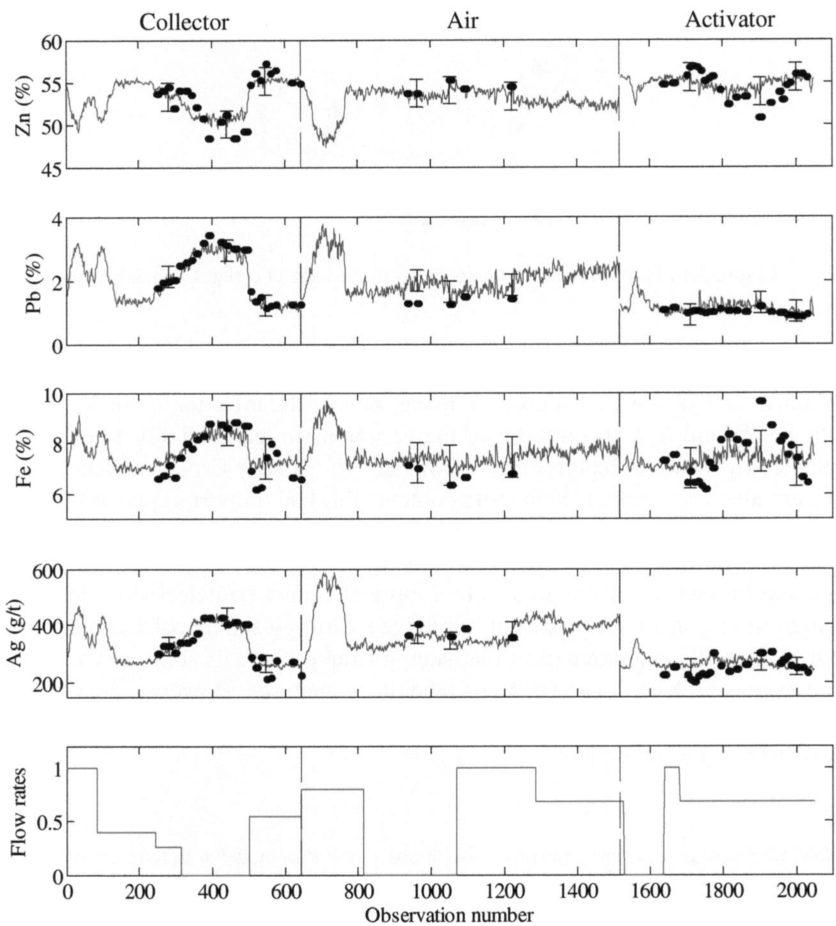

Figure 3.15 Concentrate grade prediction results obtained for the tests on collector, air, and activator flow rates at Agnico-Eagle/Laronde's zinc circuit. PLS model predictions (–), grade assays (•). Error bars correspond to the model one standard prediction error estimate ($SE_{Jackknife}$). Adapted from Duchesne *et al.* [35]

Time series of model predictions compared with laboratory grade assays are shown in Figure 3.15. The results for the three tests (collector, air, activator) are juxtaposed from left to right, respectively, and separated using black vertical dash lines. Note that time was replaced by observation number since the tests were performed on different days. In this plot, predictions obtained using each of the 2050 images are provided (gray lines). When available, assays are shown using black dots; these observations correspond to the data used to train the PLS model. Model uncertainties are also included in this figure for a few observations. Each error bar represent the values of $SE_{Jackknife}$ for each element (*i.e.*, approximate confidence intervals or reproducibility intervals on model predictions). The last plot in Figure 3.15 shows the step changes that were made on the collector, air, and activator flow rates, respectively, also juxtaposed from left to right according to each test (*i.e.*, during the collector test, variations in KAX flow rate is shown, then air flow rate in the second test, and $CuSO_4$ flow rate in the third). Each flow rate was changed one at a time, keeping all other conditions constant (as possible). The numerical values of the flow rates were scaled between 0 and 1 using their minimum and maximum values during each test. This was necessary to preserve confidential information.

In general, grade predictions are in a good agreement with the trends of the laboratory assays. As expected, step changes on the collector and activator flow rates had an important impact on the Zn grade of the froth, whereas air flow rate had very little effect. In all of these tests, it seems that the model predicted the correct trends. The results obtained at the Laronde plant were found encouraging and granted further investigation by the research consortium COREM [36]. Several COREM internal reports were also written on this subject. Some issues that were further studied in order to reduce model uncertainties are briefly discussed below.

- Froth grade was found to be heterogeneous across the surface of the contact cell (see location of sampling points in Figure 3.14), which may explain some of the model lack-of-fit. It is indeed very important to verify this issue prior to implementing such a vision sensor. This problem was found to be more acute in the contact cell than the mechanically agitated cells.
- Long-term testing was performed at Laronde to verify the robustness of the model predictions over time. This allowed the collection of additional grade data and adaptation of the model, which were found to help reduce prediction uncertainties.
- Variations in lighting conditions were also found to have an impact on grade predictions. Regular cleaning of the lighting system needs to be made. In addition, the choice of light source is also very important. Lights having a flat and stable emission spectra over their lifespan contribute to model robustness.

Finally, the results presented in this section are not valid only for zinc flotation. Several other mineral systems were also investigated [35]. Imaging of copper froth flotation (mainly chalcopyrite) in mechanically agited cells was studied at Barrick Gold/Est-Malartic plant (now closed, formely located in the Abitibi region in northwest Québec) and similar results were obtained. Batch flotation of iron ore (inverse flotation) was investigated using ground iron ore from the Iron Ore Company of Canada (Labrador City, Newfoundland, Canada). It was shown that image color features were strongly correlated with batch flotation kinetic curves and grade. The reader is referred to Duchesne *et al.* [35] for more details.

3.5.2 Flotation Froth Health Monitoring using MR-MIA

This section presents a second case study of the froth flotation process, based on the work of Liu *et al.* [37], aiming at illustrating an unsupervised image classification problem using both color and textural features. Plant operators typically use their visual appreciation of the general state of the flotation process, referred to as "froth health" in this paper, to verify whether the dosage of the various reagents (*e.g.*, collector, frother, activator, *etc.*) is appropriate for the current state of the process (*e.g.*, grade of ore feed), and to decide upon implementing control actions. For example, overdosage of collector may lead to an overloaded froth and, eventually, to bubble collapse and loss of separation capacity, generally requiring a significant amount of time to recover normal operation. On the other hand, if insufficient collector is used, the froth mineralization will be lower and loss of recovery will result. Operators usually detect the onset of these two extreme situations by their assessment of froth color and morphology, the later being essentially related with bubble size, distribution and shape [63, 64, 69, 70], also frequently called froth structure. Since froth color extraction methods have been reviewed in the previous case study (Section 3.5.1), this section focuses on the extraction of froth structural features.

Over the years, several machine vision approaches were developed for monitoring froth state in order to overcome the lack of reproducibility in human visual assessment and the limited time operators have to spend looking at the froth of any given cell or column. Indeed, it is well known that qualitative visual assessment varies from person-to-person and as a function of time, period of the day, state of mind, *etc.* On the other hand, a machine vision solution provides on-line, consistent, and quantitative measurements of froth characteristics, available to operators in the control room for several flotation units.

The most frequently used approaches for quantifying froth structure were based on image segmentation algorithms [64, 71–73] (*e.g.*, watershed [74]) and textural methods, such as gray Level co-occurrence matrix (GLCM) [70], [69, 75, 76] and Fourier transform [61]. Image segmentation may lead to very precise results for bubble size, but is computationally intensive, and sensitive to variations in lighting conditions. A new approach was proposed by Liu *et al.* [37] combining both the computational efficiency of textural methods and their robustness to lighting conditions, and the ability to estimate a distribution of bubble sizes. The estimated distributions are not as accurate as those provided by segmentation algorithms. However, robust and reproducible estimates rather than accurate bubble size distributions are required for process control purposes. The proposed approach uses both froth color and textural features, called *wavelet size signatures*, for classifying froth state. These features are obtained using the MR-MIA II algorithm discussed in Section 3.4.1.3.

Industrial data and images were also collected from the zinc flotation circuit (contact cell) of Agnico-Eagle/Laronde's plant (see Figure 3.13).

3.5.2.1 Wavelet Size Signatures

The concept of wavelet size signatures relies upon the ability of a certain wavelet to capture the shape and size of the bubbles. This is determined by the shape of the mother wavelet itself as well as the length or width covered by the wavelet as a function of decomposition level. First, some wavelets have a shape that is somewhat close to the shape of froth bubbles. For example, both the wavelet and scaling function of the symlet order 4 shown in Figure 3.16 look very similar to a bubble sliced perpendicularly to the froth surface. A peak occurs at the center of the bubble, and two valleys appear on each side of that peak. It is expected that this wavelet will correlate well with the image signals showing bubbles of different sizes.

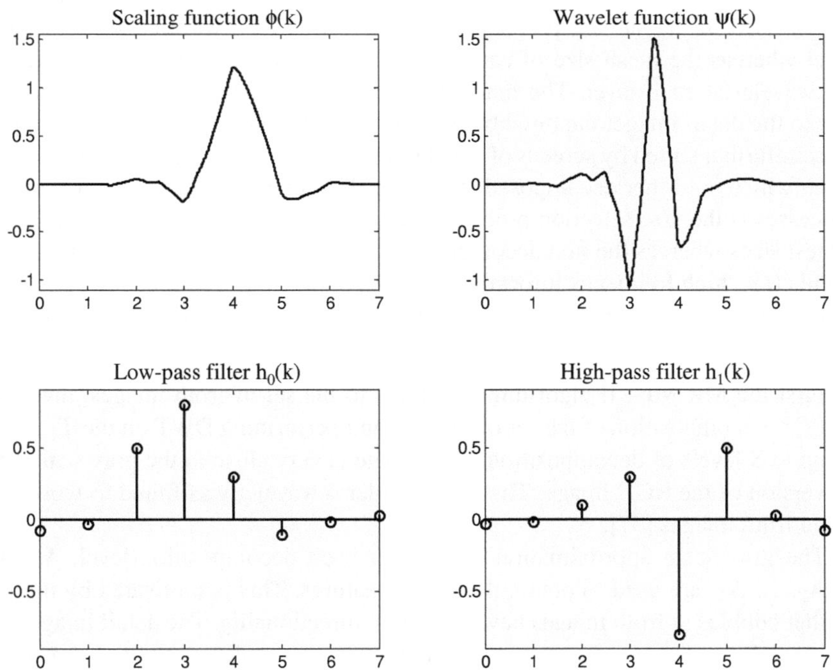

Figure 3.16 Discrete wavelet and scaling function of the symlet order 4, as well as their corresponding high-pass and low-pass filters. Obtained using Matlab R2006b Wavelet Toolbox 3.1 (MathWorks, Inc.)

To understand how the wavelet captures bubble size, it is important to recall that discrete wavelets have a finite duration, length or width. The symlet 4 shown in Figure 3.16 spans eight equally spaced discrete points, and the length covered by these on the image signal depends on the decomposition level. In the first decomposition level, the original froth image is used. When the wavelet signal is applied to that image, it will cover 8 pixels. At the second decomposition level, the dyadic sampling of DWT will make the wavelet cover a length of 16 pixels, and so on. Therefore, as the number of decomposition level increases, the wavelet will capture bubbles of larger sizes or, alternatively, lower frequency information. The correlation between the wavelet and the image signal will be maximal when the size of the bubble matches the width of the wavelet at a given decomposition level. Since the wavelet function can be represented as a high-pass filter (see Figure 3.16), bubbles with smaller diameter than the width of the wavelet will also be captured in the detail images since these are higher frequency information. Larger bubbles, which are lower frequency information, will remain in the approximation image. Therefore, detail and approximation images will capture a *range* of bubble sizes at each decomposition level and not a single size [37].

Wavelet size signatures can also be explained by analogy with experimental determination of ore particle size distribution via sieving. The initial ore sample is equivalent to the initial froth image (*i.e.*, before wavelet decomposition). Each screen used in the sieving procedure can be viewed as a wavelet decomposition level whereas the mesh size of each screen corresponds to the length covered by the wavelet at each level. The material retained on a particular screen is equivalent to the detail images captured by the wavelet function. The material passing the screen (further sorted by screens of smaller mesh) is the approximation image at this level, which is further decomposed. The main difference between these two analog processes is the size selection process. The first screen retains the materials of the largest sizes whereas the first decomposition level depletes the image with the finer details (*i.e.*, high frequency information corresponding to smaller bubbles).

Computing the wavelet size signatures involves a series of steps described below [37]:

1. First the MR-MIA II algorithm is applied to the set of froth images, involving PCA decomposition of the set of images, and performing DWT on the $\mathbf{T}_1$ image up to S levels of decomposition. This image is very close to the gray scale conversion of the RGB image. The symlet order 4 wavelet was found to work well on froth images [37].
2. The gray scale approximation images for each decomposition level, $\mathbf{A}_0$, $\mathbf{A}_1$, $\mathbf{A}_2, \ldots, \mathbf{A}_S$, are used to obtain the size signatures. This is motivated by the fact that bubbles in froth images have very little directionality. The detail images $\mathbf{D}_s^q$, $s = 1,2,\ldots,S$, capturing strongly oriented features, will not be used directly in this approach for computing size signatures. At each level of decomposition s, the approximation images are depleted from a range of smaller bubbles (captured in detail images). When applied at level $s = 1,2,\ldots,S$ (*i.e.*, from high to low frequencies) the approximation images $\mathbf{A}_s$ will contain information about bubbles larger in size than the width of the wavelet at level s. The difference between adjacent approximation images (*i.e.*, $\mathbf{A}_s$ and $\mathbf{A}_{s+1}$) will eventually provide information about the quantity of bubbles of size corresponding to level s that have been removed or captured by the detail images (*i.e.*, associated with the wavelet

function) at that level. This will allow estimation of bubble size distribution as described in the following steps.

3. A threshold is applied to each $\mathbf{A}_s$ image to remove dark pixels and valleys from the image, thus enhancing the areas covered by the bubbles observed at this level, which size is greater than the wavelet width at level s. Let A_s be the total area covered by the bubbles in approximation image $\mathbf{A}_s$. This correspond to the total number of pixels within the image remaining after the threshold was applied (*i.e.*, pixels belonging to bubbles). The fractional area covered by these bubbles is computed as follows based on the total area seen by the camera (*i.e.*, field of view) or, alternatively, the total number of pixels of the image A_T:

$$F_s = \frac{A_s}{A_T}. \tag{3.14}$$

4. To obtain the range of bubbles sizes extracted at each level, adjacent fractional areas F_S are subtracted, that is, for six decomposition levels (*i.e.*, $S = 6$), $F_1 = \frac{(A_0 - A_1)}{A_T}$, $F_2 = \frac{(A_1 - A_2)}{A_T}, \ldots, F_5 = \frac{(A_4 - A_5)}{A_T}$. The last approximation image is usually left out since it often contains very low frequency information such as variations in lighting conditions [51].

5. Finally, the number of bubbles having a size within the range captured at each level N_s is computed as shown below, assuming circular bubbles. Indeed, horizontal and vertical diameters $D_{H,s}$ and $D_{V,s}$ are assumed equal, and correspond to the width of the wavelet at level s, that is the largest bubble size captured by the wavelet function at level s.

$$N_s = \frac{A_T F_s}{\frac{\pi}{4} D_{H,s} D_{V,s}}. \tag{3.15}$$

This procedure leads to very robust estimates of bubble sizes. It is not corrupted by noise, which is captured by the detail images at the first level, and not influenced by lighting variations, which for the resolution of images considered (*i.e.*, 720×480), appear in the sixth decomposition level. When A_s and A_T are computed in terms of number of pixels, the bubble sizes are relative measures or scaled bubble diameters. These could be calibrated to provide bubble diameters in engineering units (*i.e.*, cm) by measuring the actual area covered by the field of view A_T, and computing A_s as a proportion of the field of view.

To illustrate estimation of bubble size distribution using the wavelet size signatures, a histogram of the number of bubbles of different sizes (*i.e.*, N_s, $s = 1, 2, \ldots, 5$) for three different froth images are shown in Figure 3.17, where the scaled bubble diameter ($D_{H,s}$ or $D_{V,s}$) related with wavelet width at the various decomposition levels s. The difference in bubble size distribution observed in the images is clearly distinguished in the size histogram.

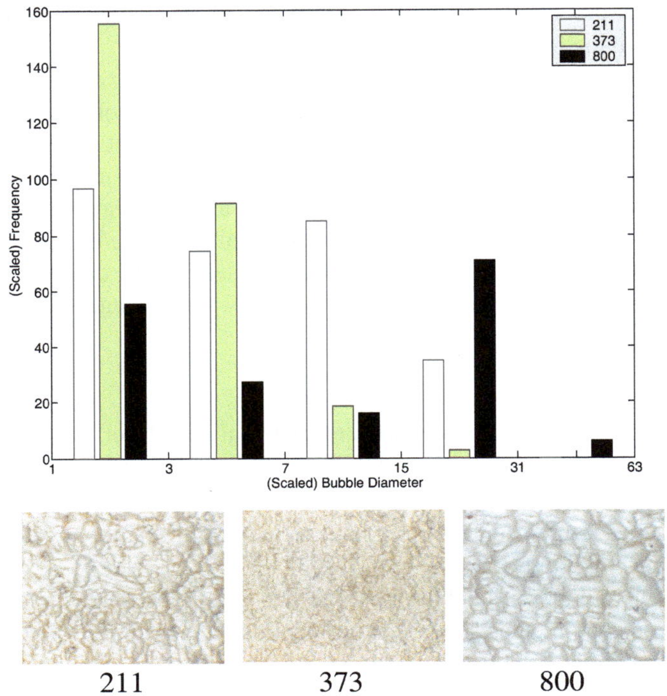

Figure 3.17 Bubble size histograms computed for three froth images. From Liu *et al.* [37]

3.5.2.2 Froth Color Features

Color features consists of black holes (BH) and clear windows (CL). Both of these are shown in Figure 3.18(a). When the froth is poorly loaded with mineral particles, bubbles size is large and regions not covered by particles starts appearing on top of bubbles. The black spots located on top of the bubbles of the froth image shown in Figure 3.18(a)(left) correspond to the so-called CL. Black holes appear in the opposite situation: when the froth is overloaded with mineral particles and bubbles are very small. The BH shown in Figure 3.18(a) (right) indicate the onset of froth collapse. The color features used in classification of froth state correspond to the number of pixels belonging to either BH or CL. The method used for computing these for each incoming image is illustrated in Figure 3.18.

A composite image made of two froth images, a poorly loaded (left) and a heavily loaded froth (right) is shown in Figure 3.18(a). Multivariate image analysis applied to this composite image (*i.e.*, PCA applied to the unfolded composite image) yields the $\mathbf{t}_1$–$\mathbf{t}_2$ score density histogram shown in Figure 3.18(b). The white mask appearing on that histogram, found by trial and error, encompasses most pixels having a black (or nearly black) color as shown in Figure 3.18(c). The pixels falling under-

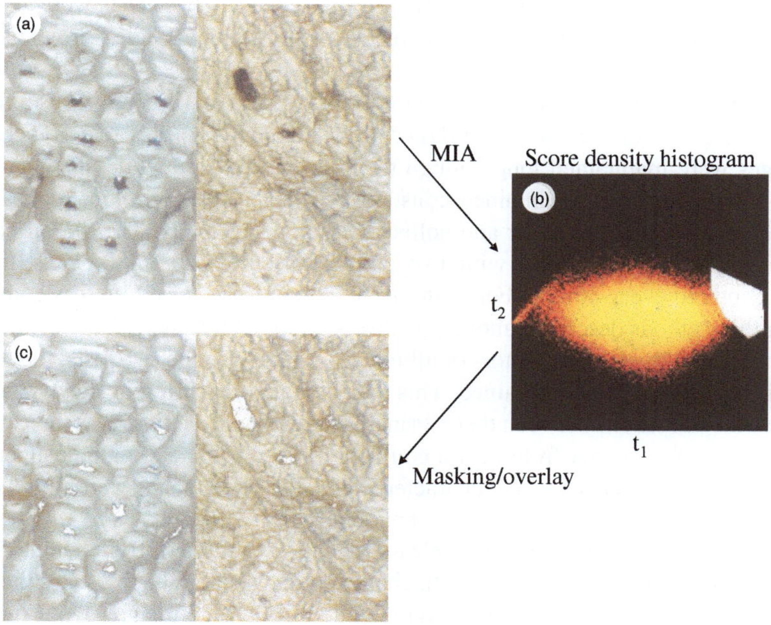

Figure 3.18 Computing the amounts of BH and CL using a composite image of a zinc froth. (a) Composite image of a lean (left) and heavily load froth (right). (b) 2-D score density histogram obtained for MIA applied to the composite image. (c) Extraction of BH and CL features using the score histogram (white mask). Adapted from Liu *et al.* [37]

neath the mask are mapped into the original composite image. This technique allows one to compute the total number of black pixels within a given froth image.

Since both CL and BH features appear black in froth images, it would be impossible to discriminate these features without additional, contextual information. These two features appear in very different froth conditions, the former when the froth is lean (*i.e.*, large bubbles) and the latter when the froth is overloaded (*i.e.*, small bubbles). Therefore, a two-class Fisher linear discriminant classifier [77] was trained on the froth wavelet size signatures discussed previously in order to assign the black pixels (*i.e.*, those falling under the white mask) to either CL or BH.

3.5.2.3 Froth Health Monitoring

The objective of froth health monitoring is to provide early detection of changes in the ore feed that would require the operators to adjust reagent dosage in order to maintain desired metallurgical performance (*i.e.*, grade and recovery). Observing such changes in ore feed would necessitate collecting images over long periods of time. Thus, it was decided to mimic inappropriate dosage of an ore feed by making designed changes on some of the reagents while the ore feed was relatively constant (*i.e.*, create over- and under-dosage for a given ore feed). These tests were also performed on the zinc flotation circuit (contact cell) of Agnico-Eagle/Laronde's plant (Figure 3.13).

Two plant tests were made on two different days 7 weeks apart. These involve steps changes in the activator ($CuSO_4$) and collector (KAX) flow rates. These step changes were maintained long enough to reach steady-state while all other operating conditions were maintained constant. A total of 487 and 456 images were collected during the activator and collector tests, respectively. The digital camera provided 24 bit RGB images with a spatial resolution of 720×480 pixels. Images were captured every minute. For each image, a feature vector of 1×7 is computed using the methods described above (*i.e.*, five size signatures, $N_1 - N_5$, BH and CL). After merging the feature vectors of all images captured in the two tests, the feature matrix $\mathbf{X_F}$ (943×7) was obtained. This data contains six different steady-states as well as the transitions between them. Principal component analysis was then applied to $\mathbf{X_F}$ to verify whether froth health could be tracked using a simple unsupervised classification of the froth visual characteristics. The steady-states were found easily distinguished in the $t_1 - t_2$ score plot shown in Figure 3.19.

The PCA score space is clearly able to track the evolution of froth health, from a poorly loaded froth (*e.g.*, image 840), characterized by large bubbles and the presence of clear windows, to a very heavily loaded froth (*e.g.*, image 413). The latter almost look like oatmeal and BH appear on the surface. Froth collapse could occur if no control actions are made to improve this condition (*i.e.*, reduce collector and/or activator dosage). Optimal operating conditions would fall somewhere in between these two extremes. Since the steady-states are clearly distinguished in between these extreme conditions (from left to right in the score plot), it would be straightforward to develop a multivariate statistical process control chart to warn operators that suboptimal conditions are detected and control actions are required.

The movements from left to right in the score space can be interpreted in terms of changes in the froth characteristics using the corresponding $p_1 - p_2$ loading plot shown in Figure 3.20. The froth images having high CL, N_4 and N_5 values (but lower values for the other variables) will fall on the left hand side on the score plot. Indeed, their p_1 loadings are negative, and since the scores are linear combinations of all variables, negative t_1 values will be obtained. Images having large amounts of CL and larger bubble size (higher values of N_4 and N_5) are associated with poorly loaded froth (see image 840 in Figure 3.19). On the other hand, the heavily loaded froths falling on the right hand side in the score plot (*i.e.*, positive t_1 values in Figure 3.19), such as image 19 and 413, have smaller bubble sizes (*i.e.*, higher values of N_1, N_2, and N_3) and more BH. This is shown in Figure 3.20 where the p_1 loadings for these variables are positive (*i.e.*, high values for these variables will push t_1 values in the positive quadrant). The excursion in the negative t_2 values (right hand side of the score plot) are associated with the presence or absence of BH on heavily loaded froths. For example, images 19 and 413 have similar bubble sizes (*i.e.*, they have similar t_1 values) but image 413 has some BH. The higher value for BH brings the t_2 score values into the bottom right quadrant. The appearance of BH indicates the onset of froth collapse.

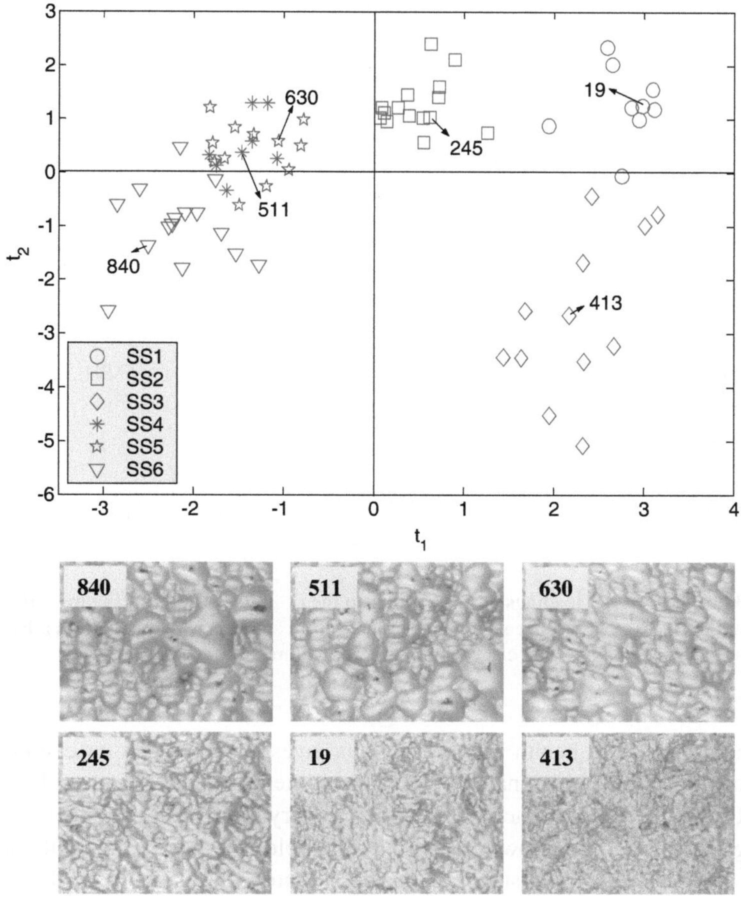

Figure 3.19 The $t_1 - t_2$ score plot of MR-MIA II features for selected steady-states images (SS1–SS6). Representative images for each steady-state are also shown at the bottom. From Liu *et al.* [37]

The continuous progression from one steady-state to the other is shown in greater detail in Figure 3.21. The transition from steady-states SS2 to SS3 is tracked using the solid line. Some representative froth images are also provided in order to follow this transition visually. This figure clearly demonstrates the ability of the proposed approach to monitor froth health, both in steady-state as well as in transitions. Note that the PCA model projection residuals (*i.e.*, SPE or DMODX) should also be used to monitor froth health. This plot would safeguard against operators reacting in a normal way to abnormal images (*i.e.* problem with the camera or lighting conditions) or other extreme conditions [37]. This plot was not included here since all the images were found to be well explained by the PCA model (*i.e.*, DMODX values were lower than limit), but is available in the paper by Liu *et al.*[37].

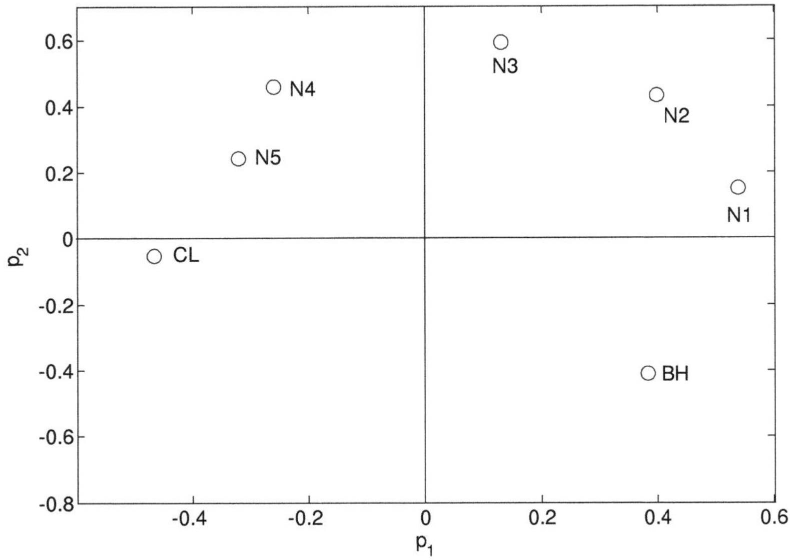

Figure 3.20 The $p_1 - p_2$ loadings corresponding to the $t_1 - t_2$ score plot of Figure 3.19. $N_1 - N_5$ correspond to the five bubble size signatures, ordered from smaller (N_1) to larger (N_5). BH and CL variables are the black holes and clear windows color features. From Liu *et al.* [37]

In summary, froth health monitoring was investigated in this case study using both wavelet bubble size signatures as well as some froth color features obtained using MR-MIA II. This approach was found to be very robust to noise and variations in lighting conditions, and allows estimation of bubble size distribution without image segmentation. Multivariate statistical process control charts could easily be established in order to warn the operator that changes need to be made to reagent dosage to maintain high metallurgical performance despite changes in ore feed properties. These charts could be made available in the control room for all the flotation cells or columns equipped with a digital camera. More appropriate and timely changes to reagent dosage would result, as well as a more efficient use of operator work time. An automatic froth health control scheme was also proposed by Liu *et al.* [59] based on the monitoring strategy presented in this case study. It essentially consists of computing the required control moves to implement reagent dosage in order to bring froth health, quantified by PCA scores (*i.e.*, $t_1 - t_2$ in Figure 3.19), back into a desired region. This control strategy uses both process data and images, and accounts for process dynamics. The control scheme is cast into an optimization problem. The reader is referred to Liu *et al.* [59] for more details on the approach.

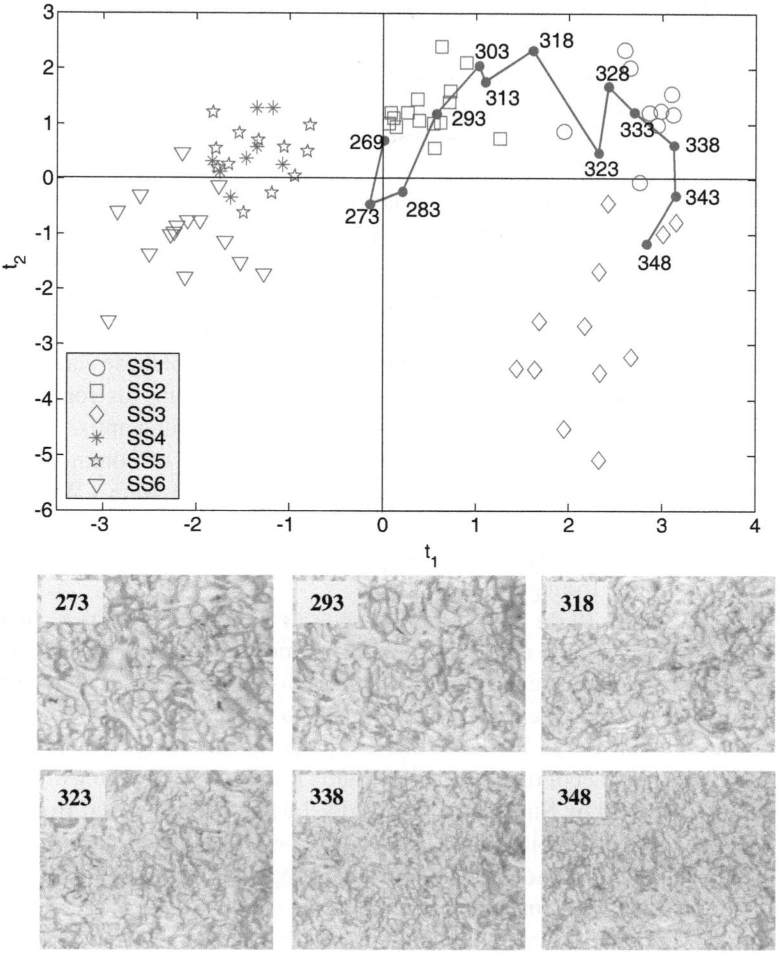

Figure 3.21 Process transition from steady-state SS2 to SS3 in the $t_1 - t_2$ score space of the PCA model. Selected images are also shown along the transition. From Liu *et al.* [37]

3.5.3 Estimation of Run-of-Mine Ore Composition on Conveyor Belts

In this last case study, the problem of estimating ROM ore composition (*i.e.*, lithological type, not elemental composition) on conveyor belts is investigated using multivariate imaging techniques. The very challenging nickel mineral system of Raglan's mine (Xstrata Nickel Canada, Inc.) is used to illustrate the methods. Each mineral is very heterogeneous, some have similar visual characteristics (*i.e.*, colors and surface texture), and the rock fragments can be dry or wet due to rain or snow. The machine vision approach therefore needs to take these issues into account in order to minimize rock type recognition errors. Hence, the proposed methodology requires a greater number of steps compared with the previous case studies due to the higher level of complexity of this machine vision problem. This section summarizes the work published by Tessier *et al.* [26, 78], and presents a supervised classification problem of image features using PLS discriminant analysis or PLS-DA.

It is well known that variations in ROM ore properties such as size, composition, and grindability strongly affect the performance of autogeneous (AG) and semi-autogeneous (SAG) grinding mills, such as power consumption, circulating load, and product size distribution [79]. The Julius Kruttschnitt Mineral Research Centre (JKMRC) has pioneered the Mine-to-Mill™[4] optimization projects to address these issues since 1998. In the past, most efforts to track and control these variations have focused on developing on-line size analysis using vision systems for troubleshooting and/or for feedforward/feedback control of crushers and grinding mills, for reducing power consumption and avoiding mill overload. A variety of commercial vision systems are currently available for that purpose, such as WipWare's WipFrag [80] and Metso's VisioRock [81]. These commercial packages use some segmentation algorithms and other traditional image processing techniques to compute ore size distributions on-line from gray scale images.

Monitoring other ROM ore properties, such as mineral species (*i.e.*, lithotype) and grindability, is a more difficult problem and most approaches developed in the past are either limited in the location where the measurement can be taken within the plant or in the size and type of the ore fragments to be analyzed. Furthermore, these approaches are generally specific to one ore property, and imaging approaches do not consider the impact of rock surface moisture. For example, tracking of ore grindability was proposed by Simkus and Dance [82] using data collected from drilling machines to infer grindability, then following truck loads using GPS technology, and finally modeling of stock piles, bins and hoppers to estimate the work index of the ore incoming the concentrator plant. State estimation techniques were also proposed for grindability [83, 84], involving detailed mechanistic models as well as input–output process data. Machine vision approaches to mineral type recognition have also been proposed for a high color contrast binary mixture of fine chalcopyrite and molybdenite particles [62] as well as for ternary mixtures of manganese, iron and alumina [85], again having very high color contrasts. A method for gold ores having different textural characteristics was proposed by Petersen *et al.* [86]. Copper ores were classified into seven lithotypes using several image features (*i.e.*, 667 color, textural and geometrical features) and neural networks . Methods for rock texture retrieval in large databases were also found in the pattern recognition literature [88],[89]. Finally, grading of crushed aggregates (*i.e.*, civil engineering materials) was proposed by Murtagh *et al.* [90] using WTA and PCA decomposition. A more extensive literature review is available in Tessier *et al.* [26, 78].

The objective of this study is to develop a machine vision approach that would classify rock types on a conveyor belt in order to estimate their proportions in the

[4] Mine-to-Mill is a registered trademark of the Julius Kruttschnitt Mineral Research Centre Isles Road, Indooroopilly, Qld Australia 4068, www.jkmrc.com.au

ROM mixtures. On-line rock composition could be help modify the sorting/blending strategy for regularizing the grindability of the ROM mixture before feeding the AG/SAG grinding mills. Furthermore, this information could be used as an input to a feedfoward controller adjusting the operating conditions of the mill (*e.g.*, throughput or water addition), and even be used to modify flotation reagent dosage proactively [81].

3.5.3.1 Mineral Characteristics and Vision System Hardware

About 3000 tons/day of ROM is extracted from an open pit and different underground mines at Raglan (Xstrata Nickel Canada, Inc.). The mine is located in the Nunavik territory of northern Quebec (Canada). The five mineral types shown in Figure 3.22 are the most frequently observed in the ROM ore. These are massive sulphide (MS), disseminated sulphide (DS), "net textured" (NT), gabbro (G), and peridotite (P). These rocks are highly variable in nickel grade and grindability (*i.e.* crushing or breaking energy) as shown in Table 3.2. Since DS and NT rocks are very similar in terms of Ni grade and grindability, they are considered as belonging to the same class in the rock recognition problem (see Table 3.2). Gabbro and P rocks are also pooled in the same class for similar reasons. Note that, in this particular case, classifying the rocks according to nickel grade or grindability is the same problem since high grade rocks are the softest, medium grade have intermediate grindability, and waste rocks are the hardest. Therefore, the classification results obtained in this work are equally valid for both properties.

Figure 3.22 Images of the five different mineral types from Raglan's mine. From Tessier *et al.* [26]

A large quantity of each rock type was obtained from Raglan's mine. These were sorted by experienced mineralurgists. Rock fragments were 100% -12 inches, but fines were removed from the analysis. Indeed, the fines are difficult to see on the conveyor belt.

Images of rocks fragments from individual classes or mixtures were obtained using a RGB digital camera (JAI CV-M7+) mounted with a 25 mm lens. This cam-

Table 3.2 Main characteristics and class assignment for the five mineral types. From Tessier *et al.* [26]

Mineral type	Impact work index	Ni grade	Density (kg/m^3)	Hardness	Class
Massive sulphide (MS)	Low	High	4000	Soft	1
Disseminated sulphide (DS)	Medium	Medium	2300	Medium	2
Net textured (NT)	Medium	Medium	2300	Medium	2
Gabbro (G)	High	Waste	3000	Hard	3
Peridotite (P)	High	Waste	2700	Hard	3

era provides 8 bit images of 1024×1376 pixels spatial resolution. The camera was installed 90 cm above the conveyor leading to a 32×24 cm field of view. The conveyor was located at COREM's pilot plant (Québec City, Canada). More details on the hardware used, including the lighting system, are available in Tessier *et al.* [26].

3.5.3.2 The Mineral Recognition Problem

Rock type identification for this mineral system is challenging for three reasons:

1. The visual characteristics of individual rock fragments are themselves very heterogeneous. Their color and surface texture patterns vary from one face to the other. This can be clearly observed in the massive sulphide and gabbro images available in Figure 3.22. Hence, the intra-class variability in the rock fragments visual features is important, at least for some of them.
2. A strong overlap in some color and surface textural features between rock fragments belonging to different classes is observed in Figure 3.23 A-B. Image A is a composite image made from larger images of each rock type. The color signature of the pixels segmented by the blue region drawn on the MS rock image was quantified using MIA. The pixels belonging to other rock types but having similar color features as MS are identified by the blue color in image B. The strong color overlap between MS, DS and NT rocks is clearly observed. Hence, interclass variability may be small between some of the classes (*i.e.* some classes may overlap in the classification problem).
3. The rock surfaces can be completely dry or wet depending on weather conditions. Surface moisture modifies the visual appearance of the fragments, their color mainly. Wet rocks are systematically darker when wet as shown in Figure 3.23 C (dry) and D (wet).

The machine vision approach will address each of these issues in order to achieve accurate rock type classification.

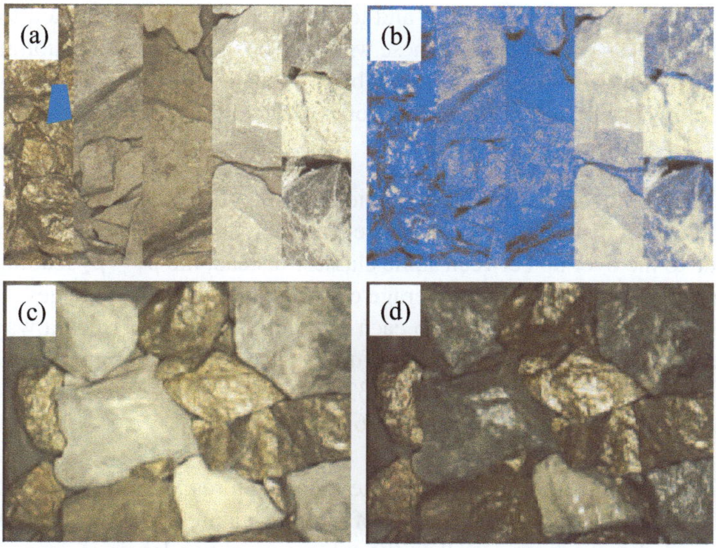

Figure 3.23 Difficulties with the investigated mineral system: (a) composite image of the five rock types including a mask selecting pixels belonging to MS. (b) Composite image overlaid with pixels (in blue) having similar color characteristics to those falling under the MS mask. (c) Dry run-of-mine ore. (d) Wet ROM ore. From Tessier *et al.* [26]

3.5.3.3 The Machine Vision Approach

The proposed machine vision approach is presented here using the framework shown in Figure 3.4.

1. **Image pre-processing.** Visual features cannot be directly extracted from images of rock mixtures (*e.g.*, Figure 3.23(c)) due to the color overlap between some rock classes (Figure 3.23(a) and (b)). Furthermore, applying texture methods to such an image would extract rock edge information and overall surface texture of the rock blend which are not relevant informations for recognizing individual rock types. The full size images of rock mixtures are therefore subdivided into smaller images which more likely belong to a single rock type. This can be accomplished using the traditional image segmentation algorithms. This would be particularly efficient when optical rock sizing is also desired. Alternatively, the full size image can be divided using a rectangular grid of the certain mesh size. This is a simpler method, but requires the subimages to be relatively small in order to make sure that most of them contain information from a single rock fragment. The latter approach was used here. The full size 1024×1376 images were divided into 512 smaller subimages of 64×43 pixels, with a corresponding field of view of 1.5 cm^2. A similar approach was used in other mining vision problems [85, 88].

2. **Feature extraction.** Both color and textural features were extracted from each subimage as described below. These will later be used to classify subimages into one of the three rock classes (see Table 3.2). The classification results will, in turn, allow estimation of the proportions of the various rock types within a mixture.

   - **Color features.** As discussed previously, pixel-by-pixel color signatures (*i.e.*, MIA color distribution features) are not efficient for discriminating rock types due to the overlap between some classes. Instead, the $p_1 - p_2$ loading vectors are used as representative average color features of the subimages. These are obtained by applying PCA to each individual subimage. Such a way of representing average image color is further discussed and compared with other methods in Yu and MacGregor [32]. The vector of color features for each subimage $[\mathbf{p}_1^\top \mathbf{p}_2^\top]$ is therefore (1×6)-D.

   - **Textural features.** Two sets of textural features were computed for each subimage. The first set consists of energy features (*i.e.* Equation 3.13) obtained after applying DWT on the subimages at two levels of decomposition. This yields six energy features since all three detail images (h, v, d) were used at both levels. To generate even more informations about the images, second order statistics obtained from the six detail images were also computed as proposed in [36, 91]. These are the energy, entropy, contrast and correlation features obtained from the GLCM [54] of the six details images. This provides 24 additional texture descriptors for each subimage, resulting in a (1×30) texture vector after combining the two sets of textural features.

3. **Feature reduction and analysis.** After extracting and combining the color and textural features for all subimages, the following matrices are obtained for one full size rock mixture image: $\mathbf{X_F}$ (512×36) and $\mathbf{Y}$ (512×3). The latter corresponds to binary numbers (*i.e.*, 0–1) assigning each subimage to a rock class. If the ith subimage belongs to class 1, then $\mathbf{Y}(i,:) = [1 \quad 0 \quad 0]$, if it rather belongs to class 2, $\mathbf{Y}(i,:) = [0 \quad 1 \quad 0]$, and so on. This correspondence between subimages and rock classes is obviously only known *a priori* for the training dataset.

   - **Discriminant PLS.** To reduce the 36-D feature space and to enhance class separation, made necessary due to overlap, a supervised latent variable classification technique (*i.e.*, PLS-DA) was then applied to ($\mathbf{X_F}$, $\mathbf{Y}$). The PLS-DA model extracts those linear combinations in $\mathbf{X_F}$ that are the most highly correlated with the classes or, alternatively, the information that maximizes class discrimination. This operation also partially removes the systematic color variations caused by surface moisture. Indeed, moisture will be made independent of class information by design since all three rock classes will contain dry and wet rocks. Hence, the PLS-DA model will not capture the effect of moisture. This should increase the robustness of the rock classification model to weather conditions.

- **Establishing class boundaries.** The final step involves establishing class boundaries of each rock type within the latent variable space of the PLS-DA model (*i.e.*, in the score space of that model). A wide range of techniques are available for this purpose, from Fisher's linear discriminant analysis [77] to nonlinear classification algorithms such as self-organizing maps [92], based on artificial neural networks, or support vector machines [40, 93, 94]. The latter will be used in this case study since it is efficient with overlapping classes.

3.5.3.4 Training and Validation of the Mineral Type Classification Model

A database of rock images was first collected for training and validation of the rock type classification model. The available rock loads for each type were sequentially spread on a conveyor belt. Then, between 50 and 60 images were captured for each rock type in order to cover the variability of appearance of each type. After collecting a dry rock image, and before moving the conveyor, the rocks contained within the field of view were watered and a second image was captured (*i.e.*, the wet rock image). Therefore, a wet rock image corresponds to each dry rock image. Repeating this procedure for each rock type led to a set of about 500 images with almost half of them showing wet rocks.

The training dataset was made of 10 composite images representing the variability of each rock type in dry or wet conditions. Each composite image of a particular rock type was formed with 10 smaller images randomly selected from the larger database discussed above. The size of the smaller images was selected such that the composite images have the same size as the original images in the large database (*i.e.*, 1024×1376 pixels). Furthermore, once a smaller image was selected for one of the five dry composite images, its wet counterpart was also selected for the corresponding wet composite image. These training images are available in [78].

The procedure described in Section 3.5.3.3 was then applied to the 10 composite images of the training set, one for each rock type in dry and wet condition. That is, each image was further divided into 512 subimages after which color and textural features were extracted for each. The resulting feature and class matrices, $\mathbf{X_F}$ and $\mathbf{Y}$, were (4316×36) and (4316×3)-D, respectively, after removing some outlier images showing the conveyor belt only. The soft/high grade rocks (MS) were assigned to class 1, the medium hardness/grade rocks (DS and NT) were assigned to class 2 and, finally, the hard/waste rocks (G and P) were assigned to class 3. These components PLS-DA model was built on the training dataset, together explaining 44% of the variance of $\mathbf{X_F}$. That is, about half of the information contained in the feature matrix is correlated with the classes and is used for class discrimination. The $t_1 - t_2$ score scatter plot of the PLS-DA model showing the clusters corresponding to each class are shown in Figure 3.24. Note that in these score plots, each dot corresponds to a multivariate summary of the visual features of one subimage.

If the ROM ore consists of only soft/high grade and hard/waste rocks (*i.e.*, class 1 and 3), then the classification problem would be easier as shown in Figure 3.24(a). A simple linear classification problem aiming at finding the location of a straight line

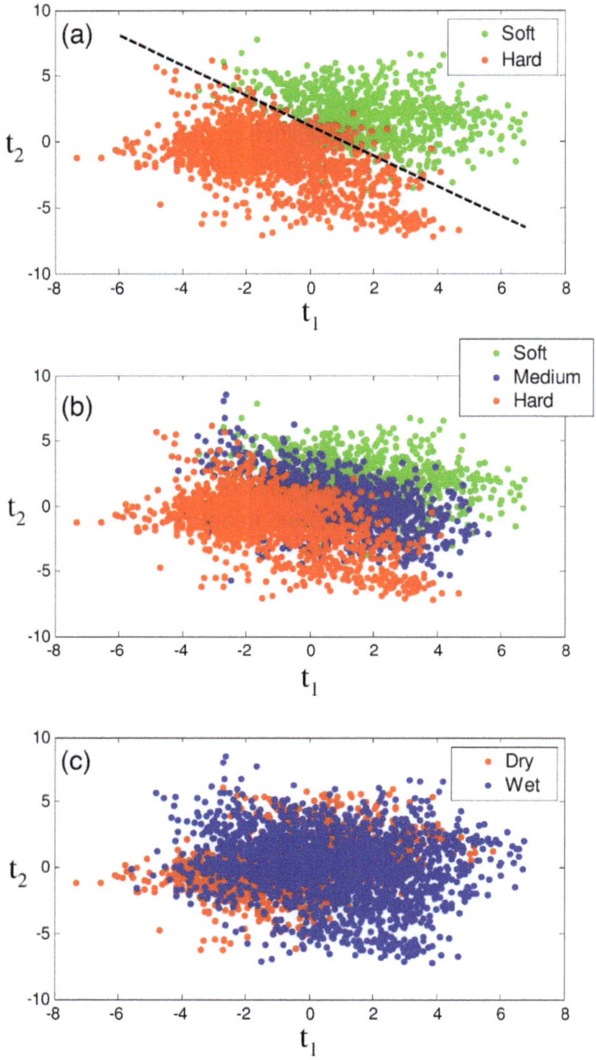

Figure 3.24 Scatter plots of the first two latent variables ($t_1 - t_2$) of the PLS-DA model for the training dataset. (a) Soft and hard classes. (b) All three classes. (c) Distribution of dry and wet subimages. From Tessier *et al.* [26]

in the $t_1 - t_2$ latent variable space that minimizes classification errors could be used for automatic identification of soft and hard rocks (*i.e.* green and red dots, respectively). However, when the medium hardness/grade rocks (class 2) are introduced the classification problem becomes more difficult as shown in Figure 3.24(b). The medium rock class (*i.e.*, blue dots) significantly overlap with the other two classes, and the contribution of the third component (not shown) to class discrimination is marginal. This justifies the need for more sophisticated methods for establishing class boundaries in the presence of significant class overlap. Support vector machines, a statistical learning technique [40, 93, 94], was used in this study for segmenting class boundaries. It essentially projects the data into a higher dimensional space where simple linear classification is possible. This is particularly useful when clusters overlap. A complete description of SVM is outside the scope of this chapter. However, the reader is referred to the following publications for more details [40, 93, 94]. A total of three 3-class SVM models were trained for establishing class boundaries in the latent variable space of the PLS-DA model (*i.e.* Figure 3.24(b)). The parameters of these models as well as their performance in training are described in [26, 78]. The performance of the complete machine vision approach will be discuss later in this section.

The last score plot in Figure 3.24(c) shows the clustering pattern of the subimages from all three rock classes, but dry and wet images were identified using red and blue dots, respectively. Both clusters overlap almost completely, which supports the assumption that PLS-DA does not treat dry and wet images differently, but focuses only on rock class discrimination based on rock hardness/grade. This should improve the robustness of the classification model to systematic variations in rock visual appearance introduced by surface moisture condition, which is irrelevant for classifying rocks in hardness/grade categories.

To validate the rock type classification model a second series of 10 composite images was created, five for dry and five for wet rocks. One composite image of dry rocks is shown in Figure 3.25(a). These composite images again consist of 10 smaller images but, instead of showing different pictures of a single rock type as for the training phase, they include smaller images belonging to all three rock classes. The advantages of this approach, compared with using pre-defined rock mixtures on the conveyor belt (*i.e.*, Figure 3.25 (b)), is that the location of each rock type within the composite image is perfectly known, and it is also easier to control the composition of the rock mixture. For example, if two smaller images of MS rocks appear in a composite image, then MS proportion in the mixture is 20%. For the dry rock composite images, five different proportions of the three rock types were simulated by deciding upon the number of smaller images of each rock type to include in the composite image. Then, the required number of smaller images of each rock type were randomly selected from the large image database discussed earlier in this section. The five wet composite images were made of the wet couterpart of the dry images. Again, a correspondence between wet and dry images was maintained for easier comparison. It is also important to mention that none of the smaller images selected in the validation step were used in the training dataset.

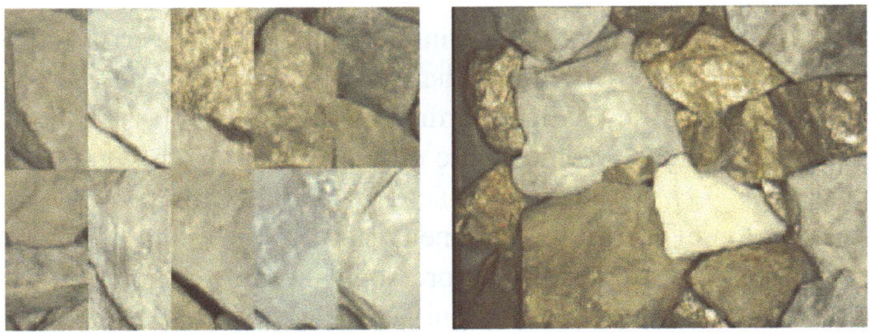

(a) (b)

Figure 3.25 Typical rock mixture images. (a) Composite. (b) As would appear on a conveyor belt. From Tessier *et al.* [26]

After applying the machine vision approach to the 10 validation composite images, the results presented in Table 3.3 were obtained: this shows a comparison between true and estimated surface composition. The latter is defined as the percentage of the field of view of the image occupied by each of the three rock classes. As expected, the results are more accurate for dry rocks than for wet rocks. Even if the PLS-DA model partially removed systematic variations associated with surface moisture, wet rock images have much less contrast. Furthermore, most classification errors come from the overlap of the medium hardness/grade rock class with the other two. On average, the classification model provides satisfactory results as shown at the bottom of Table 3.3 (see mean as well as correlation between true and estimated compositions). Results are very good for dry rocks but classification performance degrades for wet rocks.

3.5.3.5 Mineral Type Recognition on a Conveyor Belt

A simulated conveyor belt application was performed on more realistic rock mixture images (Figure 3.25(b)) to complete the proof of concept of the proposed machine vision approach. To maximize the coverage of rock mixtures, the ternary mixture design shown in Figure 3.26 was used. A total of 17 mixtures of the three rock types were selected. Among this number, 13 involve the three rock classes, and five others consist of each of the individual rock species. These mixtures are shown schematically in Figure 3.26 (a) whereas the corresponding proportions are provided in the table, Figure 3.26 (b). Note that the proportions used in this design are weight percentages.

Once the amounts (*i.e.*, weights) of each rock type required by any given experiment were gathered, the loads were manually mixed and spread on the conveyor belt. Each load covered from 8 to 15 feet of conveyor length. As the belt moved forward, several images of the rock mixtures were captured. The number of images collected for each mixture is also given in Figure 3.26(b). Here again, after capturing a dry rock mixture image, and before moving the belt, the rocks were watered and a wet image was collected. Correspondence between dry and wet rock images

Table 3.3 Validation results for the 10 composite images. From Tessier *et al.* [26]

		True surface composition (%)			Estimated surface composition (%)		
Image	Ore	Soft	Medium	Hard	Soft	Medium	Hard
1	Dry	20	50	30	19.0	50.5	30.5
2	Dry	20	60	20	25.6	56.9	17.5
3	Dry	20	30	50	12.1	24.9	63.0
4	Dry	20	40	40	21.3	34.7	44.0
5	Dry	20	30	50	22.0	28.9	49.1
6	Wet	20	50	30	23.6	61.9	14.8
7	Wet	20	60	20	29.2	67.4	3.4
8	Wet	20	30	50	22.8	34.6	42.6
9	Wet	30	40	30	39.3	47.2	13.5
10	Wet	20	30	50	29.2	39.4	31.4
Mean	Overall	21	42	37	24.4	44.6	31.0
	Dry	20	42	38	20.0	39.2	40.8
	Wet	22	42	36	28.8	50.0	21.1
Correlation between true and estimated				Dry	0.73	0.90	0.84
				Wet	0.65	0.47	0.63

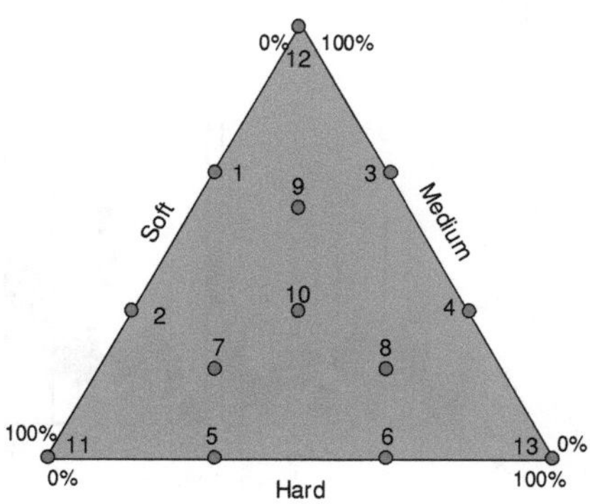

(a)

Experiment #	Weight % Soft	Weight % Medium	Weight % Hard	Number of images
1	33	66	0	26
2	66	33	0	18
3	0	66	33	19
4	0	33	66	27
5	66	0	33	31
6	33	0	66	30
7	60	20	20	23
8	20	20	60	26
9	20	60	20	23
10	33	33	33	30
11 (Pure MS)	100	0	0	24
12	0	100	0	17
13	0	0	100	42
14 (Pure DS)	0	100	0	75
15 (Pure NT)	0	100	0	52
16 (Pure G)	0	0	100	50
17 (Pure P)	0	0	100	46

(b)

Figure 3.26 Design of experiment for conveyor belt application. From Tessier *et al.* [26]

was preserved in this final application. In the end, about 560 images were collected including both dry and wet rocks. The machine vision approach was applied to each image and the rock classification results are provided in Figure 3.27, which presents a comparison between the true and estimated weight compositions (solid line and dots, respectively). To convert surface composition obtained by imaging into weight composition, the estimated surface composition were weighted according to the density of each mineral (see Table 3.2).

Very promising results were obtained, particularly for dry rock mixtures. The trends are followed quite well, although some confusion between medium and hard

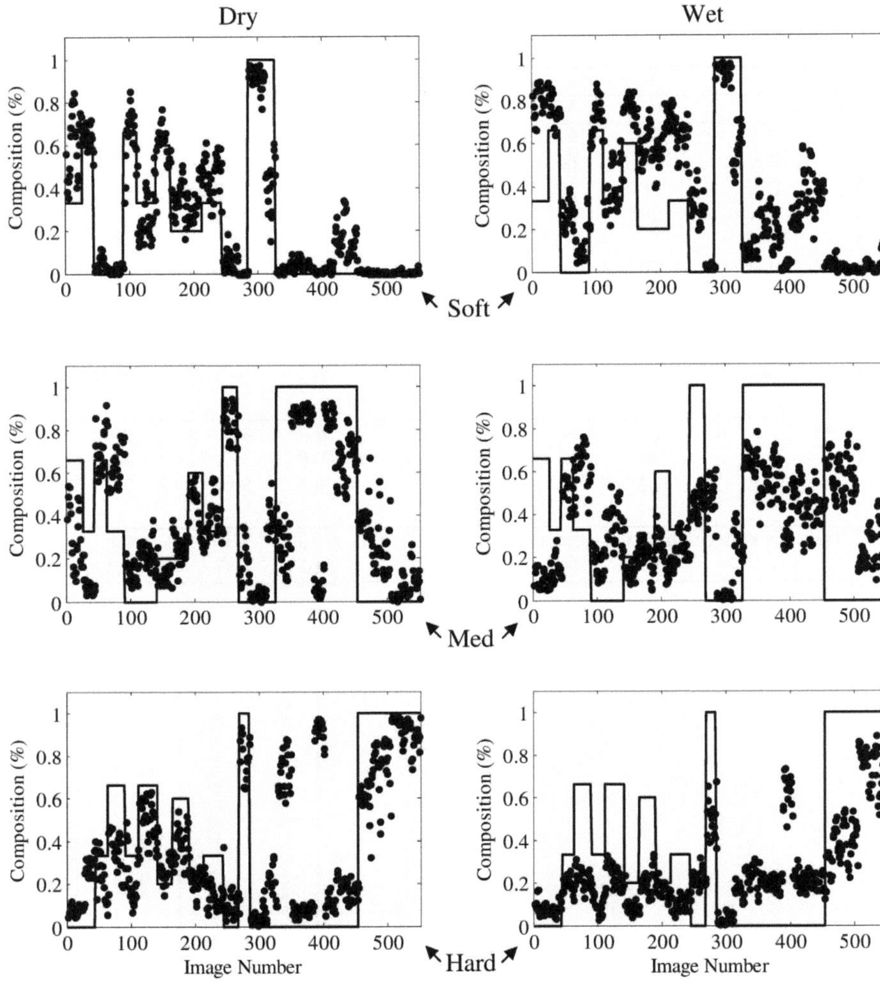

Figure 3.27 Estimation of rock composition for the simulated on-line conveyor belt application. Dots: model predictions. Solid lines: true weight compositions. Columns: dry and wet rocks. Rows: soft, medium and hard rocks, from top to bottom. From Tessier *et al.* [26]

rocks occur for images in the range 350–400. For wet rocks, specular reflection due to wet surfaces seems to be confounded with the shiny surface of MS rocks. This may explain why MS composition (soft rocks) is overestimated by 20–30% in the first half of the test. This problem is less important in the second half of the test where either 100% or 0% MS is present on the conveyor. Adding imaging hardware to remove specular reflection combined with contrast enhancement methods may help improve classification results for wet rock mixtures. Since this was one of the first attempt to directly take into account surface moisture in the rock type classification problem, further investigations of these issues are required. Overall, the results were found to be very promising and support future work on this subject.

3.6 Conclusions

Multivariate imaging techniques were presented in this chapter and applied to the development of three novel on-line vision sensors for use in the mineral processing field. These sensors aim at: (1) estimating flotation froth mineral grade; (2) froth health monitoring for appropriate reagent dosage; and (3) on-line estimation of ROM ore lithotype composition on conveyor belts. Both MIA, based on well known PCA and PLS, and MRA, relying on wavelet decomposition, have been presented and used for spectral/textural analysis of process images (*i.e.*, flotation froth and ROM rock mixtures). These techniques were shown to be very efficient for extracting the stochastic features of process images relevant for monitoring and/or predicting process conditions or product quality. As opposed to traditional automatic inspection problems in the manufacturing industries, the relevant features of images collected in process industries are usually unknown *a priori*. A machine vision framework based on multivariate methods was also presented for identifying these relevant image features, and for using them in mineral process modeling, monitoring, and advanced control.

The first case study served to illustrate a multivariate image regression problem using spectral image information only. It was shown that predictions of zinc mineral grade obtained using RGB froth images were in a good agreement with the trends of the laboratory assays. The long-term robustness of such a sensor to lighting conditions, mineral feed and other environmental conditions still needs to be demonstrated, but the results obtained at the Laronde plant were found encouraging and granted further investigation by the research consortium COREM (unpublished work). Among other issues to be investigated are (1) the optimal positioning of the camera for a given flotation cell or column to make sure that images are representative of the entire froth surface, (2) selecting the best flotation unit(s) within a bank on which to install the camera. Fast on-line froth grade measurements could be integrated into a grade control loop manipulating reagent dosage when mineral feed and other disturbances occur. It could also be used to decide how to distribute the reagents across the flotation bank for optimal use of chemicals.

An unsupervised classification problem using both spectral and textural image features was proposed as the second case study, again using the froth flotation process. The concept of on-line "froth health" was presented for detecting inappropriate reagent dosage for a given ore feed composition. A new bubble sizing algorithm based on wavelet decomposition (*i.e.*, wavelet size signatures) was discussed. It is robust to noise and to variations in lighting conditions, and does not require direct image segmentation. After combining with some froth color features (*i.e.*, clear windows and black holes), a PCA monitoring scheme was developed and was shown to be able to detect the onset of froth collapse, related to over-dosage of reagents with respect to a given ore feed, as well as the opposite situation where under-dosage leads to poor mineral recovery. Providing such information to operators in the control room could result in more stable operation, better mineral recovery and more efficient use of operator's attention. Straightforward modifications to the proposed approach could also yield froth grade predictions as discussed in the first case study.

The last case study presents the very difficult problem of estimating ROM ore lithotype composition on conveyor belts based on dry and wet rock fragments of highly heterogeneous visual appearance, but also showing significant similarities between some rock types. This problem was selected to illustrate supervised classification of both spectral and textural features, requiring several steps due to the level of complexity of the rock identification problem. On-line classification of rock fragments according to their grade or grindability could be useful for modifying the sorting/blending strategy before feeding AG/SAG grinding mills, for feedfoward control of the mill's operating conditions (*e.g.*, throughput or water addition) and for proactive adjustments to flotation reagent dosage. Very promising results were obtained, particularly for dry rock mixtures. For wet rocks, specular reflection due to wet surfaces and lower color contrast led to greater classification errors. Since this was one of the first attempt to directly take into account surface moisture in the rock type classification problem, further investigations on these issues are required. In particular, adding imaging hardware to remove specular reflection combined with contrast enhancement methods may help improve classification results for wet rock mixtures. Overall, the results were found to be very promising and suggest future work on this subject.

The multivariate imaging methods presented in this chapter have also been used with great success in a wide variety of fields ranging from remote sensing [25, 29, 95], combustion, pyrometallurgy, and the steel industry [34, 38, 39], forest products and pulp and paper [30, 31], snackfood industry [32, 33], plastics [52, 96, 97], and even medical imaging [29, 98]. In most process cases, the vision sensors were used to develop new advanced control schemes that could not be implemented using standard process instrumentation. The author hopes that this chapter will help stir some interest in the mineral processing research community. Fast, robust and reliable vision sensors are still required to improve mineral process control, stability and productivity.

Acknowledgements Financial support from the Natural Sciences and Engineering Research Council of Canada (NSERC) and from COREM is greatly acknowledged. Thanks to the personnel of Agnico-Eagle/Laronde and Xstrata Nickel Canada for providing the necessary materials and access to their plants as well as for sharing their process knowledge. Special thanks go to Dr June J. Liu and Professor John F. MacGregor for very insightful discussions on multivariate imaging and for providing some of the materials used in this chapter. Finally, the author would like to thank Gianni Bartolacci (COREM) for close collaboration in most of the case studies presented in this chapter.

References

[1] Kourti T (2002) Process analysis and abnormal aituation detection: from theory to practice. IEEE Control Syst Mag 22:10–25

[2] Miletic I, Quinn S, Dudzic M, Vaculik V, Champagne M (2004) An industrial perspective on implementing on-line applications of multivariate statistics. J Process Control 14:821–836

[3] MacGregor JF, Yu H, Muñoz SG, Flores-Cerrillo J (2005) Data-based latent variable methods for process analysis, monitoring and control. Comput Chem Eng 29:1217–1223

[4] Kourti T (2005) Application of latent variable methods to process control and multivariate statistical process control in industry. Int J Adapt Control Signal Process 19:213–246

[5] Cutmore NG, Liu Y, Middleton AG (1997) Ore charactetisation and sorting. Miner Eng 10:421–426

[6] Zeng Y, Zheng M, Forssberg E (1993) Monitoring jaw crushing parameters via vibration signal measurement. Int J Miner Process 39:199–208

[7] Zeng Y, Forssberg KSE (1995) Multivariate statistical analysis of vibration signals from industrial scale ball grinding. Miner Eng 8:389–399

[8] Bergh LG, Yianatos JB, León A (2005) Multivariate projection methods applied to flotation columns. Miner Eng 18:721–723

[9] Gardner S, Le Roux NJ, Aldrich C (2005) Process data visualisation with biplots. Miner Eng 18:955–968

[10] Groenewald JW de V, Coetzer LP, Aldrich C (2006) Statistical monitoring of a grinding circuit: an industrial case study. Miner Eng 19:1138–1148

[11] Hodouin D, MacGregor JF, Hou M, Franklin M (1993) Multivariate statistical analysis of mineral processing plant data. CIM Bull 86(975):23–34

[12] Gonzalez GD, Orchard M, Cerda JL, Casali A, Vallebuona G (2003) Local models for soft-sensors in a rougher flotation bank. Miner Eng 16:441–453

[13] Liu J (2004) Machine vision for process industries: monitoring, control, and optimization of visual quality of processes and products. PhD thesis, Department of Chemical Engineering, McMaster University, Hamilton, Ontario, Canada

[14] Marshall AD, Martin RR (1992) Computer Vision, Models and Inspection. World Scientific Publishing, Singapore

[15] Wold S, Esbensen K, Geladi P (1987) Principal component analysis. Chemom Intell Lab Syst 2:37–52

[16] Wold S (1978) Cross-validatory estimation of the number of components in factor and principal components models. Technometrics 20:397–405

[17] Geladi P, Kowalski BR (1986) Partial least-squares regression: a tutorial. Anal Chim Acta 185:1–17

[18] Geladi P (1988) Notes on the history and nature of partial least squares (PLS) modelling. J Chemom 2:231–246

[19] Höskuldsson A (1988) PLS regression methods. J Chemom 2:211–228

[20] Burnham AJ, Viveros R, MacGregor JF (1996) Frameworks for latent variable multivariate regression. J Chemom 10:31–45

[21] Eriksson L, Johansson E, Kettaneh-Wold N, Trygg J, Wikstrom C, Wold S (2006) Multivariate and Megavariate Data Analysis: Part I – Basic Principles and Applications. Umetrics Academy

[22] Valle S, Li W, Qin SJ (1999) Selection of the number of principal components: the variance of reconstruction error criterion with a comparison to other methods. Ind Eng Chem Res 38:4389–4401

[23] Duchesne C, MacGregor JF (2001) Jackknife and bootstrap methods in the identification of dynamic models. J Process Control 11:553–564

[24] Nomikos P, MacGregor JF (1994) Monitoring of batch processes using multi-way principal component analysis. AIChE J 40:1361–1375

[25] Bharati MH, MacGregor JF (1998) Multivariate image analysis for real-time process monitoring and control. Ind Eng Chem Res 37:4715–4724
[26] Tessier J, Duchesne C, Bartolacci G. (2007) A machine vision approach to on-line estimation of run-of-mine ore composition on conveyor belts. Miner Eng 20:1129–1144
[27] Gonzalez RC, Woods RE (1993) Digital Image Processing. Addison-Wesley, Reading, MA
[28] Esbensen KH, Geladi P (1989) Strategy of multivariate image analysis (MIA). Chemom Intell Lab Syst 7:67–86
[29] Geladi P, Grahn H (1996) Multivariate Image Analysis, Wiley, New-York
[30] Bharati MH, MacGregor JF, Tropper W (2003) Softwood lumber grading through on-line multivariate image analysis, Ind Eng Chem Res 42:5345–5353
[31] Bharati MH, MacGregor JF, Champagne M (2004) Using near-infrared multivariate image regression techniques to predict pulp properties. TAPPI J 3:1–7
[32] Yu H, MacGregor JF (2003) Multivariate image analysis and regression for prediction of coating content and distribution in the production of snack foods. Chemom Intell Lab Syst 67:125–144
[33] Yu H, MacGregor JF, Haarsma G, Bourg W (2003) Digital imaging for on-line monitoring and control of industrial snack food processes. Ind Eng Chem Res 42:3036–3044
[34] Yu H, MacGregor JF (2004) Monitoring flames in an industrial boiler using multivariate image analysis. AIChE J 50:1474–1483
[35] Duchesne C, Bouajila A, Bartolacci G, Pelletier P, Breau Y, Fournier J, Girard D (2003) Application of multivariate Image analysis (MIA) to predict concentrate grade in froth flotation processes. In: Proceedings of the 35th Annual CMP Meeting, Ottawa (Ontario), Canada, January 21–23, 511–526
[36] Bartolacci G Pelletier P, Tessier J, Duchesne C, Bossé PA, Fournier J (2006) Application of numerical image analysis to process diagnosis and physical parameter measurement in mineral processes - Part I: Flotation control based on froth textural characteristics, Miner Eng 19:734–747
[37] Liu JJ, MacGregor JF, Duchesne C, Bartolacci G (2005) Flotation froth monitoring using multi-resolutional multivariate image analysis. Miner Eng 18:65–76
[38] Szatvanyi G, Duchesne C, Bartolacci G (2006) Multivariate image analysis of flames for product quality and combustion control in rotary kilns. Ind Eng Chem Res 45:4706–4715
[39] Graham KJ, Krishnapisharody K, Irons GA, MacGregor JF (2007) Ladle eye area measurement using multivariate image analysis. Can Metall Q 46:397–405
[40] Liu JJ, Bharati MH, Dunn KG, MacGregor JF (2005) Automatic masking in multivariate image analysis using support vector machines. Chemom Intell Lab Syst 79:42–54
[41] Mallat S (1989) A theory for multiresolution signal decomposition: the wavelet representation. IEEE Trans Pattern Recog Machine Intell 11:674–693
[42] Mallat S (1989) Multiresolution approximations and wavelet orthonormal bases of L2(R). Trans Am Math Soc 315:69–87
[43] Mallat S (1989) Multifrequency channel decompositions of images and wavelet models. IEEE Trans Acoust, Speech, Signal Process 37:2091–2110
[44] Mallat S (1999) A Wavelet Tour of Signal Processing, 2nd edition. Academic Press, San Diego, California USA
[45] Meyer Y (1990) Ondelettes et Opérateurs. Hermann, Paris
[46] Vetterli M, Kovačević J (1995) Wavelets and Subband Coding, Prentice Hall PTR, Englewood Cliffs, New Jersey
[47] Burke Hubbard B (1998) The World According to Wavelets – The story of a Mathematical Technique in the Making, 2nd edition, AK Peters Ltd, Natick, Massachusetts
[48] Daubechies I (1990) The wavelet transform, time-frequency localization and signal analysis. IEEE Trans Inform Theory 36:961–1005
[49] Daubechies I (1998) Orthonormal bases of compactly supported wavelets. Commun Pure Appl Math 41:909–996
[50] Chang T, Kuo CCJ (1993) Texture analysis and classification with tree-Structured wavelet transform. IEEE Trans Image Process 2:429–441

[51] Bharati MH, Liu, J, MacGregor JF (2004) Image texture analysis: methods and comparisons. Chemom Intell Lab Syst 72:57–71
[52] Gosselin R, Rodrigue D, González-Núñez R, Duchesne C (2009) Potential of hyperspectral imaging for quality control of polymer blend films. Ind Eng Chem Res 48:3033–3042
[53] Van de Wouwer G (1998) Wavelets for Multiscale Texture Analysis. PhD thesis, University of Antwerp, Antwerp, Belgium
[54] Haralick RM, Dinstein I (1973) Textural features for image classification. IEEE Trans Syst Man Cybern 3:610–621
[55] Esbensen KH, Kvaal K, Hjelmen KH (1996) The AMT approach in chemometrics first forays. J Chemom 10:569–590
[56] Liu JJ, MacGregor JF (2007) On the extraction of spectral and spatial information from images. Chemom Intell Lab Syst 85:119–130
[57] Kaspar MH, Ray WH (1993) Dynamic PLS modeling for process control. Chem Eng Sci 48:3447–3461
[58] Lakshminarayanan S, Shah SL, Nandakumar K (1997) Modeling and control of multivariate processes: dynamic PLS approach. AIChE J 43:2307–2322
[59] Liu JJ, MacGregor JF (2008) Froth-based modeling and control of flotation processes. Miner Eng 21:642–651
[60] Gebhardt JE, Tolley WK, Ahn JH (1993) Colour measurement of minerals and mineralised froth. Miner Metall Process 10:96–99
[61] Moolman DW, Aldrich C, Van Deventer JSJ, Stange WW (1994) Digital image processing as a tool for on-line monitoring of froth in flotation plants. Miner Eng 7:1149–1164
[62] Oestreich JM, Tolley WK, Rice DA (1995) The development of a color sensor system to measure mineral compositions. Miner Eng 8:31–39
[63] Hätönen J, Hyötyniemi H, Miettunen J, Carlsson LE (1999) Using image information and partial least squares method to estimate mineral concentrations in mineral flotation. In: Proceedings of the Second International Conference on Processing and Manufacturing of Materials (IPMM'99), Honolulu, Hawaii, July 10-15 1:459–464
[64] Bonifazi G, Serranti S, Volpe F, Zucco R. (1999) Characterisation of flotation froth color and structure by machine vision. In: Geovision 99, May 6–7, Liege, Belgium
[65] Bonifazi G, Massacci P, Meloni A (2000) Prediction of complex sulphide flotation performances by a combined 3D fractal and colour analysis of the froths. Miner Eng 13:737–746
[66] Quenouille M. (1949) Approximate tests of correlation in time series. J Royal Statist Soc B11:18–44
[67] Tukey JW (1958) Bias and confidence in not quite large samples. Ann Math Statist 29:614
[68] Efron B (1979) Bootstrap methods: another look and the jackknife. Ann Statist 7:1–26
[69] Moolman DW, Eksteen JJ, Aldrich C, Van Deventer JSJ (1996) The significance of flotation froth appearance for machine vision control. Int J Miner Process 48:135–158
[70] Hargrave JM, Hall ST (1997) Diagnosis of concentrate grade and mass flowrate in tin flotation from colour and surface texture analysis. Miner Eng 10:613–621
[71] Sadr-Kazemi N, Cilliers JJ (1997) An image processing algorithm for measurement of flotation froth bubble size and shape distribution. Miner Eng 10:1075–1083
[72] Cipriano A, Guarini M, Vidal R, Soto A, Sepúlveda C, Mery D, Briseno H (1998) A real time visual sensor for supervision of flotation cells. Miner Eng 11:489–499
[73] Wang W, Bergholm F, Yang B (2003) Froth delineation based on image classification. Miner Eng 16:1183–1192
[74] Beucher S, Meyer F (1993) The morphological approach to segmentation: the watershed transformation. In: Dougherty E (Ed), Mathematical Morphology in Image Processing, Marcel Dekker, New York, 433–481
[75] Aldrich C, Moolman DW, Eksteen JJ, Van Deventer JSJ (1995) Characterization of flotation processes with self-organizing neural nets. Chem Eng Commun 139:25–39
[76] Nguyen KK, Thornton AJ (1995) The application of texture based image analysis techniques in froth flotation. In: Proceedings of DICTA95, Brisbane, Australia 613–618

[77] Duda RO, Hart PE, Stork DG (2000) Pattern Classification. 2nd Edition, Wiley-Interscience, New York
[78] Tessier J (2006) Détermination de la composition de l'alimentation des circuits de broyage par analyse d'images multivariées. MSc thesis, Department of Chemical Engineering, Université Laval, Québec, Québec, Canada
[79] Napier-Munn T, Morrell S, Morrison R, Kojovic T (1996) Mineral comminution circuits: their operation and optimisation. JKMRC monograph, The University of Queensland, Brisbane
[80] Maerz NH (2001) Automated online optical sizing analysis. In: Proceedings of the Third International Conference on Autogeneous and Semiautogeneous Grinding Technology (SAG'01), Vancouver, Canada, September 30-October 3 2:250–269
[81] Guyot O, Monredon T, LaRosa D, Broussaud A (2004) VisioRock, an integrated vision technology for advanced control of comminution circuits. Miner Eng 17:1227–1235
[82] Simkus R, Dance A (1998) Tracking hardness and size: measuring and monitoring ROM ore properties at Highland Valley Copper. In: Proceedings of the Mine to Mill 1998, Brisbane, Australia, October 11–14
[83] Herbst JA, Pate WT (1999) Object components for comminution system softsensor design. Powder Technol 105:424–429
[84] Schroder AJ (2000) Towards automated control of milling operations. In: IIR Conference-Crushing and Grinding Technologies for Mineral Extraction. Perth, Australia, 18/19 May 2000
[85] Singh V, Mohan Rao S (2005) Application of image processing and radial basis neural network techniques for ore sorting and ore classification. Miner Eng 18:1412–1420
[86] Petersen KRP, Aldrich C, Van Deventer JSJ (1998) Analysis of ore particles based on textural pattern recognition. Miner Eng 11:959–977
[87] Casali A, Gonzalez G, Vallebuona G, Perez C, Vargas R (2001) Grindability soft-sensor based on lithological composition and on-line measurements. Miner Eng 14:689–700
[88] Lepistö L, Kunttu I, Visa A (2005) Color-based classification of natural rock images using classifier combinations. In: Proceedings of the SCIA 2005, Joensuu, Finland, June 19-22 2005, 901–909
[89] Paclik P, Verzakov S, Duin RPW (2005) Improving the maximum-likelihood co-occurrence classifier: a study on classification of inhomogeneous rock images. In: Proceedings of the SCIA 2005, Joensuu, Finland, June 19–22 2005 998–1088
[90] Murtagh F, Qiao X, Crookes D, Walsh P, Muhammed Basheer PA, Long A, Starck JL (2005) A machine vision approach to the grading of crushed aggregate. Machine Vision Appl 16:229–235
[91] Van de Wouwer G, Scheunders P, Van Dyck D (1999) Statistical Texture Characterization from Discrete Wavelet Representations. IEEE T Image Process 8:592–598
[92] Haykin S (1999) Neural Networks - A Comprehensive Foundation. 2nd Edition, Prentice-Hall
[93] Schölkpof B, Smola AJ (2002) Learning with Kernels: Support Vector Machines, Regularization, Optimization, and Beyond. The MIT Press, Cambridge, MA
[94] Jemwa GT, Aldrich C (2006) Kernel-based fault diagnosis on mineral processing plants. Miner Eng 19:1149–1162
[95] Esbensen KH, Edwards G, Eldridge NR (1993) Multivariate image analysis in forestry applications involving high resolution airborne imagery. In: 8th Scandinavian Conference on Image Analysis, NOBIM, Tromsø, 953–963
[96] Liu JJ, MacGregor JF (2005) Modeling and optimization of product appearance: application to injection-molded plastic panels. Ind Eng Chem Res 44:4687–4696
[97] Gosselin R, Rodrigue D, Duchesne C (2008) On-line prediction of crystallinity spatial distribution across polymer films using NIR spectral imaging and chemometrics methods. Can J Chem Eng 86:869–878
[98] Bruwer MJ, MacGregor JF, Noseworthy MD (2007) Dynamic contrast-enhanced MRI diagnostics in oncology via principal component analysis. J Chemom 22:708–716

Chapter 4
Soft Sensing

Guillermo D. González

Abstract This chapter addresses different aspects of soft sensors and their importance, especially for mineral processing plants. It considers current state-of-the-art soft sensors and the future outlook of the industry, taking into account different classes of linear and nonlinear in-the-parameter soft sensor models, their design, and soft sensor shell systems for providing a robust operation in industrial environments. A geometrical approach to soft sensor modeling is given for nonlinear but linear in-the-parameter models. Inclusion of phenomenology in linear in-the-parameter models is highlighted. The chapter includes a review and examples from the mineral processing industry.

4.1 Introduction

This chapter is concerned with the problem of availability of information provided by measurements in plants either when sensors become unavailable or when they have not been installed. The solution to this problem is approached by means of soft sensors, *i.e.*, systems replacing the missing measurement (primary measurement) with an estimation or prediction of the missing measurement based on the other measurements (secondary measurements) correlated with the primary measurement. A detailed analysis of soft sensors for industrial applications in general is given by Fortuna *et al.* [1] where different topics and industrial applications are covered. A summary of soft sensor topics may be found in [2]. However in [1], cases of soft sensors used in mineral processing are not included. In this chapter, the focus is on soft sensors in mineral processing plants, which due to their particular characteristics require special attention to certain topics. In addition, by covering

Guillermo D. González
Department of Electrical Engineering, Universidad de Chile, Santiago, Chile, e-mail: gugonzal@ing.uchile.cl

application of soft sensors to this industry it is expected that gaps are filled in the available reviews in papers and in books on soft sensors.

In order to fulfill the objectives stated for a plant (circuit, system, unit, *etc.*) it is essential to be able to extract information about the plant's condition, in order either to drive it to a given operating point, or to a desired trajectory, or to such condition that a performance measure is optimized. Such required information is obtained directly or indirectly through sensors that measure plant variables. If any of these measurements becomes unavailable, technical and economic objectives will not be met because the plant's control, either automatic or manual, will be impaired and the plant products will fail to meet specifications.

The sensor measurement may become unavailable because an installed sensor system fails or its operation is interrupted for maintenance or repairs. A sensor for a given measurement may also not be installed at all, *e.g.*, because it is not available on the market, or due to the high cost of the sensor system and its maintenance. A soft sensor may also provide measurement estimations when an actual sensor is shared among several measuring points (*e.g.*, X-ray copper grade analyzer) or, in general, it can fill in measurements when sampling is infrequent.

An unmeasured variable, such as an unmeasured disturbance, prevents the system from benefiting from information necessary for better plant performance. Through changing an unmeasured disturbance into a measured disturbance, a sensor or a soft sensor makes it possible to use such a measurement in the control of the plant (*e.g.*, by means of feedforward control or other model-based control schemes). As a result, it is sometimes possible to considerably improve the plant's operation. This is particularly important when the unmeasured disturbance has a relatively large effect on the controlled variables, as in some mineral processing plants.

On the contrary, when the sensor or soft sensor is not available, the use of advanced control strategies would become necessary. For example, if the controlled variable is highly correlated with a random unmeasured disturbance, either some form of stochastic control may be required, or a less efficient control system must be used. But it is a well-known fact that complicated control strategies tend to be vulnerable and of reduced robustness. As a result, it often happens that after several problems come up, the control strategy is abandoned and control reverts to one that is less elaborate, but surely less efficient. All this may be avoided if the sensor measurement is available or if conditions are such that a soft sensor may be designed to estimate the missing unmeasured disturbance.

Among the main unmeasured disturbances in mineral processing are lithology, mineralization, liberation, grindability, and concentration of reactants in flotation pulps. The absence of sensors for these plant variables implies that the soft sensor must not only be designed using off-line measurement samples that are analyzed in a laboratory, but also be updated using off-line samples.

There is clearly a question of economics in deciding whether a sensor's back up should be to a spare sensor or to a soft sensor. Justification for a soft sensor depends, on the one hand, on the relative cost of a real back-up sensor. For example, as a back up for a wattmeter, a spare wattmeter may be best, which can easily be plugged in to replace the missing one. On the other hand, for backing up other complex sensors, a soft sensor may be a best choice. Such would be the case in the replacement of a particle size analyzer used in the mineral processing industry where an undue period of time and effort may be required to replace the missing sensor. However, if a soft sensor is available, it may be used during this period for manual or automatic control. A soft sensor may even be an option to a real sensor if certain conditions are met, mainly if only infrequent updating of the soft sensor model using laboratory samples is required.

The relevant relations between the primary measurement and the secondary measurements or functions of them may be expressed in a model. Its output is an estimation or prediction of the missing primary measurement and constitutes a soft sensor for the primary measurement (Figure 4.1). In this case the core of a soft sensor is a soft sensor model [1, 3], which must be chosen from a class of models, *e.g.*, from ARX, NARX (nonlinear but linear-in-parameter (LIP) models), ARMAX, NARNAX, fuzzy models, state-state/output models (Kalman filter, extended Kalman filter), fuzzy models (*e.g.*, Takagi and Sugeno models), neural network models, radial basis function network models, support vector machine regression models, models based on principal component analysis (PCA), projection to latent structure (PLS) and clustering, and models incorporating combinations of them.

Alternatively, an approach is to use features extracted from measurements of secondary variables, including static or dynamic images captured by vision sensors (video cameras). These features are used to generate the soft sensor output for the primary measurement.

The principal modeling aspects of softsensors with emphasis on mineral processing is undertaken in Section 4.2. A geometric approach to soft sensor modeling in the special but important case of nonlinear but LIP models is found in Section 4.3.

Modeling is not a simple task when it comes to mineral processing, compared with mechanical, electrical, and other systems. Mineral processes tend to be quite nonlinear, multivariable, and subject to stochastic unmeasured disturbances (*e.g.*, liberation, mineralogy, lithology), which have a considerable influence on the variables (*e.g.* particle size, concentrate and tailings grade) that are important to the automatic or manual control of the mineral processing unit, circuit or plant, and therefore to its economic performance. This fact complicates matters because obviously the unmeasured disturbances – influential as they may be on the primary measurement – cannot be used as secondary measurements.

Development of an accurate phenomenological model turns out to be too complex or even impossible, so accurate local relatively simple models around operating points are desired (*e.g.*, linear models such as ARX models). But such simple models cannot in general be a representation of the portion of the plant being modeled over a broad operating region and time span. Therefore up-dating of the model is required as the operating point shifts during plant operation due to changes in manipulated variables, measured and unmeasured disturbances, and because of changes of plant parameters.

A solution to the problem of the conflicting model requirements of simplicity versus coverage may be afforded by nonlinear but LIP models, such as NARX models [1, 4]. This class of model may have a much larger operating region coverage than linear models (*e.g.*, ARX, ARMAX) and at the same time share the linear regression possibilities of linear models.

Soft sensor design must be approached not only considering the design of the soft sensor model. In addition a soft sensor shell providing robustness for the primary measurement estimation is needed, mainly in case secondary measurements may themselves become degraded and even unavailable and also because of changes in operating points not dealt with by the soft sensor model. The main functions of the soft sensor shell system are signal conditioning and monitoring, determining that the primary sensor has failed in some sense (including loss of calibration), on-line determination of which soft sensor model of a possible set of models is best for replacing the unavailable actual sensor, adaptation of the soft sensor models to changing plant conditions, providing adequate excitation of plant inputs for on-line estimation of the model parameters, *etc.* If conditions previous to sensor replacement by the corresponding soft sensor are favorable, it may become convenient to choose the model parameters using their optimal prediction rather than fixed values at the replacement instant. The soft sensor shell and its main functions are treated in Section 4.4.

Section 4.3 contains a geometric approach to the subject of soft sensor modeling in the case of linear models or nonlinear models that are LIP. A global approach to modeling LIP models is taken with respect to the kinds of primary and secondary variables, whether they are represented by deterministic time sequences, discrete time stochastic processes (*i.e.*, random sequences), or continuous time functions and stochastic processes. The issue of using time averages instead of expected values is considered, together with conditions that allow the averages to be valid representations of expected values. This is important because the use of expected values is generally the appropriate way to approach the problem of model structure determination and parameter estimation. However, probability distributions are mostly not available for determining expected values, so they must be estimated by their corresponding time averages.

An additional problem to be considered when a sensor in a control loop is replaced by a soft sensor is the effect that this change may have on the performance of the control, since now the control loop contains the additional system constituted by the soft sensor model. This fact may cause deterioration of control and even instability as the plant operating point changes unless the model is updated fast enough (Section 4.2).

The use of soft sensors has become wide-spread in different industries, particularly in the mineral processing industry. A sample of the soft sensors currently used in this industry and of the different classes of models that are being used is given in Tables 4.1 to 4.4, which show that soft sensors have been developed for particle size in grinding circuits, density or solids concentration, grindability, work index, operational work index, lithology, grades in flotation cells, weight and flow. This sample is not intended to be exhaustive, but its purpose is to cover the scope and most of the approaches concerning soft sensor design and operation with regard to the mineral processing industry. For each soft sensor, this set of tables contains the reference citation, the class of model used, its main features, and its application in the industry or to tests performed using data obtained from industrial plants or from simulated plants. Examples of soft sensor design for industrial ore concentrators are given in Section 4.4.

In the future, we expect that the developments of soft sensors will be driven by the development of new data processing and sensor technologies in general, and of new equipment in the mineral processing industry.

Nanotechnologies will permit the development of networks of a great number of miniature sensors profusely located almost anywhere in the processing units and connected to many miniature high-speed processors through newly developing wireless communications systems. Paralleling these developments, these new sensors may have their corresponding soft sensors based on these nanotechnologies. Also, the development of soft sensors for which no sensors have been developed may also be favored. It is not difficult to imagine also that new data processing and communications technologies may also have a favorable impact on soft sensor shell systems [5].

Thus, the new technologies point to an omnipresence of multiple sensors and soft sensors in the processing units together with the pervasiveness of miniature signal processing computers.

It is also reasonable to expect that the development of new equipment based on new technologies may require new sensors and soft sensors. In order to improve the efficiency of the comminution processes new equipment is being introduced in the mineral processing industry, *e.g.*, the HPGR mill (high pressure grinding rolls mill) [6] and the IsaMill [7]. Concerning flotation, a potentially increasing use of the Jameson cell is also to be mentioned, where this flotation technology produces an intense flotation recovery process using robust equipment [8].

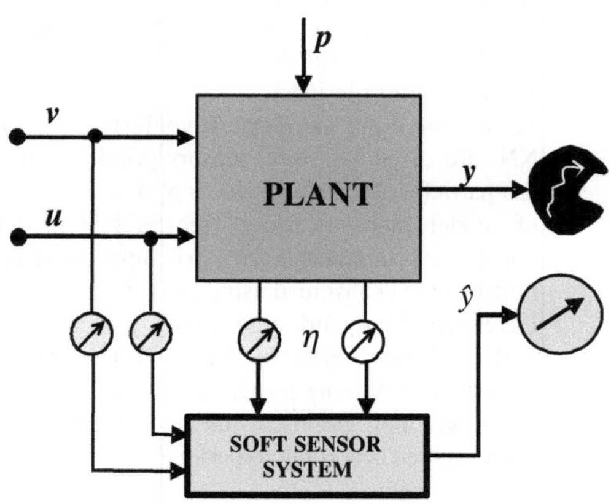

Figure 4.1 Soft sensor system output $\hat{y}$ replacing failed sensor signal y using a model having as inputs correlated secondary measurements: commands u, measured disturbances v, and other plant outputs η. Although the unmeasured disturbances p affect the plant output y, they cannot be directly used as inputs to the model

Table 4.1 Examples of soft sensors for particle size in grinding circuits

Reference	Model class	Main features	Primary variable, applications, tests
González *et al.*, 1985 [9]	ARX	Static particle size estimator with dynamic compensator.	+65 # for grinding circuit consisting of a rod mill and a ball mill.
González *et al.*, 1995 [10]	NARX	Stepwise regression. Comparison between gray and black box NARX soft sensor models. Effects due to improper sampling period.	Tests using data from industrial grinding circuit for estimating or predicting % +65# in circuit product. Sampling period effects tested in simulated simple plant.
Casali *et al.*, 1998 [11]	NARX	Stepwise regression. Robustness to secondary measurements failures using a set of candidate soft sensor models. Gray models using composite candidate regressors determined by phenomenological model.	Tests using data from industrial grinding circuit for estimating or predicting % +65# in circuit product. Stepwise regression selects mainly phenomenological regressors instead of direct secondary measurements.
Du *et al.*, 1997 [12]	NN	NN soft sensor for particle size in product of grinding circuit considering adaptation to changing conditions. Use of PCA to simplify the SS model in order to improve on-line adaptation performance.	Tests using simulated industrial grinding circuit. SS estimates passing 53 μm particle size.
Sbarbaro *et al.*, 2008 [13]	NARX	Adaptation of soft sensor model parameters to changing conditions considering parameter constraints using the error projection (EP) algorithm. Regressors involving inputs in LIP NARMAX model are convolutions.	Tests using data from industrial grinding circuit for estimating or predicting % +65# in circuit product. Comparison of EP and RLS algorithms.
Sbarbaro *et al.*, 2001 [14]	NN	Feedforward and autoassociative NN are used as soft sensors for particle size. After selection of model input variables (secondary measurements) NN parameters are determined using the Levenberg–Marquardt algorithm	Tests using data from industrial grinding circuit for estimating or predicting % +65#. Comparison of NN models used concerning input noise and number of parameters.
Sun *et al.*, 2008 [15]	SVM	Particle size soft sensor model designed by combining SVM, identification, and empirical methods weighted according to performance of each.	Tests using particle size data form industrial grinding circuit. One hour sampling time over 23 days.

Notation: LIP = Linear-in-parameters; NARX = Nonlinear ARX; NN = Neural Network; SS = Soft Sensor; PCA = Principal Component Analysis; SVM= Support Vector Machine.

Table 4.2 Examples of soft sensors for density or solids concentration in ore concentrators

Reference	Model class	Main features	Applications, tests
Casali *et al.*, 1995 [16]	NARX	Stepwise regression used to select SS inputs from set of candidate components. NARX Gray models through introduction of composite components having physical significance, which are selected instead of direct simple measurements. Robustness is obtained by on-line selection of a SS model from a set of models, using an availability index	Tests in pulp density using data from an industrial grinding plant. The SS model is selected on-line from a set of four candidate models.
Casali *et al.*, 1998 [17]	NARX	Solids concentration soft sensor at hydrocyclone overflow. Gray models through introduction of composite components having physical significance. Selection by stepwise regression	Test using data from industrial grinding circuit.
Chuk *et al.*, 2005 [18]	Dynamic and static empirical models	SS based on transfer function model empirical relations for froth density estimation in column flotation used in conjunction with expert fuzzy system to prevent froth collapse.	Tests by simulation using a model developed for column flotation.
González *et al.*, 1993 [19]	ARX	Effects of choosing frozen, average, and optimally predicted soft sensor parameters at the time of replacing a sensor by the soft sensor.	Tests for pulp density (solids concentration) using data from an industrial grinding circuit show improved performance in case of predicted parameters.
González *et al.*, 1995 [10]	NARX	Stepwise regression. Comparison between gray and black box (linear) soft sensor models. Comparison between models with original measurements and models including phenomenological components.	Hydrocyclone overflow particle size sensor and feed pulp density (solids concentration) soft sensor tested using data from industrial grinding circuit.

Notation: LIP = Linear-in-parameters; NARX = Nonlinear ARX; NN = Neural Network; SS = Soft Sensor; PCA = Principal Component Analysis.

Table 4.3 Example of soft sensors for mill loading, work index, grindability in grinding circuits, weight

Reference	Model class	Main features	Applications, tests
Amestica *et al.*, 1993 [20]	State model	Extended Kalman filter for nonlinear reduced dimension state model based on phenomenological knowledge. Combined state and parameter estimation. Main states: total ore, fine ore, and water mill contents. Estimated parameters (transformed into states) are grinding rate, water and ore discharge rates.	Tests by simulation using data from a detailed dynamical model of a pilot semiautogenous mill having 27 size fractions states. Considerable improvement is found with model based on nonlinear extended Kalman filter compared with linear Kalman filter. Tests by simulation using the developed model.
Apelt, 2002 [21]	Various	Thirty-six dimension state estimation of SAG mill contents and parameters: ore grindability, mill discharge grate parameters turned into states. There are 27 size fraction states.	Tests using data from industrial SAG circuit give acceptable results if a bulk ore discharge flow is considered to be measured, and almost perfect estimation if a 27 size fraction ore discharge is assumed to be measured.
González *et al.*, 2008 [22]	Extraction of features of measured variables	On-line detection of grindability index based on concatenated wavelet transform variances of three measured variables. Reduced dimension dynamic model of SAG circuit is developed and used for testing	Tests by simulation using dynamic SAG circuit state model adjusted to data from a large industrial grinding circuit show important improvements due to concatenation.
Casali *et al.*, 2001 [23]	Extraction of features of measured variables (images) and NN	On-line ore work index SS based on lithological ore composition determined from image analysis of ore in conveyor belt using a NN classifier. Grindability determined on the basis of estimated lithology.	Laboratory tests using ores from a grinding plant have given good results.
González *et al.*, 2000 [24]	Gray (ARX plus empirical)	Global operational work index SS for a grinding circuit using empirical models for operational work index and ARX for p_{80}. Sensitivity analysis is used to determine that mean f_{80} may be used as empirical model input.	Tests using ore samples from mine site and data from grinding circuit.

Notation: LIP = Linear-in-parameters; NARMAX = Nonlinear ARMAX; NN = Neural Network; SVM = Support Vector Machine SS = Soft Sensor; PCA = Principal Component Analysis; SAG = Semi-autogenous. Continued on next page.

Table 4.3 (*continued*)

Reference	Model class	Main features	Applications, tests
Perez *et al.*, 1999 [25]	Extraction of features of measured variables (images) and genetic classification	Estimation of ore lithology by classifier using genetic algorithm.	Laboratory tests using monthly rock samples from mine site. Seven lithological classes are considered and 130 features are extracted from each rock sample out of a database consisting of 760 digital images.
Espinoza, *et al.*, 1995 [26]	Clustering of operating points	Clusters formed represent different types of overload conditions. Distances to cluster centers give early warning of overload.	Clusters are formed using data from an industrial ball mill grinding circuit. If distance to an overload cluster is below a threshold corrective actions are taken according to the cluster characteristics.
Pan *et al.*, 2003 [27]	NN	Industrial weightometer soft sensor including monitoring system for data validation, failure detection, replacement of failing weightometer and alarming. Among inputs are particle size distribution and power draw.	Several models are developed and tested in an industrial diamond processing plant.
Herbst *et al.*, 1989 [28]	State model	Kalman filter for combined state and parameter estimation in SAG and AG mill circuits. States are total ore and water mill hold-ups, ball hold-up, liner weight. Estimated parameters are ore grindability, liner wear rate and ball wear rate.	Design and tests using milling model. Estimated variables are used in a model-based control and expert control systems designed using the models.
Herbst *et al.*, 1996 [29]	State model	SAG mill charge volumes estimation using Kalman filter.	Installed in industrial grinding circuits.
Notation: LIP = Linear-in-parameters; NARMAX = Nonlinear ARMAX; NN = Neural Network; SVM = Support Vector Machine SS = Soft Sensor; PCA = Principal Component Analysis; RMS = root mean square; SAG = Semi-autogenous.			

Table 4.4 Examples of soft sensors for grades and density in flotation

Reference	Model class	Main features	Applications, tests
Chuk *et al.*, 2005 [18]	Dynamic and static empirical models	SS based on transfer function model empirical relations for froth density estimation in column flotation used in conjunction with expert fuzzy system to prevent froth collapse.	Tests by simulation.
Espinoza *et al.*, 1995 [26]	ARX	Stepwise regression. Comparison of concentrate grade soft sensors with and without clustering.	Tests of rougher flotation grade using data from an industrial grinding plant. Clustered SS gives remarkably better RMS and correlation.
González *et al.*, 1994 [30]	NARX	Concentrate grade SS. Stepwise regression. Gray models. Comparison between soft-sensor models (large prediction span) and control models (low prediction span and forcible inclusion of some manipulated variables).	Tests using data from an industrial rougher flotation plant and from a model.
González *et al.*, 2003 [31]	Various	Local soft sensor models using various models stemming from a general model structure: ARMAX, NARX NARMAX, cluster, neural networks, fuzzy Takagi–Sugeno, and combinations of them. Comparisons. Data validation for soft-sensor design.	Test using concentrate grade data from industrial rougher flotation plant .
Hyötyniemi, 2000 [32]	Extraction of features of measured variables (images)	SS giving froth characteristics on the basis of froth image analysis, including characteristics dynamics.	Installed in flotation plant and used in expert control loops.
Notation: LIP = Linear-in-parameters; NARMAX = Nonlinear ARMAX; NN = Neural Network; SVM = Support Vector Machine SS = Soft Sensor; PCA = Principal Component Analysis; RMS = root mean square;			

Table 4.5 Soft sensors in general

Reference	Model class	Main features	Applications
Basso *et al.*, 2002 [4]	NARX	Identification of gas turbine using NARX model.	NARX model of an industrial gas turbine, developed using plant data, shows better performance than ARX models.
Crisafulli, 1996 [33]	State model	Infrequent sampling. Use of phenomenological knowledge. Kalman filter.	Fiber rate in sugar cane mill.
Dunia, 1996 [34]	PCA	Missing measurement reconstruction (soft sensing) for sensors within a set of sensors. Types of faults. Fault detection. Identification of failed sensor through sensor validity index. Effect of filtering. Separating sensor fault detection from plant changes.	Data from boiler process with a nine sensor set.
Fortuna *et al.*, 2007 [1]	Various	Book containing treatment of different topics concerning soft sensors: justification, SS models, measurement selection, sensor fault detection, *etc.*	Examples of soft sensors in industrial plants, except mineral processing plants.
González, 1999 [3]	Various	Paper containing a review of soft sensors including topics in SS model design, different model classes (especially LIP or NARX), clustering, effect on control loop due to replacement of sensors by SS, the need for a SS administration system.	Examples of soft sensors in mineral processing plants and other plants. Control.
Control Eng. Europe, 2001 [2]	Various	Examples on benefits provided by soft sensors in several industrial plants. The need for data preprocessing and adaptation.	Example of commercial packages for SS development. Soft sensor basics is addressed. Examples showing SS displays in industrial environment.
González *et al.*, 1996 [35]	Transfer function	Effects of replacing actual sensors by soft sensors in control loops.	Tests using simulation.
Gonzlez, 1992 [36]	ARX	Design problem when soft sensor replaces sensor in control loop.	Study using model and simulation.

Notation: LIP = Linear-in-parameters; NARX = Nonlinear ARX; NN = Neural Network; SVM = Support Vector Machine SS = Soft Sensor; PCA = Principal Component Analysis; RMS = root mean square; LS =least square; RBF = radial basis function

Table 4.5 (*continued*)

Reference	Model class	Main features	Applications
Wang, 1996 [37]	RBF NN	Soft sensor based on radial basis function (RBF) NN. Data is subject to clustering and a RBF NN is developed for each cluster. The SS output is given by the sum of the several RBF NN weighted by the respective membership functions to each cluster.	Tests in high purity distillation column.
Wendt, 1999 [38]	Gray (empirical + phenomenological)	Soft sensor for flow in open channels having an overshot (submerged) gate, using as inputs to a gray box model upstream and downstream levels and gate angle.	Tests in open channel point to the selection of one of three proposed models.
Herbst, 1999 [39]	State	Generic soft object oriented development tool for soft sensor development includes on-line model updating.	Applications to industrial SAG and ball milling circuits.
Zhang, 2006 [40]	LS-SVM	SS model design using least square regression and SVM. Nonlinear regression transformed to linear regression but in higher dimension space. Cost function includes a term for penalizing model complexity. A simple training algorithm is given.	Tests using data of gasoline absorbing rate in cracking unit of a refinery. Better results are reported than those obtained if NN SS model is used, including less data needed in training.
Dunia *et al.*, 1996 [34]	PCA	Identification of sensor faults using sensor validity index.	Tests using boiler process data.
King, 2004 [41]	—	Problems arising in soft sensors for the hydrocarbon processing industry are analyzed, including warnings of wrong interpretations and procedures. The subject may be extended to other industries.	Several graphs are given based on industrial results to support the different subjects considered.
Barrera, 1996 [42]	—	System for handling SS in industrial environment. Includes: several soft sensor models per primary measurement, selected according to analysis of secondary measurement quality and availability; parameter estimation while actual sensors are available; monitoring of soft sensor errors and decision to update parameter of models using appropriate excitation signals. Flexible configuration using text strings.	Tested on particle size estimation in an industrial grinding plant.
Notation: LIP = Linear-in-parameters; NARX = Nonlinear ARX; NN = Neural Network; SVM = Support Vector Machine SS = Soft Sensor; PCA = Principal Component Analysis; RMS = root mean square; LS =least squares; RBF = radial basis function; SAG = Semi-autogenous.			

4.2 Soft Sensor Models

The task of determining soft sensor models is the same as the general modeling procedure in which model identification plays a central role. Some important issues affecting the development of soft sensor models are covered in this section, including an alternative way, by analyzing features of measured plant variables. Other important matters are found in the books by Fortuna *et al.* [1] and Ljung [44]. Due to the particular characteristics of mineral processing units circuits and plants some special issues are included in this section, such as the introduction of process phenomenology in models which are LIP.

Modeling in mineral processing is generally more difficult than in other industries. Indeed, some processes here are highly nonlinear (semiautogenous grinding, flotation) and subject to unmeasured disturbances having a heavy influence on the controlled variables (*e.g.*, grindability, concentrate and tailing grades). In addition, there is a lack of sensors for on-line measurements (*e.g.*, grindability, mineralogy). For example, Figure 4.2 shows the important effects of unmeasured disturbances in the case of a concentrate grade soft sensor used in the process of adjusting a flotation bank phenomenological model to plant data [43]. Since a phenomenological model would be too complex to use as a soft sensor model or to be included in a model-based control system, simpler local models have to be developed around operating points. Therefore, it is to be expected that these models need to be up-dated as the process operating point shifts. As a consequence, the models are valid only during limited time spans and these soft sensor models must undergo on-line adaptation and in some cases off-line adaptation. A compromise must be made between simple soft sensor models possibly requiring frequent but faster adaptation (*e.g.*, ARX models [1, 44, 45]) and complex models requiring a less frequent but a slower adaptation process (*e.g.*, neural networks, and NARMAX models [1, 45]). In between there are models such as NARX models [1, 45] which are LIP so that adaptation may be relatively fast, even though they may be non-linear with respect to the plant measurements. A different approach to soft sensor design is based on the extraction of features from plant measurements, including images captured by video cameras (*e.g.*, of flotation froth).

The modeling process in the case of soft sensors may be defined paralleling the definition of identification given by Lotfi Zadeh (Figure 4.3): modeling is the determination of which model M from a given class of models is the best representation of the modeled primary measurements for a given class of signals and according to a given criterion (Figure 4.3), on the basis of only measured variables. Examples of class of models are difference and differential input–output equations, continuous and discrete time state–state/output equations, neural networks, wavelet models and fuzzy models. The criterion often used is the mean square error of some error measure – such as the one step ahead prediction error or the mean square equation error (between the plant output and the model output given the values of the measurements contained in the right hand side of the model equation). The class of signals is concerned with the characteristics of the variables involved in the modeling process, such as their rate of change (or frequency contents) so that the relevant plant dynamics may be captured by the model (*e.g.*, through persistent excitation).

The classes of models may be broadly classified as [44, 45]:

- Phenomenological models or white box models: Models derived from first principles for which all parameters are known. For example, in the case of a mechanical system formed by the interconnection of a mass a spring and a damper with known mass, spring constant and damper constant.
- Empirical models or black box models: Models where there is almost no knowledge of the phenomenology or first principles governing the modeled system. Such is the case of hydrocyclone models. This class of models is very important in mineral processing, since phenomenological models are too complex. In many cases they are combined with phenomenological properties to produce gray models.
- Gray models are models that have characteristics of both white and black box models. Here the phenomenological part usually consists of dynamic balances of masses, volumes, energy, *etc.*, while the empirical part is usually concerned with internal processes that are difficult or impractical to model using first principles (*e.g.*, particle breakage, classification in hydrocyclones and SAG mills, mill power draw). Many models of mineral processing units belong to this class.

4.2.1 Optimality Criterion

The criterion to decide which model is the best is usually based on the one step ahead prediction error [1, 44]

$$e(t,\kappa) = y(t) - \hat{y}(t|t-1,\kappa), \tag{4.1}$$

between the plant output $y(t)$ and its one step ahead prediction output given by the model, where κ includes all the model parameters. The model fitness is measured by the mean square prediction error

$$J = E\{[y(t) - \hat{y}(t|t-1,\kappa)]^2\} = E\{e(t,\kappa)^2\}, \tag{4.2}$$

so the best model is the one for which (4.2) is minimum. For a given model structure the parameter vector κ giving the best model is found by vector $\hat{\kappa}$ which minimizes

$$J(\hat{\kappa}) = E\{e(t,\hat{\kappa})^2\}. \tag{4.3}$$

In the case of structure determination J also incorporates the effect of the number of model parameters, *i.e.*, the dimension of κ, in relation to the length of the data sets T (see (4.74)). Since the required probability functions are rarely available the minimum of (4.3) must be estimated by minimizing the corresponding time average

$$V = \frac{1}{T}\sum_{i=1}^{T}[y(t) - \hat{y}(t|t-1,\hat{\kappa})]^2 = \frac{1}{T}\sum_{i=1}^{T}e(t,\hat{\kappa})^2. \tag{4.4}$$

The fact that $y(t)$ and $\hat{y}(t|t-1,\kappa)$ are in general stochastic processes implies that V is a random variable having an expected value and a variance depending on T. For the estimate V to be useful it should be unbiased, *i.e.*, $J = E\{V\}$, and its variance should tend to zero as T tends to infinity. This situation propagates to the estimation of the model parameters (see Section 4.3.2).

Applying the necessary condition for the minimization of (4.3)

$$\frac{\partial}{\partial\hat{\kappa}}E\{e(t,\hat{\kappa})^2\} = E\{\frac{\partial}{\partial\hat{\kappa}}e(t,\hat{\kappa})^2\} = 0, \quad E\{e(t,\hat{\kappa})\frac{\partial}{\partial\hat{\kappa}}e(t,\hat{\kappa})\} = 0. \tag{4.5}$$

From (4.1),

$$\frac{\partial}{\partial\hat{\kappa}}e(t,\hat{\kappa}) = \frac{\partial}{\partial\hat{\kappa}}[y(t) - \hat{y}(t|t-1,\hat{\kappa})] = -\frac{\partial\hat{y}(t|t-1,\hat{\kappa})}{\partial\hat{\kappa}} \tag{4.6}$$

since, obviously, the plant output $y(t)$ does not depend on the model parameters $\hat{\kappa}$. Then from (4.1),(4.5) and (4.6) the necessary condition for determining the optimal $\hat{\kappa}$ is

$$E\{[y(t) - \hat{y}(t|t-1,\hat{\kappa})]\frac{\partial\hat{y}(t|t-1,\hat{\kappa})}{\partial\hat{\kappa}}\} = 0. \tag{4.7}$$

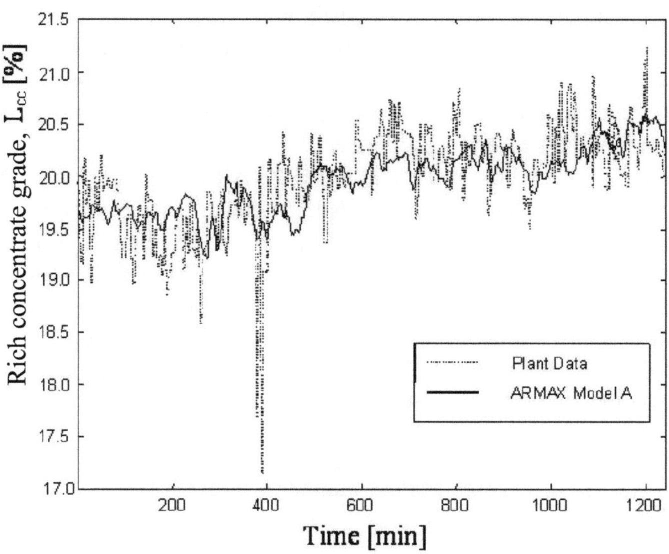

Figure 4.2 Concentrate grade sensor measurement (dotted line) and soft sensor output (solid line), where differences may be explained by the effect of important unmeasured disturbances

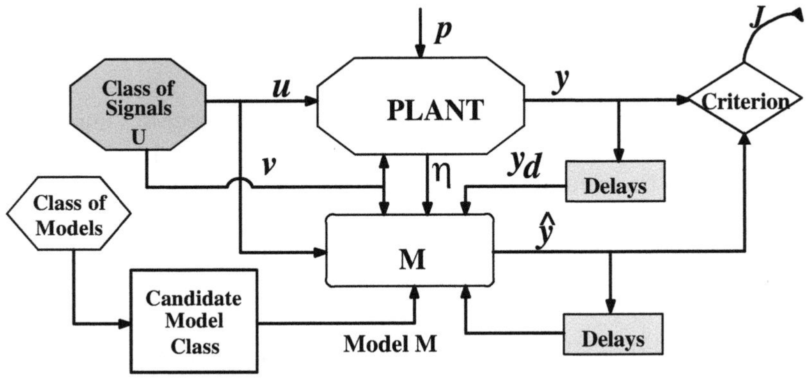

Figure 4.3 A general approach to modeling. Vector u contains the manipulated variables, v is the vector of measured disturbances, p is the vector of unmeasured disturbances, y is the plant output to be modeled, η are other measured plant outputs and $\hat{y}$ is the model output

4.2.2 A General Class of Black Box Models

Black box and gray box models are especially important for soft sensors in mineral processing plants because the complexities of the involved processes entail very intricate phenomenological models. As a consequence it is extremely difficult to identify these models accurately and reliably with plant data. Simpler back box models, on the other hand may be much more easily identified, but they usually require updating as the operation point shifts. However, gray models combining phenomenology with empirical or black box models afford a possible and useful compromise between simplicity and the need for too frequent updating. A unified approach to black box modeling is given by Sjöberg *et al.* [45] and Bakshi *et al.* [46] which covers all classes of black box models: FIR, ARX, ARMAX, OE, NFIR, NARMAX, NOE, nonlinear LIP models such as NARX models, wavelet models, radial basis function models, neural network models, Takagi and Sugeno, PCA and PLS based models, and other model structures. As an example in [31] ARX, NARX, PLS, Takagi and Sugeno, and wavelet models are used in the design of soft sensors for the concentrate grade of a rougher flotation bank of an industrial copper concentrator.

What follows is based on the general class of black box models proposed by Sjöberg *et al.* [45]. Let this general model be

$$\hat{y}(t|t-1,\kappa) = \sum_{k=1}^{m} \theta_k \varphi_k(\xi(t), \mu_k), \tag{4.8}$$

where μ_k is a row vector of parameters for function φ_k and

$$\xi(t) = [\xi_1(t)\xi_2(t)\dots\xi_q(t)]^T \tag{4.9}$$

is a vector of the selected inputs to the model, which for $k > 1$ may contain commands or manipulated variables $u(t-k)$, measured disturbances $v(t-k)$, delayed primary variable outputs $y(t-k)$, other plant outputs $\eta(t-k)$, delayed predictions $\hat{y}(t|t-1-k,\hat{\kappa})$, and prediction errors (residuals) $e(t-k,\hat{\kappa})$ [1, 45]. The selection of basis functions φ_k in the model structure determination phase is covered in Section 4.2.4.

For the case of the soft sensor models covered in this Chapter, in (4.8) vector κ contains all the model parameters to be determined, *i.e.*, θ and μ:

$$\kappa = [\theta^T \mu^T]^T, \quad \theta = [\theta_1 \theta_2 \ldots \theta_m]^T, \quad \mu = [\mu_1 \mu_2 \ldots \mu_m]^T. \tag{4.10}$$

In static models and in dynamic models where present values $\xi_j(t)$ appear, (4.8) is just the estimated output obtained and (4.1) becomes the modeling error.

Specific black box models turn out to be special cases of this general form, including neural network models. Let

$$\varphi(t,\mu) = [\varphi_1(\xi(t),\mu_1)\varphi_2(\xi(t),\mu_2)\ldots\varphi_m(\xi(t),\mu_m)]^T. \tag{4.11}$$

Equation 4.8 may be written, using (4.10) and (4.11),

$$\hat{y}(t|t-1,\kappa) = \varphi^T(\xi(t),\mu)\theta \tag{4.12}$$

For example, if the φ_k are radial basis functions or wavelets, parameters in μ are related to scaling and delays or shifts [45]. The following abbreviated notation shall be used:

$$\varphi_k(\xi(t),\mu_k) = \varphi_k(t,\mu). \tag{4.13}$$

Figure 4.4 has a schematic representation of (4.8).

If in (4.11) functions φ_k do not contain model parameters, or if their parameters are fixed and known, let this fact be reflected in the notation

$$\varphi_k(\xi(t),\mu_k) = \varphi_k(\xi(t)) = \varphi_k(t), \{k = 1,...,m\}, \tag{4.14}$$

$$\hat{y}(t|t-1,\theta) = \varphi^T(\xi(t))\theta, \tag{4.15}$$

and in this case $\kappa = \theta$. It must be noted that here $\hat{y}(i|i-1,\theta)$, for $i < t$ is not a model input ξ_r because it is a function of parameter vector θ, implying that some of the φ_k would also be functions of θ. In addition, let the set (4.14) be a linearly independent set. Then the set of functions (4.14) constitute a basis B_m for $\hat{y}(t|t-1,\theta)$ as seen in Section 4.3. Linear combinations of the basis functions given by (4.12) span a linear space to which the LIP model output $\hat{y}(t|t-1,\theta)$ belongs, since

$$\hat{y}(t|t-1,\theta) = \theta_1\varphi_1(t) + \theta_2\varphi_2(t) + \ldots + \theta_m\varphi_m(t) = \varphi^T(\xi(t))\theta. \tag{4.16}$$

Basis functions φ_k provide a means for introducing nonlinearities in a model which is LIP, since they may be nonlinear combinations of the ξ_j , *e.g.*, of direct plant measurements contained in ξ. Basis functions φ_k may have physical signifi-

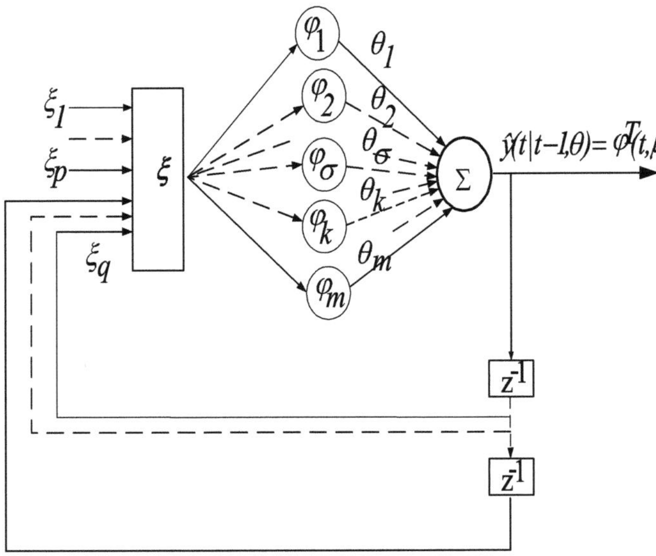

Figure 4.4 Example of general black box or gray box model structure (4.8), where $\hat{y}$ is the model output, ξ is a vector of selected model inputs, possibly including delayed versions of $\hat{y}$, and parameters θ_k define a linear combination of functions $\varphi_k = \varphi_k(t, \mu)$

cance or be terms of expansions, as in truncated wavelet, Fourier, and Taylor series expansions.

There is a general expression for a one step ahead predictor $\hat{y}(t|t-1)$ for an assumed discrete time plant model of the form

$$y(t) = G(z^{-1})u(t) + H(z^{-1})w(t), \tag{4.17}$$

where $w(t)$ is a white noise random sequence uncorrelated with $u(t)$, and $G(z^{-1})$ and $H(z^{-1})$ are fractions of polynomials in the delay operator z^{-1} . This predictor [1, 44] is given by

$$H(z^{-1})\hat{y}(t|t-1) = G(z^{-1})u(t) + (H(z^{-1}) - 1)y(t). \tag{4.18}$$

In the ARMAX case

$$A(z^{-1})y(t) = B(z^{-1})u(t) + C(z^{-1})w(t), \tag{4.19}$$

where A, B, and C are polynomials in the delay operator z^{-1} [1, 44],

$$y(t) = \frac{B(z^{-1})}{A(z^{-1})}u(t) + \frac{C(z^{-1})}{A(z^{-1})}w(t), \tag{4.20}$$

so

$$G(z^{-1}) = \frac{B(z^{-1})}{A(z^{-1})} \qquad H(z^{-1}) = \frac{C(z^{-1})}{A(z^{-1})}. \tag{4.21}$$

From (4.18) and (4.21) the one step ahead predictor for the ARMAX case is given by

$$C(z^{-1})\hat{y}(t|t-1) = B(z^{-1})u(t) + (C(z^{-1}) - A(z^{-1}))y(t). \tag{4.22}$$

For convergence, asymptotic stability of (4.22) is required, for which a necessary condition is that the roots of $C(z^{-1})$ have absolute values that are less than one.

For the ARX model class $C(z^{-1}) = 1$ in (4.22), so

$$\hat{y}(t|t-1) = B(z^{-1})u(t) + (1 - A(z^{-1}))y(t). \tag{4.23}$$

In the case of soft sensors there are usually several inputs $u_1, u_2, ...u_r$. Let then

$$u(t) = [u_1(t)\ \ u_2(t) \dots u_r(t)]^T \qquad G(z^{-1}) = [G_1(z^{-1})\ \ G_2(z^{-1}) \dots G_r(z^{-1})] \tag{4.24}$$

$$G(z^{-1})u(t) = G_1(z^{-1})u_1(t) + G_2(z^{-1})u_2(t) + \dots G_r(z^{-1})u_r(t). \tag{4.25}$$

Following the same procedure used by Ljung [44] to derive the one step ahead prediction for one input, given by (4.18), it may easily be shown that in this multiple input case the one step ahead prediction has the same form, *i.e.*,

$$H(z^{-1})\hat{y}(t|t-1) = G(z^{-1})u(t) + (H(z^{-1}) - 1))y(t), \tag{4.26}$$

where vectors $u(t)$ and $G(z^{-1})$ are defined by (4.24).

Example 1. Let the assumed plant model be the ARMAX model

$$\begin{aligned} y(t) + a_1 y(t-1) + a_2 y(t-2) + a_3 y(t-3) = {} & b_1 u(t-1) + w(t) + c_1 w(t-1) \\ & + c_2 w(t-2), \end{aligned} \tag{4.27}$$

where w is a white noise sequence uncorrelated with u. Therefore the noise in (4.27) is colored (non white). Then

$$A(z^{-1}) = 1 + a_1 z^{-1} + a_2 z^{-2} + a_3 z^{-3},\ B(z^{-1}) = b_1 z^{-1},\ C(z^{-1}) = 1 + c_1 z^{-1} + c_2 z^{-2} \tag{4.28}$$

and from (4.22),

$$\begin{aligned} \hat{y}(t|t-1) = {} & -c_1 \hat{y}(t-1|t-2) - c_2 \hat{y}(t-2|t-3) + b_1 u(t-1) \\ & -(a_1 - c_1)y(t-1) - (a_2 - c_2)y(t-2) - a_3 y(t-3). \end{aligned} \tag{4.29}$$

Let

$$\theta = [c_1\ \ c_2\ \ (a_1 - c_1)\ \ (a_2 - c_2)\ \ a_3\ \ b_1]^T. \tag{4.30}$$

and

$$\varphi(t) = [-\hat{y}(t-1|t-2)\ \ -\hat{y}(t-2|t-3)\ \ -y(t-1)\ \ -y(t-2)\ \ -y(t-3)\ \ u(t-1)]^T. \tag{4.31}$$

Since the one step ahead predictions depend on the model parameters, the explicit notation

$$\hat{y}(t|t-1) = \varphi^T(t,\theta)\theta \tag{4.32}$$

has been used, since (4.32) is nonlinear in the parameter vector θ .

In the case of the ARX class of models, from (4.23), or letting $c_1 = c_2 = c_3 = 0$ in (4.29),

$$\hat{y}(t|t-1) = b_1 u(t-1) - a_1 y(t-1) - a_2 y(t-2) - a_3 y(t-3), \tag{4.33}$$

so

$$\varphi(t) = [-y(t-1) \;\; -y(t-2) \;\; -y(t-3) \;\; u(t-1)]^T \tag{4.34}$$

does not depend on θ . Then

$$\hat{y}(t|t-1) = \varphi^T(t)\theta \tag{4.35}$$

is linear in the parameter vector

$$\theta = [a_1 \;\; a_2 \;\; a_3 \;\; b_1]^T. \tag{4.36}$$

Example 2. Let the assumed plant model plant model be of the NARX class

$$\begin{aligned} y(t) &= -a_1 y(t-1) - a_2 y(t-2) - a_3 y(t-3) + b_1 u(t-1) + b_2 u^2(t-1) \\ &\quad + b_3 \frac{u(t-1)}{3 - \sqrt{u(t-1)}} + b_4 l(t) + w(t). \end{aligned} \tag{4.37}$$

Here $l(t)$ is a function having a real value 1 for all discrete times t that must be considered as an additional input to account for the constant term b_4.

Let

$$\begin{aligned} u_1(t-1) &= u(t-1) \quad && u_2(t-1) = u^2(t-1) \\ u_3(t-1) &= \frac{u(t-1)}{3-\sqrt{u(t-1)}} \quad && u_4(t-1) = l(t). \end{aligned} \tag{4.38}$$

Replacing (4.38) in (4.37) gives

$$\begin{aligned} y(t) &= -a_1 y(t-1) - a_2 y(t-2) - a_3 y(t-3) + b_1 u_1(t-1) + b_2 u_2(t-1) \\ &\quad + b_3 u_3(t-1) + b_4 u_4(t-1) + w(t), \end{aligned} \tag{4.39}$$

i.e., a single output ARX class model with four inputs.

$$y(t) = \frac{b_1 u_1(t-1) + b_2 u_2(t-1) + b_3 u_3(t-1) + b_4 u_4(t-1)}{A(z^{-1})} + \frac{w(t)}{A(z^{-1})}, \tag{4.40}$$

where $A(z^{-1}) = 1 + a_1 z^{-1} + a_2 z^{-2} + a_3 z^{-3}$.

Hence,

$$G_1(z^{-1}) = \frac{b_1 z^{-1}}{A(z^{-1})} \quad G_2(z^{-1}) = \frac{b_2 z^{-1}}{A(z^{-1})} \quad G_3(z^{-1}) = \frac{b_3 z^{-1}}{A(z^{-1})} \quad G_4(z^{-1}) = \frac{b_4 z^{-1}}{A(z^{-1})} \tag{4.41}$$

and

$$H(z^{-1}) = \frac{1}{A(z^{-1})}. \tag{4.42}$$

From (4.26), (4.41), and (4.42),

$$\frac{1}{A(z^{-1})}\hat{y}(t|t-1) = G(z^{-1})^T u(t) + [\frac{1}{A(z^{-1})} - 1]y(t), \tag{4.43}$$

$$\begin{aligned} \hat{y}(t|t-1) = & -a_1 y(t-1) - a_2 y(t-2) - a_3 y(t-3) + b_1 u_1(t-1) \\ & + b_2 u_2(t-1) + b_3 u_3(t-1) + b_4 u_4(t-1) \end{aligned} \tag{4.44}$$

$$\begin{aligned} \hat{y}(t|t-1) = & -a_1 y(t-1) - a_2 y(t-2) - a_3 y(t-3) + b_1 u(t-1) \\ & + b_2 u^2(t-1) + b_3 \frac{u(t-1)}{3 - \sqrt{u(t-1)}} + b_4 l(t). \end{aligned} \tag{4.45}$$

Hence,

$$\varphi(t) = [-y(t-1) \;\; -y(t-2) \;\; -y(t-3) \;\; u(t-1) \;\; u^2(t-1) \frac{u(t-1)}{3-\sqrt{u(t-1)}} \;\; l(t)]^T \tag{4.46}$$

and $\hat{y}(t|t-1) = \varphi^T(t)\theta$ is LIP vector

$$\theta = [a_1 \;\; a_2 \;\; a_3 \;\; b_1 \;\; b_2 \;\; b_3 \;\; b_4]^T \tag{4.47}$$

even though it is nonlinear in the plant input $u(t)$.

Example 3. Let the assumed plant model be the same as that of Example 2, except that parameter b_5 is unknown, and must be estimated along with the other six parameters. Then

$$\begin{aligned} y(t) = & -a_1 y(t-1) - a_2 y(t-2) - a_3 y(t-3) + b_1 u(t-1) + b_2 u^2(t-1) \\ & + b_3 \frac{u(t-1)}{b_5 - \sqrt{u(t-1)}} + b_4 l(t) + w(t) \end{aligned} \tag{4.48}$$

and in view of (4.48)

$$\varphi(t,\mu) = [y(t-1)\;\; y(t-2)\;\; y(t-3)\;\; u(t-1)\;\; u^2(t-1)\frac{u(t-1)}{b_5 - \sqrt{u(t-1)}}\;\; l(t)]^T \tag{4.49}$$

so $\varphi(t,\mu)$ depends on parameter b_5. Then this is a NLIP (nonlinear in the parameters) model and, in general, a NARX model [1].

Example 4. In the case of the single layer perceptron $\varphi_k(t,\mu_k)$ contains parameters (weights) in vector β_k, and the bias γ_k:

$$\varphi_k = f(d^T \beta_k + \gamma_k), \tag{4.50}$$

where f is, *e.g.*, a sigmoid function (Figure 4.4). This model is NLIP.

4.2.2.1 Optimal Parameter Vector

The optimal parameter vector is obtained through minimization of functional (4.2) using the necessary condition (4.7)

$$E\{[y(t) - \hat{y}(t|t-1,\hat{\kappa})]\frac{\partial \hat{y}(t|t-1,\hat{\kappa})}{\partial \hat{\kappa}}\} = 0. \tag{4.51}$$

From (4.7)

$$\frac{\partial \hat{y}(t|t-1,\hat{\kappa})}{\partial \hat{\kappa}} = \sum_{k=1}^{m} \frac{\partial \hat{\theta}_k \varphi_k(\xi(t),\hat{\mu}_k)}{\partial \hat{\kappa}}. \tag{4.52}$$

For LIP models (*e.g.*, ARX, NARX),

$$\hat{y}(t|t-1,\hat{\kappa}) = \hat{y}(t|t-1,\hat{\theta}) = \sum_{k=1}^{m} \hat{\theta}_k \varphi_k(t) = \varphi^T(t)\hat{\theta} \tag{4.53}$$

and

$$\frac{\partial \hat{y}(t|t-1,\hat{\kappa})}{\partial \hat{\kappa}} = \frac{\partial \hat{y}(t|t-1,\hat{\kappa})}{\partial \hat{\theta}} = \varphi^T(t), \tag{4.54}$$

since no parameters appear in $\varphi(t)$.

From (4.7), (4.53) and (4.54)

$$E\{[y(t) - \varphi^T(t)\hat{\theta}]\varphi^T(t)\} = 0,\;\; E\{\varphi^T(t)y(t)\} = E\{\varphi^T(t)\hat{\theta}\varphi^T(t)\}, \tag{4.55}$$

$$E\{\varphi(t)y(t)\} = E\{\varphi(t)\varphi^T(t)\}\hat{\theta},\;\; \hat{\theta} = E\{\varphi(t)\varphi^T(t)\}^{-1}E\{\varphi(t)y(t)\}. \tag{4.56}$$

Assuming $\varphi(t)$ is a wide sense stationary stochastic vector, *i.e.*, its elements are stochastic processes (random sequences) [47], the correlation matrices in (4.56) are independent of time t and may be written

$$R_{\varphi\varphi} = E\{\varphi\varphi^T\},\quad R_{\varphi y} = E\{\varphi y\}. \tag{4.57}$$

If $R_{\varphi\varphi}$ is nonsingular (a condition achieved through persistent excitation [44]), an explicit solution

$$\hat{\theta} = R_{\varphi\varphi}^{-1} R_{\varphi y} \tag{4.58}$$

is obtained for the optimal vector $\hat{\theta}$ of LIP models, such as ARX and NARX models (see (4.34) in Example 1, and (4.46) in Example 2). In Section 4.4 the same result is obtained (see (4.110)) in a general inner product space by projecting the primary measurement $y(t)$ onto the space spanned by the base $B_m = \{\varphi_1, \varphi_2, \ldots, \varphi_m\}$ to which $\hat{y}(t|t-1)$ belongs, and specializing the results to the case of a random variable inner product space.

But, again, the problem is that the optimal model vector is given in terms of expected values, and in the large majority of cases the probability functions needed to compute these expected values are not available. Therefore, they must be estimated using time averages.

$$\hat{R}_{\varphi\varphi} = \frac{1}{T} \sum_{i=t-T+1}^{t} \varphi(i)\varphi^T(i) \approx E\{\varphi\varphi^T\}, \tag{4.59}$$

$$\hat{R}_{\varphi y} = \frac{1}{T} \sum_{i=t-T+1}^{t} \varphi(i)y(i) \approx E\{\varphi y\}, \tag{4.60}$$

$$\hat{\hat{\theta}} = \hat{R}_{\varphi\varphi}^{-1} \hat{R}_{\varphi y}. \tag{4.61}$$

In Section 4.3.2 it can be seen that $\hat{R}_{\varphi\varphi}$ is a random matrix and $\hat{R}_{\varphi y}$ is a random vector, so $\hat{\hat{\theta}}$ is a random vector having an expected value and a covariance matrix. Hence the estimator $\hat{\hat{\theta}}$ for a given set of measurements is just one of all possible outcomes that could have resulted. It should be that realizations of this estimator are distributed around $\hat{\theta}$ and that variances of the parameter estimations are small. Required properties for these estimators through time averages then are unbiasedness and ergodicity or consistency, so that variances decrease to zero as length T of time window $[t-T+1, t]$ increases to infinity and $\hat{\hat{\theta}}$ converges in the mean to $\hat{\theta}$. Ljung [44] shows that estimator $\hat{\hat{\theta}}$ satisfies these conditions.

If the model is NLIP, *e.g.*, a NARMAX model (see Examples 3 and 4), the optimal parameter vector must be obtained using algorithmic methods [48] to directly minimize the time average (4.3), *e.g.*, using the System Identification Toolbox of MATLAB® [49], or algorithms used in neural network optimal parameter (weight) determination [12, 14].

From (4.135) in Section 4.3

$$\hat{\hat{\theta}} = (X^T X)^{-1} X^T Y, \tag{4.62}$$

where

$$X = \begin{bmatrix} \varphi^T(t-T+1) \\ \vdots \\ \varphi^T(t) \end{bmatrix}, Y = \begin{bmatrix} y(t-T+1) \\ \vdots \\ y(t) \end{bmatrix}, \tag{4.63}$$

so the estimator is given in terms of matrix operations instead of sums, which is very convenient in computer programming. The covariance matrix of estimator $\hat{\theta}$ is given by [44]

$$Cov(\hat{\hat{\theta}}) = \lambda_0 \left[\sum_{i=t-T+1}^{t} \varphi(i)\varphi^T(i) \right] = \lambda_0 (X^T X)^{-1}, \tag{4.64}$$

where an unbiased estimator of λ_0 is

$$\hat{\lambda}_T = \frac{1}{T-m} \sum_{i=t-T+1}^{t} (y(i) - \varphi^T(i)\hat{\hat{\theta}}) \tag{4.65}$$

and m is the dimension of φ. The diagonal of $Cov(\hat{\hat{\theta}})$ are the variances of the parameter estimations.

4.2.2.2 Gray Models

The phenomenological part of gray models usually consists of dynamic balances of masses, volumes, energy, *etc.*, as in the case of particle masses for different size intervals in a mill. The empirical part is usually concerned with some internal processes governing the relation between masses, volumes, *etc.*, for example, in a mill model [50] the transfer of ore particles from one size range to another. Figure 4.5 gives a block diagram of this state model. Other examples of Gray models for soft sensors are given in [11, 16, 17] and in [10, 19, 31]. In these cases nonlinear combinations of secondary variables having phenomenological meaning are used in the definition of the basis functions φ_k, but the models are LIP, hence NARX models. An example is given in detail for the case of a +65# particle size soft sensor in Section 4.3. One of the models determined there is

$$f_{65}(t|t-1) = \theta_0 f_{65}(t-1) + \theta_1 J_{BM}(t-1)S_p^3(t) + \theta_2 J_{BM}(t-3)S_p(t-3) + \theta_3 l(t). \tag{4.66}$$

It can be seen that it is a NARX model, LIP but nonlinear in plant measurements: mill power draw J_{BM} and solids concentration in the hydrocyclone feed flow S_p.

4.2.3 The Use of the Model as a Soft Sensor

Once the model has been determined it may be used as a soft sensor. Since the soft sensor is replacing the actual sensor, no measurements are available for measure-

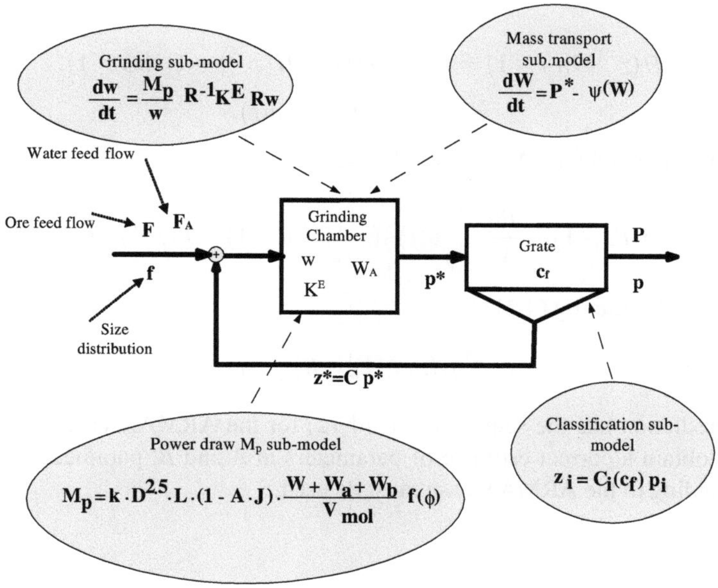

Figure 4.5 Gray state model for a SAG mill in which dynamic mass balances are combined with empirical (black box) relations [50]

ments involving the delayed versions $y(t-k)$ of the primary variable prior to the present time t. Instead, the previous estimated measurements are used. For example let, the one step ahead prediction be the one found in Example 1. Then replacing

$$y(t-k) = \hat{y}(t-k|t-k-1) \tag{4.67}$$

the soft sensor output is given by

$$\begin{aligned}\hat{y}(t|t-1) = & -c_1\hat{y}(t-1|t-2) - c_2\hat{y}(t-2|t-3) + b_1u(t-1) \\ & -(a_1-c_1)y(t-1|t-2) - (a_2-c_2)y(t-2|t-3) - a_3y(t-3|t-4).\end{aligned} \tag{4.68}$$

Collecting terms,

$$\hat{y}(t|t-1) = -a_1\hat{y}(t-1|t-2) - a_2\hat{y}(t-2|t-3) - a_3\hat{y}(t-3|t-4) + b_1u(t-1). \tag{4.69}$$

This is a difference equation for the soft sensor output $\hat{y}(t)$ driven by input u. The solution depends on initial conditions, *e.g.*, $\hat{y}(1), \hat{y}(0)$. These initial conditions could be arbitrary, but a better choice could be some reasonable value derived from knowledge of the plant, ideally their expected values. In any case, if (4.69) is asymptotically stable, the effect of the initial conditions vanish after a transient. This requires that the roots of $z^3 + a_1z^2 + a_2z + a_3 = 0$ must be inside the unit circle in the complex plane [44].

Considering now the general case (4.18) and using (4.67),

$$H(z^{-1})\hat{y}(t|t-1) = G(z^{-1})u(t) + (H(z^{-1})-1)\hat{y}(t|t-1), \tag{4.70}$$

$$\hat{y}(t|t-1) = G(z^{-1})u(t). \tag{4.71}$$

For example in the ARMAX case, by (4.19)

$$\hat{y}(t|t-1) = \frac{B(z^{-1})}{A(z^{-1})}u(t)A(z^{-1})\hat{y}(t|t-1) = B(z^{-1})u(t). \tag{4.72}$$

In the ARX case, by (4.21)

$$A(z^{-1})\hat{y}(t|t-1) = B(z^{-1})u(t), \tag{4.73}$$

so the prediction has the same form as (4.72) for the ARMAX case. However, in order to obtain a correct estimate of parameters in A and B, parameter estimation corresponding to the ARMAX case must be used.

4.2.4 Development of Soft Sensor Models

The process of finding the best model for a soft sensor may be separated into two stages:

- structure determination;
- parameter estimation.

Prior to using the data in model determination it must be analyzed and subjected to certain processing, as indicated in Section 4.3 and by Fortuna *et al.* [1] and Ljung [44] (see also [31]).

4.2.4.1 Model Structure

The determination of model structure is concerned with finding a form for the soft sensor model. In particular in the case of ARX and NARX class of models this involves finding the most significant basis functions φ_k correlated with the primary measurement which will be used in the model. In the neural network case, the structure comprises the type of network, the network topology in general, the most significant measurements to be used as inputs, the number of layers, the selection of the activation functions, and the number of nodes in each layer. Composite components may also be considered as suitable inputs for a neural network. A schematic diagram of the determination of model structure is shown in Figure 4.6. The selection of plant measurements is determined by the measurements required by the selected components or basis functions.

A model having a given structure may be valid for a large operating region of a process, but its parameters may need to be updated as the operating point undergoes changes. It may also happen that models with different structures may be the best choice when the operating point moves to other operating regions, so the model must be changed. For a smooth transition between two regions clustering and membership functions may be used, as described below under Clustering. There are various indexes defined for the training stage of structure determination.

In order to reduce the complexity of the model to be determined the number of parameters or the number of bases is penalized in relation to the length of the training record. Examples are the FPE (final prediction error) and AIC (Akaike's information criterion) indexes:

$$FPE = \frac{1+m/T}{1-m/T} \quad AIC = \log\left[V(1+2\frac{m}{T}),\right] \tag{4.74}$$

where T is the length of the data sets, m is the number of model parameters and V is the estimation of the mean square prediction error or the modeling error (4.2) estimated by (4.4) [44]. Another index that has been used in connection with reducing complexity of neural network models is the Lipschitz number to select the inputs, and a weighted sum of the modeling error and a term penalizing the sum of the squares of the parameters [14]. Other indexes used for model structure determination are given by Fortuna *et al.* [1].

All these indexes tend to avoid overfitting, which may cause the mean square output error to be very small for the training data set but very large for validation (generalization) data sets. To ensure that this situation does not occur, in addition the structure determination should be done with one set of data called the training set and tested with a different set, the test or validation set.

4.2.4.2 Clustering

The advantages of dealing with a simple, parsimonious model may sometimes be achieved by means of separating the operation region into clusters, each of which has a good simple model. Then if the operating point is identified as belonging to a given cluster, the corresponding model is used. Such a procedure has been used in building a soft sensor for concentrate grade measurement in an industrial rougher flotation plant [26]. Much better results were obtained with clustering than with the case of a single model determined for the complete operating range.

In the design of a soft sensor for the rich copper concentrate grade g_{cc} of an industrial flotation rougher flotation plant two clusters, C_1 and C_2 were employed using fuzzy C-means clustering [31, 51]. For each cluster a soft sensor model was determined. From the relevant membership functions $\mu_1(g_{cc})$ and $\mu_2(g_{cc})$ relative membership functions were defined as

$$\psi_1(g_{cc}) = \frac{\mu_1(g_{cc})}{\mu_1(g_{cc})+\mu_2(g_{cc})} \quad \psi_2(g_{cc}) = \frac{\mu_2(g_{cc})}{\mu_1(g_{cc})+\mu_2(g_{cc})}. \tag{4.75}$$

The total soft sensor output was then obtained by a weighted combination of the outputs of soft sensors for clusters C_1 and C_2 using the relative membership functions as weights. Figure 4.7 shows the evolution of these relative membership functions for the data in Figures 4.2 and 4.8. The clustering approach was also used for this plant to design a Takagi and Sugeno model [31].

Clustering has also been used for a soft sensor designed to model an event and the proximity to the occurrence of such event [26]. In this case the objective was to give early warning for the event defined by the overload of a ball mill in an industrial grinding plant. C-means clustering was used to group the plant's operating points. Later developments led to retain two clusters representing two different types of overload. The distances from the operating point to each of these two clusters, computed on-line, give an indication of how far the operation was from reaching one of the two overload conditions. Different corrective actions could then be taken according to the distance to the nearest cluster.

4.2.4.3 Stepwise Regression

This is one of the methods used to determine which are the most significant components (bases) associated with the soft sensor model primary measurement. One version of stepwise regression begins by selecting that component from a set of candidate components which is most closely correlated with the variable to be modeled, *i.e.*, the primary variable. In this way a first partial model is determined. The residue of this model is then correlated with the remaining candidate components. In each stage that follows, the component to be included is the particular component that gives the largest partial correlation with the previous step residue, calculated after a multilinear regression is performed with the previously selected components. With this procedure, the model structure increases through the addition of one component at a time, such that the added component is the one that contributes the greatest improvement to the goodness of fit to the model. Before each new component is included in the model, though, it is tested for its statistical significance. Each model parameter estimate has an estimated standard deviation. The ratio between this standard deviation and the coefficient value, is used to decide the inclusion of every new component. If for any component this ratio exceeds a given threshold then the corresponding component is not included, or excluded if it had been previously included. This procedure is repeated until no component from the list of candidate components is either deleted or included in the model being determined. Alternatively the procedure stops if inclusion of new candidates does not cause a significant improvement according to an F-test [52].

MATLAB® stepwise or stepwisefit functions may be used to perform stepwise regressions. It should be mentioned that stepwise models are locally optimal, but may not be globally optimal (Mathworks™[1], stepwisefit) [53]. Another MATLAB® function for model structure determination is lsselect [54] with which models with

[1] Mathworks is a registered trademark of The MathWorks, Inc., 3 Apple Hill Drive, Natick, MA 01760-2098, USA, http://www.mathworks.com

all possible subsets of candidate components are systematically determined in order to choose the best one. It also has options for the forward inclusion or forward regression method [48] and various criteria for assessing the best model.

Good results have been obtained by Casali *et al.* [11, 16, 17] and by Gonzalez *et al.* [10, 31] using stepwise regression in the design of soft sensors for grinding circuits and flotation concentrate grades.

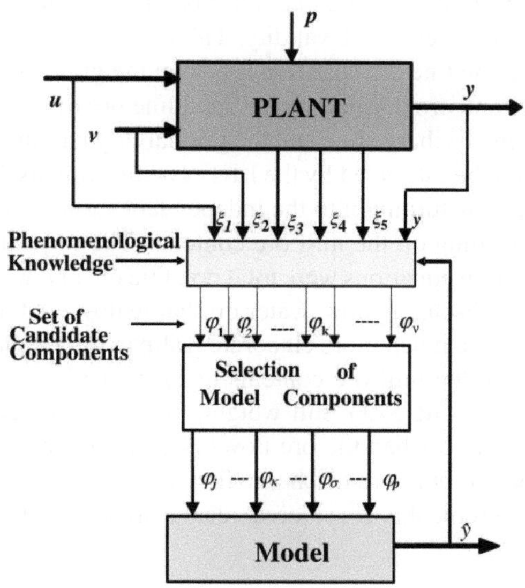

Figure 4.6 Example of a structure determination for a black box model, where the candidate components are measured variables or functions of them, derived from phenomenological knowledge about the plant. The manipulated variables are in vector u. Vector v contains the measured disturbances and vector p the unmeasured disturbances

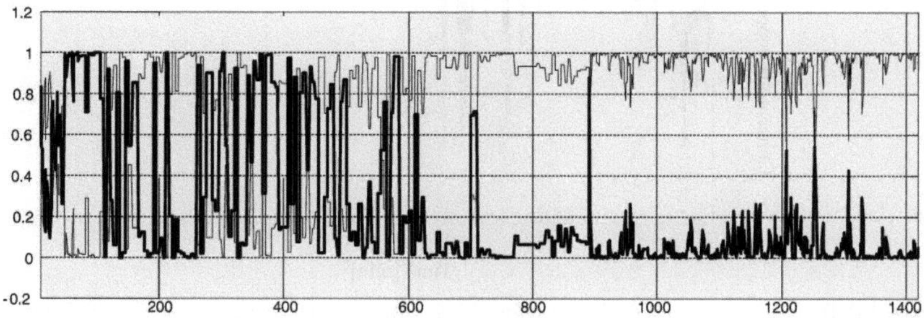

Figure 4.7 Evolution of the relative membership functions ψ_1 and ψ_2 used to combine the outputs of models 1 and 2 corresponding to clusters C_1 and C_2 in the design of a concentrate grade soft sensor for a rougher flotation bank (from [31])

4.2.4.4 Soft Sensors Based on the Kalman Filter

Soft sensor models using Kalman filters have been used when the phenomenological knowledge about the plant allows a convenient state-state/output model to be built [20, 21, 29, 39]. A soft sensor has been tested using data from a pilot semiautogenous grinding mill in which the ore contents comprise 27 size fractions which are states [20], with aid from the phenomenological model [20, 55]. In this case a low dimension extended (nonlinear) Kalman filter (EKF) was used, since a linear state model had a very small region of validity. The estimated states were water, total mill ore content and the fine ore (of sizes less than the grate opening) mill content, grinding rate, water and ore discharge rates, and fine ore discharge flow. The water discharge rate, the ore discharge rate and the grindability parameters were converted to states so they could be estimated by the EKF. The fine ore discharge flow was also converted into a state by turning into the independent variable of a first order equation with input depending on the mill ore contents. The measurements considered for the state/state-output equations were total ore, fine ore and water feed flows, total mill weight, total ore discharge flow, water discharge flow, and mill power draw.

Apelt *et al.* [21] use a much more elaborate Kalman filter equation with 36 states of which 27 represent the mill ore contents by size fractions (as in [20]), and the rest are the ball charge, the SAG mill weight, the water content, and parameters. Good results are reported when the ore flows per each of the 27 ore size intervals are measured instead of bulk ore discharge flow measurement. If only the bulk ore discharge flow measurement is considered, results are acceptable for the most part.

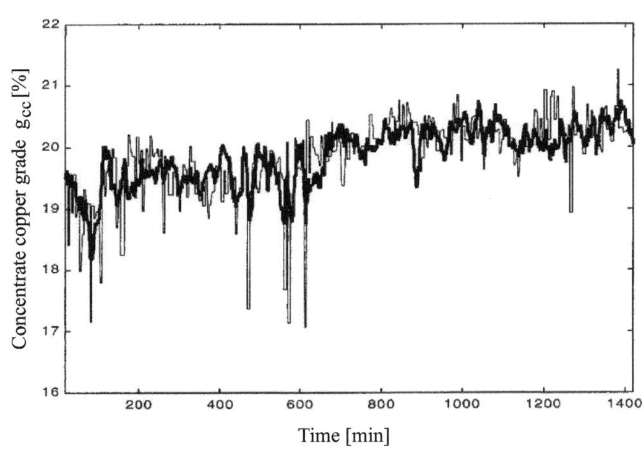

Figure 4.8 Test of soft sensor designed using PLS. Concentrate grade sensor measurement (fine line) and soft sensor output (broad line)

4.2.4.5 Principal Components

In the principal component modeling methods an $N \cdot m$ sample measurement matrix X is formed in which each row is a vector $x(j)^T$ containing the m measurements $x_1(j), x_2(j), \ldots, x_m(j)$ performed in the plant at sample times $j = 1, \ldots., N$. Then each column of matrix X may be considered a realization of a random sequence () in the time window corresponding to $j = 1, \ldots., N$. The m eigenvalues λ_p and eigenvectors p_i of the symmetrical matrix $X^T X$ are found. Since $X^T X$ is symmetric, all eigenvectors are orthogonal to each other, and are normalized to Euclidean length 1. Letting V be the matrix whose columns are the eigenvectors p_i, it turns out that VV^T = I. The columns of vector $Z = XV$ are the principal components $z_1(j), z_2(j), \ldots, z_m(j)$ corresponding to measurements $x_1(j), x_2(j), \ldots, x_m(j)$. Principal components are mutually orthogonal, *i.e.*, $z_i(j)^T z_k(j) = 0$ for $i \neq k$ and span the principal component space. Furthermore $z_i(j)^T z_i(j) = \|z_i(j)\|^2 = \lambda_i$, [56–58]. A subset of principal components that explains most of the data variation is selected, *e.g.*, by choosing those having the largest eigenvalues. In this way a space of sometimes considerably reduced dimensions is obtained which facilitates modeling, fault detection, and the solution of other problems.

Principal components have been used in the estimation of a grindability index by Gonzalez *et al.*, [22]. Here an original dimension 48 space has been reduced to an 8 dimension space in which Fisher discrimination analysis [57] was made simpler for the identification a grindability index, which corresponds to the soft sensor output.

Using a PCA method a missing measurement may be reconstructed [34], [59]. But more than one measurement may be reconstructed, depending on the dimension of the principal component space and the number q of sensors. Hence several soft sensors are implicit in the PCA modeling method and the set of secondary measurements for any of them depends on which of the q measurements are missing. Here the model structure is contained in the principal component space and is determined from the sample correlation matrix $X^T X$.

In principal component regression [56, 58] a regression model is determined by regressing the soft sensor variable on the principal components of the secondary measurements or basis functions. Here matrix X is similar to the one in PCA, but contains only the secondary measurements or basis functions used to model the soft sensor.

Projection to latent structures, an extension of PCA, has been used in the design of a soft sensor for concentrate grade in a rougher flotation plant. Figure 4.8 shows a test of the soft sensor with the same data used in Figure 4.2 [31].

4.2.4.6 Neural Networks

Neural network models may be considered as general adaptable nonlinear functions generators whose parameters (weights) may be determined so as to minimize indexes reflecting the error between the model output and the plant measurement, following the general modeling approach [1, 12, 14, 27, 31]. Being nonlinear, neural network models tend to be valid in larger operating regions, but updating takes longer [12].

4.2.4.7 Support Vector Machines (SVM)

Support vector machines have recently started being used [60] for designing soft sensors for the mineral processing industry. Sun *et al.*[15] have used SVM in the design of a particle size soft sensor at the hydrocyclones overflow in an industrial grinding circuit.

4.2.5 On-line Parameter Estimation

Once a model structure has been determined, the model parameters (weights) are estimated for an operating region. For this purposes gradient techniques are used to minimize a function of an error by measuring the discrepancy between the plant measurement and the estimated measurement obtained using the model (*e.g.*, the mean square prediction error or the output error). For certain cases such as ARX and NARX models explicit solutions are available and no search for the optimum is necessary (Section 4.2.2).

As already pointed out, model parameters may have to be updated as the operating point moves within a region – *e.g.*, defined by clusters – even if the structure remains valid. Parameter estimation may be performed on-line or off-line, depending on whether there is an installed operating sensor for the primary measurement or not. If this sensor is not installed, sampling and laboratory measurements would be required. In the off-line case the model should be valid in a large operation region so that the model parameters do not need to be updated frequently, because of the operational and cost matters involved.

A model based on the slower than appropriate sampling rate will not give satisfactory results, since it is derived using samples that do not represent the actual evolution of the variable involved. Instead, a model must be built using a sampling rate that is fast enough for the sampling theorem to be satisfied [61]. If there are other correlated secondary measurements that are sampled at an appropriately fast rate there is a solution. A model is built based on the fast sampling rate, notwithstanding which its parameters (weights) may be found using the infrequent samples provided by the measurement of the modeled primary variable as well as the fast rate samples of the correlated (secondary) measurements. For example, an instrumentation system for measuring grades in a flotation plant is used in a time sharing mode to assay 14 samples piped into an X-ray analyzer. If the analyzer takes 30 s to measure each mineral sample, the sampling rate for any of the 14 measured grades is 7 min. In a similar case there is a laboratory analysis performed infrequently on the modeled variable but due to the dynamics of the process a faster rate is required for good manual or automatic control [33, 62].

When the sensor becomes unavailable at t_f there are three choices for the soft sensor parameters. On the one hand the values that had been estimated up to time t_f may be used (frozen parameters). But a better choice might be to use the average of the estimated parameters or, better still, a prediction of the future parameter values using a parameter evolution sub-model, whenever appropriate conditions are met and the sub-model parameters may be estimated while the soft sensor is working. The sub-model is then used to give an optimal prediction of the future value of the soft sensor model parameters for $t > t_f$. This optimal prediction for a given parameter is its conditional expectation given its value prior to the time of failure t_f [19]. As an example a first order sub-model for each parameter of a soft sensor for particle size measurement in an industrial grinding plant has been used. The prediction evolves exponentially from the initial value at t_f to its unconditional expected value for $t >> t_f$. As should be expected, the frozen parameter choice gives better prediction for times near the fault, while the unconditional expected value parameters give a better prediction for $t >> t_f$ [19].

4.2.5.1 Region of Validity

Once a sensor fails, the parameter updating must cease, since it is no longer possible to determine the error between the soft sensor output and the actual measurement. The soft sensor signal then begins to perform its job, providing a virtual measurement of the missing one, using the secondary measurements or composite components as bases. But plant characteristics may change during the period after the sensor failure, *e.g.*, because the operation point shifts due to disturbances or control actions. If the model involves a simplification of reality, as is usually the case, it may be expected that, except in very simple instances, the model will likely not be able to represent the part of the actual plant to be modeled for all possible operating points or operating trajectories. This is clearly the situation when for the sake of having a simple model, a class of linear models is selected containing the candidates that represent a nonlinear plant. As the operating point changes, the optimal parameter set should change, so a frequent updating of the model parameters may be required. But since the primary sensor signal is no longer available this updating is no longer possible, and the soft sensor error becomes unacceptable since its region of validity has been surpassed. One way to palliate this performance degradation is to increase the region of validity of the soft-sensor model by incorporating more structure into it, *e.g.*, using NARX or NARMAX models having physically significant components (*i.e.*, gray models),[10, 11, 16, 17, 31] or sufficiently complex neural networks. In any case, on-line adaptation requires a compromise to be made between simple soft sensor models requiring frequent but fast adaptation (*e.g.* ARX models) and complex models requiring a less frequent but slower adaptation process, *e.g.*, neural networks of increasing complexity [12]. In between are those models that although nonlinear are LIP (*e.g.*, NARX models), and hence may undergo faster adaptation even though they may be nonlinear in the secondary measurements. Another way is to use clustering, as explained above.

4.2.5.2 Soft Sensor Models and Plant models

From the point of view of soft sensor models (and in many other cases: control system design, fault isolation and detection systems) it is very convenient to have a dynamic or static model of the plant. If a model is available the soft sensor's performance may be evaluated for different operating conditions. But since plant models in the mineral processing industry are usually quite complex, in most cases a model which is qualitatively good but only approximate from the quantitative point of view is sufficient to design and test the soft sensor's basic behavior. The final adjustment of the soft sensor may then be performed with plant data using the best models determined. For example, plant models have been used to design and test soft sensors in [12, 18, 20, 28]. Figure 4.25 in Section 4.4.2 shows the SAG mill circuit for which a dynamic model able to handle two types of ore was developed, and used for designing a grindability index soft sensor [22]. In Figure 4.9 the response of the model circuit variables to a step change in fresh ore feed flow is shown.

4.2.6 Soft Sensors Designed From Measurement Features

A different way to approach soft sensor design is by analyzing characteristics or features extracted from measured variables, such as variances and wavelet transforms, possibly after undergoing transformations using PCA and discriminant analysis. Such is the case of the on-line estimation of the grindability index based on concatenated wavelet transform variances. Tests done by simulation using a dynamic SAG circuit state model of a large industrial grinding circuit show good results, as seen in [22] and in Section 4.4.2. See also Figures 4.24 and 4.27.

Another example is the estimation of ore lithological composition by a classifier using genetic algorithms on the results of image processing of the ore carried on a conveyor belt. Seven lithological classes are considered and 130 features are extracted from each rock sample out of a database consisting of 760 digital images [23, 25]. Calibration was done in the laboratory using ore samples obtained at a mine site .

Hyötyniemi *et al.* [32] have used image analysis of flotation froth to extract features such as speed of froth movement, color, area, aspect ratio, perimeter, roundness and transparency of bubbles in the froth. These features are used for categorizing the froth into classes, and for flotation control by means of an expert control system.

4.2.7 Soft Sensors in Control Loops

One of the most important factors determining a suitable control strategy for a plant is the availability of sensors to measure not only controlled and manipulated vari-

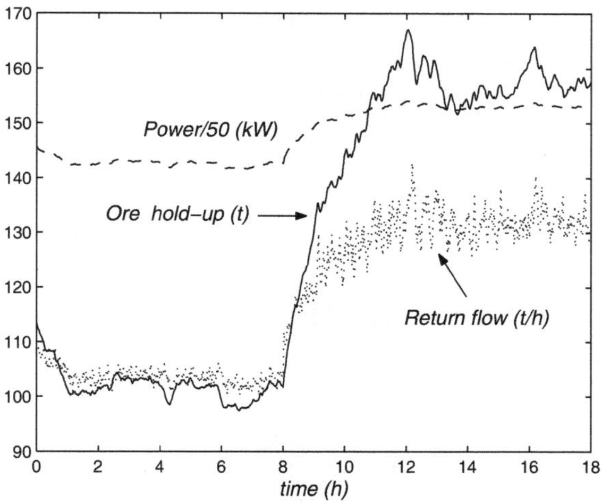

Figure 4.9 Model response of ore hold-up, mill power draw and return ore flow from screen when fresh feed ore flow changes from 1.150 to 1.265 ton/h at $t = 8h$ [22]

ables, but also disturbances. In general, control strategies become simpler as more measurements of plant variables become available. For example, if a fast acting random disturbance is measured, it may be used in a relatively simple feedforward control scheme to compensate its effect upon the controlled variable. But some sensors used in mineral processing plants fail, or have to be removed for maintenance or repairs relatively more frequently than in other industries. The options are:

1. switch to a complex control strategy which does not include the signal of the now unavailable disturbance measurement;
2. try to somehow continue using a manual or automatic control without the important information provided by the failed sensor;
3. use an estimation or prediction of measurements delivered by a soft sensor, based on other correlated measurements, and thus maintain the satisfactory simple control scheme [3].

Therefore soft sensors provide a means for graceful degradation of the control performance when sensors fail. But the partial modeling involved in soft sensor design may cause unexpected control performance when soft sensors are used in automatic control loops because of two circumstances: (a) change of the plant-controller model structure introduced because of the soft sensor model; and (b) parameter coupling between the unmodeled and modeled part of the plant because of (i) inherent coupling of input/output model parameters due to changes in parameters of the state/state-output plant model, and (ii) a disturbance causing changes in parameters of both the modeled and unmodeled part of the plant. Among the components selected by the structure determination method measurements usually appear that are not necessarily inputs to the plant (controls or measured disturbances), but other measurements of plant variables from sensors which are close to the real sensor to be replaced. In this way, only a part of the plant might be modeled, so that a control loop using the soft sensor contains a modeled and an unmodeled part of the plant. Coupling between the modeled and unmodeled part of the plant may lead to off-specification control performance – even instability – unless the fact that a soft sensor is eventually going to replace the actual sensor is considered in the design of the control loop [3, 35, 36].

4.3 Geometrical View of Modeling for Linear in the Parameter and Nonlinear Models

The purpose of this section is to provide geometrical insight for modeling and identification problems. A general approach appears to be possible for linear and nonlinear models which are LIP or that may be transformed into LIP models. In such case, this approach covers a broad class of variables involved in modeling, *e.g.*, random variables, stochastic processes (random functions or random sequences [47]), continuous functions, and integrable functions. In particular, the case of discrete time random sequences is considered since they represent the preferential treatment for soft sensor models adopted in the literature and the corresponding applications [1]. Due to the stochastic characteristics of the variables involved, results turn out to be in terms of expected values. As a consequence, the problem of estimating them by time averages of plant measurements appears, since the probability functions for calculating expected values are usually not available. The modeling problem and the estimation of expected values come together in the problem of finding the optimal parameter vector for a model which is linear in the parameters, although it may be nonlinear with respect to measured plant variables.

4.3.1 Modeling From the Inner Product Spaces Point of View

A number of modeling problems may be approached using the properties of inner product spaces, of which the familiar Euclidean space $\Re^n$ is just a particular case. Inner product spaces are linear spaces endowed with an inner product. Other linear spaces are also special cases of inner product spaces, *e.g.*, spaces of random variables, spaces of random functions (stochastic processes), spaces of random sequences, spaces of continuous functions, spaces of integrable functions [63, 64]. If an inner product space has the property of completeness it is called a Hilbert Space, in which problems concerning limits may be covered.

The fact that inner product spaces share the same axioms with Euclidean spaces allows one to resort to geometrical images in the solution of problems, in particular, in the case of modeling. Indeed, the Projection Theorem provides this geometrical insight for finding a model for a variable y (primary variable in the case of soft sensors) in terms of a finite number of basis functions $\varphi_1, \varphi_2, ..., \varphi_m$, which are generally nonlinear functions of the plant measured variables (secondary variables). The model is simply obtained by projecting y into the space spanned by the linear combination of these bases. This is so whether y is a vector in a Euclidean space, a function in an inner product function space, a random variable in an inner product random variable space, a random sequence in a random sequence inner product space, *etc.*

For the purposes pursued in this chapter an inner product space $\aleph$ is a linear space in which a real valued function called inner product is defined on any two of its elements x, y. The inner product is denoted as $\langle x,y \rangle$, so

$$\langle x,y \rangle : \aleph \rightarrow \Re. \tag{4.76}$$

The axioms that an inner product must satisfy are the same as the ones for an Euclidean space. This is why other inner product spaces "look like" a Euclidean space.

4.3.1.1 Inner Product Axioms

The four axioms an inner product must satisfy are [63, 64]

$$\begin{array}{l} (IP1)\ \langle x+z,y \rangle = \langle x,y \rangle + \langle z,y \rangle, \\ (IP2)\ \langle \alpha x,y \rangle = \alpha \langle x,y \rangle,\ \ \alpha \in \Re, \\ (IP3)\ \langle x,y \rangle = \langle y,x \rangle, \\ (IP4)\ \langle x,\ x \rangle > 0,\ \ x \neq 0, \langle x,\ x \rangle = 0 \Longleftrightarrow x = 0. \end{array} \tag{4.77}$$

It may easily be seen that in the Euclidean space

$$\langle x,y \rangle = x^T y, \tag{4.78}$$

also known as dot product, is an inner product, because it satisfies these four axioms.

Two elements x, y in an inner product space are said to be orthogonal if $\langle x,y \rangle = 0$ and the notation $x \perp y$ is also used in general. Orthogonality defined as $x^T y = 0$ in a Euclidean space is a particular case.

Just as in Euclidean spaces, the "size" of an element x is its norm , and

$$||x||^2 = \langle x,x \rangle. \tag{4.79}$$

This is also a general property of inner products and this norm is called the norm induced by the inner product. The norm assigns to each element a real number according to (4.79). The four axioms for the norm in a normed linear space – and also in an inner product space, where the norm is the induced norm – are [63, 64]

$$\begin{array}{l} (N1)\ ||x|| \geq 0, \\ (N2)\ ||x+y|| \leq ||x|| + ||y||, \\ (N3)\ ||\alpha x|| = \alpha ||x||, \\ (N4)\ ||x|| = 0 \Longleftrightarrow x = 0. \end{array} \tag{4.80}$$

It is also easily and directly seen that in a Euclidean space $\|x\|^2 = \langle x,x\rangle = x^T x$ satisfies the four norm axioms.

Table 4.6 shows real inner products and corresponding induced norms in some inner product spaces. It is relatively easy to prove that these inner products satisfy axioms $IP1$ to $IP3$, but proving $IP4$ may require in some cases – such as random variable inner product spaces – deeper mathematics (*e.g.*, using the Lebesgue integral).

Table 4.6 Inner product spaces

Space		$\langle x,y\rangle$	$\|x\|^2$
Euclidean	$X = \Re^n$	$x \cdot y = x^T y$	$x^T x$
Continuous real functions of real argument	$X = C[a,b]$	$\int_a^b x(t)y(t)dt$	$\int_a^b x(t)^2 dt$
Finite real valued sequences	$X = \{x_1,\ldots,x_n\}$	$\sum_1^n y_k x_k$	$\sum_1^n x_k^2$
Random variables (*) with finite expected values and variance	$X : \Omega \to \Re$	$E\{xy\}$	$E\{x^2\}$
Real stochastic processes (**)	$X : \Omega \to \Im^{(**)}$	$E\{xy\}$	$E\{x^2\}$

(*) Ω is the outcome set in a probability space $\{\Omega, F, P\}$ [47, 64].
(**) $\Im$ is a function space (or sequence space in case of discrete time), *e.g.*, L_2 or l_2 spaces [64].

4.3.1.2 Projection on an Element

In an inner product space, let the projection of y on x be defined as

$$Pr(y|x) = \frac{\langle y,x\rangle}{\|x\|^2} x = \langle y,x_u\rangle x_u, \tag{4.81}$$

where

$$x_u = \frac{x}{\|x\|} \tag{4.82}$$

has a unit norm. Since the norm is scalar, by $IP2$ $1/\|x\|$ may be incorporated into the inner product. This is a generalization of the projection of a vector y onto a vector x in Euclidean space, *i.e.*,

$$\hat{y} = Pr(y|x) = (y^T x_u)x_u. \tag{4.83}$$

It almost immediately follows that in the general case $y - \hat{y} \perp x$, since

$$\langle y - \hat{y}, x\rangle = \langle y,x\rangle - \langle \hat{y},x\rangle \tag{4.84}$$

and

$$\langle \hat{y}, x \rangle = \langle \hat{y}, \|x\| x_u \rangle = \|x\| \langle \hat{y}, x_u \rangle = \|x\| \langle \langle y, x_u \rangle - x_u, x_u \rangle. \tag{4.85}$$

By replacing (4.81) in (4.85), and by (4.82)

$$\|x\| \langle y, x_u \rangle \langle x_u, x_u \rangle = \|x\| \langle y, x_u \rangle = \langle y, \|x\| x_u \rangle = \langle y, x \rangle. \tag{4.86}$$

Hence, by (4.84), (4.85) and (4.86) $\langle y - \hat{y}, x \rangle = 0$.

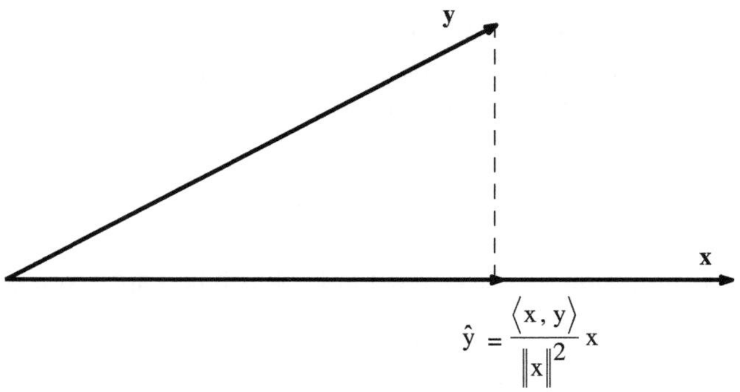

Figure 4.10 Projection $\hat{y}$ of y onto x, which may be vectors in Euclidean space, or elements in other inner product spaces such as those in Table 4.6

A general Pythagoras theorem holds in inner product spaces. Let $x \perp y$, *i.e.*, $\langle y, x \rangle = 0$, and let $z = x + y$. Then

$$\|z\|^2 = \langle z, z \rangle = \langle x+y, x+y \rangle = \langle x, x \rangle + \langle x, y \rangle + \langle y, x \rangle + \langle y, y \rangle = \|x\|^2 + \|y\|^2 \tag{4.87}$$

Then the Pythagoras theorem in Euclidean spaces is just a particular case of the Pythagoras theorem in inner product spaces (and in Hilbert spaces in general).

4.3.1.3 Projection on a Subspace

Let $B_m = \{\varphi_1, \varphi_2, \ldots, \varphi_m\}$ be a base for a subspace $V_m \subset \aleph$ of the inner product space $\aleph$. Therefore, any element $h \in \aleph$ may be written as the linear combination

$$h = \sum_{k=1}^{m} \alpha_k \varphi_k. \tag{4.88}$$

Let

$$B_{mon} = \{\zeta_1, \zeta_2, \ldots, \zeta_m\} \tag{4.89}$$

be an orthonormal basis in V_m , so $\langle \zeta_j, \zeta_k \rangle = 0$ for $k \neq j$ and $\langle \zeta_k, \zeta_k \rangle = \|\zeta_k\|^2 = 1$ for $j = k$. Then the projection of y into the subspace V_m is defined as

$$\hat{y} = Pr(y|V_m) = \sum_{k=1}^{m} \langle y, \zeta_k \rangle \zeta_k. \tag{4.90}$$

It turns out that $\varepsilon = y - \hat{y}$ is orthogonal to subspace V_m, *i.e.*, ε is orthogonal to any element in V_m, $\varepsilon \perp V_m$, in particular to each base ζ_k (Figure 4.11).

Projection Theorem. Let $\aleph$ be a Hilbert space and let B_{mon} be an orthonormal basis for a closed subspace $V_m \subset \aleph$. Then for $y \in \aleph$ the element $\hat{y}$ for which $\|y - \hat{y}\| = \|\varepsilon\|$ is minimum is the projection of y into V_m, *i.e.* [64, 65],

$$\hat{y} = \sum_{k=1}^{m} \langle y, \zeta_k \rangle \zeta_k. \tag{4.91}$$

In other words, projection $\hat{y}$ is the element in V_m which best approximates y.

Moreover,

$$\varepsilon = (\hat{y} - y) \perp V_m. \tag{4.92}$$

If basis B_m is not orthonormal, then by a Gram–Schmitt procedure [64] it may be turned into an orthonormal basis and (4.92) still holds, since the Gram–Schmitt procedure produces just a particular change of coordinates (Figure 4.11). There is a clear similarity with the case of Euclidean spaces, which is to be expected since the axioms for inner products and norms are the same.

Let $y \in \aleph$ and let $B_m = \{\varphi_1, \varphi_2, \ldots, \varphi_m\}$ be a basis – not necessarily orthonormal – for subspace V_m, and let g be any element in V_m so it may be written as a linear combination of bases φ_k (Figure 4.11).

Then, the element in V_m which best approximates y (in the sense that $\|\varepsilon\| = \|y - \hat{y}\|$ is minimum) is orthogonal to V_m:

$$\langle y - \hat{y}, g \rangle = 0 \qquad \forall g \in V_m. \tag{4.93}$$

In particular, in the case of the basis elements φ_k,

$$\langle y - \hat{y}, \varphi_k \rangle = 0 \qquad k = 1, \ldots, m. \tag{4.94}$$

This property is illustrated in Figure 4.12 for $m = 2$. Let $\hat{y}$ be the projection of y into the space V_2 spanned by bases φ_1 and φ_2, and let $\hat{y}_a$ be any other element in V_2. Then, by (4.92) ε is orthogonal to V_2 and hence to every element in V_2, in particular to $\hat{y} - \hat{y}_a$. By the general Pythagoras Theorem

$$\|\varepsilon_a\|^2 = \|\varepsilon\|^2 + \|\hat{y}_a - \hat{y}\|^2, \tag{4.95}$$

so $\|\varepsilon\|^2 \leq \|\varepsilon_a\|^2$, *i.e.*, $\|\varepsilon\| \leq \|\varepsilon_a\|$, and $\|\varepsilon\|$ is minimum with respect to any element in V_2

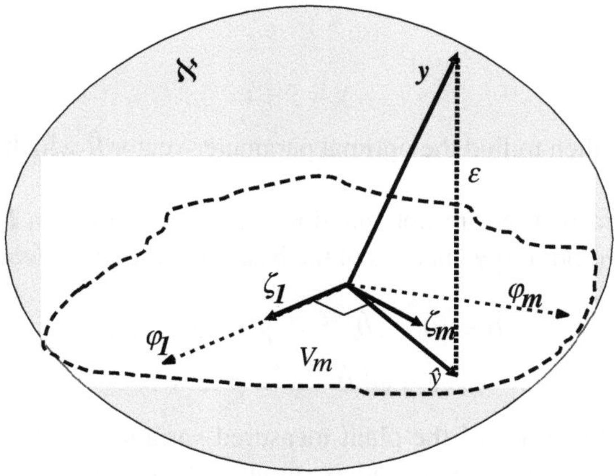

Figure 4.11 Projection of y into space V_m spanned by base B_m or the corresponding orthonormal base B_{mon}

In particular in a random variable space $\|\varepsilon\| = E(\varepsilon^2)$, so minimizing $\|\varepsilon\|^2$ implies minimizing the mean square error between y (*e.g.*, the primary variable to be estimated by a soft sensor) and $\hat{y}$ (*e.g.*, the soft sensor output).

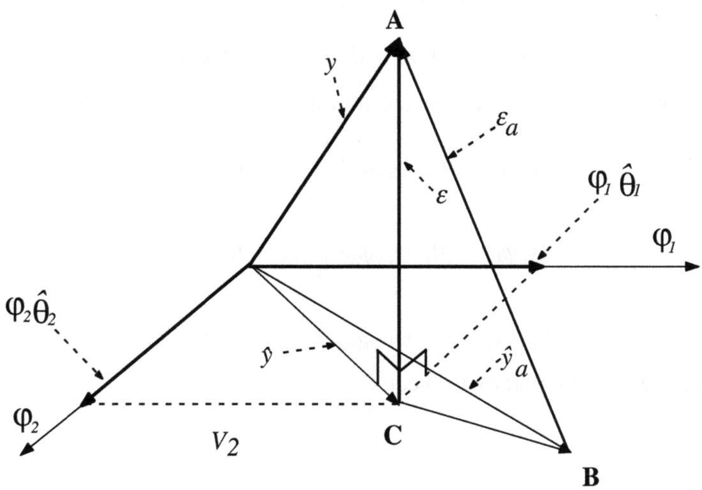

Figure 4.12 Norm $\|\varepsilon\|$ is minimum if ε is orthogonal to the space V_2 spanned by bases φ_1 and φ_2

The problem at hand is to find the best approximation $\hat{y}$ of y. Since $\hat{y}$ belongs to V_m it may be expressed as

$$\hat{y} = \hat{\theta}_1 \varphi_1 + \hat{\theta}_2 \varphi_2 + \ldots \hat{\theta}_m \varphi_m \tag{4.96}$$

and

$$y = \hat{y} + \varepsilon. \tag{4.97}$$

The problem is then to find the optimal parameter vector $\hat{\theta}$ which minimizes $\|\varepsilon\| = E(\varepsilon^2)$.

It is assumed that φ_k are not functions of $\hat{\theta}$ so (4.96) is a LIP model. If not, changes of $\hat{\theta}$ would imply changes in the bases and the space would not be linear.

Letting

$$\hat{\theta} = [\hat{\theta}_1 \ldots \hat{\theta}_m]^T \quad \varphi = [\varphi_1 \ldots \varphi_m]^T, \tag{4.98}$$

$$\hat{y} = \varphi^T \hat{\theta} \quad y = \varphi^T \hat{\theta} + \varepsilon. \tag{4.99}$$

The φ_k are functions of the plant measured variables. For example, in a plant having as inputs $u_1(t)$ and $u_2(t)$ and as output $y(t)$ some of the bases might be

$$\varphi_j(t) = -y(t-3) \;\; \varphi_l(t) = u_1(t-1)u_2(t-1) \;\; \varphi_r(t) = \sqrt{\log(u_2(t-3))}. \tag{4.100}$$

See also (4.45) in Example 2 in Section 4.2.

Clearly this form makes it possible to introduce nonlinear terms representing plant phenomenology in LIP models. This is an important property in the case of soft sensors.

Then, from (4.94) and (4.96):

$$\langle y - \hat{y}, \varphi_k \rangle = \langle y - (\hat{\theta}_1 \varphi_1 + \hat{\theta}_2 \varphi_2 + \ldots + \hat{\theta}_m \varphi_m), \varphi_k \rangle = 0; \tag{4.101}$$

$$\langle y, \varphi_k \rangle = \langle \hat{\theta}_1 \varphi_1 + \hat{\theta}_2 \varphi_2 + \ldots + \hat{\theta}_m \varphi_m, \varphi_k \rangle. \tag{4.102}$$

Therefore, for $k = 1, \ldots, m$, there is a system of m equations with m unknown $\hat{\theta}_k$ given by

$$\begin{aligned}
\langle y, \varphi_1 \rangle &= \hat{\theta}_1 \langle \varphi_1, \varphi_1 \rangle + \ldots + \hat{\theta}_m \langle \varphi_m, \varphi_1 \rangle, \\
&\vdots \\
\langle y, \varphi_k \rangle &= \hat{\theta}_1 \langle \varphi_1, \varphi_k \rangle + \ldots + \hat{\theta}_m \langle \varphi_m, \varphi_k \rangle, \\
&\vdots \\
\langle y, \varphi_m \rangle &= \hat{\theta}_1 \langle \varphi_1, \varphi_m \rangle + \ldots + \hat{\theta}_m \langle \varphi_m, \varphi_m \rangle.
\end{aligned} \tag{4.103}$$

Hence the optimal parameter vector,

$$\hat{\theta} = [\hat{\theta}_1 \ldots \hat{\theta}_k \ldots \hat{\theta}_m]^T, \tag{4.104}$$

which minimizes the squared error norm $\|e\|^2 = \|y - \hat{y}\|^2$ is given by

$$P_{xx} \hat{\theta} = P_{yx}, \tag{4.105}$$

where

$$P_{xx} = \begin{bmatrix} \langle\varphi_1,\varphi_1\rangle & \cdots & \langle\varphi_m,\varphi_1\rangle \\ \vdots & & \vdots \\ \langle\varphi_1,\varphi_k\rangle & \cdots & \langle\varphi_m,\varphi_k\rangle \\ \vdots & & \vdots \\ \langle\varphi_1,\varphi_m\rangle & \cdots & \langle\varphi_m,\varphi_m\rangle \end{bmatrix} \quad P_{yx} = \begin{bmatrix} \langle y,\varphi_1\rangle \\ \vdots \\ \langle y,\varphi_k\rangle \\ \vdots \\ \langle y,\varphi_m\rangle \end{bmatrix}. \tag{4.106}$$

If P_{xx}^{-1} exists, then

$$\hat{\theta} = P_{xx}^{-1} P_{yx}. \tag{4.107}$$

For example, let the φ_k and y be random variables, in a random variable inner product space $\aleph$, where the inner product is $\langle x_j,\varphi_k\rangle = E\{x_j\varphi_k\}$. Let

$$\varphi = [\varphi_1 \cdots \varphi_m]^T \tag{4.108}$$

be a random vector whose elements are the random variables φ_i. Then

$$P_{xx} = \begin{bmatrix} \langle\varphi_1,\varphi_1\rangle & \cdots & \langle\varphi_m,\varphi_1\rangle \\ \vdots & & \vdots \\ \langle\varphi_1,\varphi_k\rangle & \cdots & \langle\varphi_m,\varphi_k\rangle \\ \vdots & & \vdots \\ \langle\varphi_1,\varphi_m\rangle & \cdots & \langle\varphi_m,\varphi_m\rangle \end{bmatrix} = \begin{bmatrix} E\{\varphi_1^2\} & \cdots & E\{\varphi_m\varphi_1\} \\ \vdots & & \vdots \\ E\{\varphi_1\varphi_k\} & \cdots & E\{\varphi_m\varphi_k\} \\ \vdots & & \vdots \\ E\{\varphi_1\varphi_m\} & \cdots & E\{\varphi_m^2\} \end{bmatrix} \tag{4.109}$$

and $P_{xx} = E\{\varphi\varphi^T\} = R_{\varphi\varphi}$ is the correlation matrix of random vector φ, while $P_{xy} = E\{y\varphi\} = R_{\{}y\varphi\}$ is the correlation vector between the random variable y and the random vector φ. The optimal parameter vector then is

$$\hat{\theta} = E\{\varphi\varphi^T\}^{-1}E\{\varphi y\}. \tag{4.110}$$

This is the same result (4.58) obtained in Section 4.2.2 for the optimum parameter vector $\hat{\theta}$ in the case of LIP models when the necessary condition for optimality was used in the case of random variables. However, here the optimal $\hat{\theta}$ has been found by particularizing the general result (4.107) to the case of random variables.

In most practical cases, the probability functions needed to compute expected values are not known, so parameter vector given by (4.110) cannot be found. However, under certain assumptions (*i.e.*, wide sense stationarity, ergodicity [47]) expected values may be estimated using time averages. In such cases the estimation $\hat{\hat{\theta}}$ of $\hat{\theta}$ becomes a random vector having an expected value and a covariance matrix, as shown in Section 4.3.2.

Let it be assumed that the LIP plant model is

$$y = \varphi^T\theta + w, \tag{4.111}$$

where w is orthogonal to all elements of φ . Then, from (4.99),

$$\varepsilon = \varphi^T(\theta - \hat{\theta}) + w. \tag{4.112}$$

But since $\|\varepsilon\|$ is minimum, by the Projection Theorem ε is orthogonal to the space V_m spanned by the elements of φ^T (*e.g.*, see Figures 4.11 and 4.12) so

$$\varepsilon = \varphi_1(\theta_1 - \hat{\theta}_1) + \cdots + \varphi_k(\theta_k - \hat{\theta}_k) + \ldots + \varphi_m(\theta - \hat{\theta}_m), \tag{4.113}$$

$$\langle \varepsilon, \varphi \rangle = ((\theta_1 - \hat{\theta}_1))\langle \varphi_k, \varphi_1 \rangle + \cdots + (\theta_n - \hat{\theta}_) + \langle \varphi_k, \varphi_m \rangle = 0, \tag{4.114}$$

for $k = 1,2,...,m$, since ε is orthogonal to φ_k. A system of m equations is then formed which in matrix form is

$$\underbrace{\begin{bmatrix} \langle \varphi_1, \varphi_1 \rangle & \cdots & \langle \varphi_m, \varphi_1 \rangle \\ \vdots & & \vdots \\ \langle \varphi_1, \varphi_k \rangle & \cdots & \langle \varphi_m, \varphi_k \rangle \\ \vdots & & \vdots \\ \langle \varphi_1, \varphi_m \rangle & \cdots & \langle \varphi_m, \varphi_m \rangle \end{bmatrix}}_{P_{xx}} (\theta - \hat{\theta}) = 0. \tag{4.115}$$

Hence, if P_{xx} is nonsingular, a conditions already required in (4.107) for finding $\hat{\theta}$, then the only solution is $(\theta - \hat{\theta}) = 0$. Thus, $\theta = \hat{\theta}$ and from (4.112) the error $\varepsilon = w$ in this general LIP case. Figure 4.13 is an illustration of this case in an inner product space.

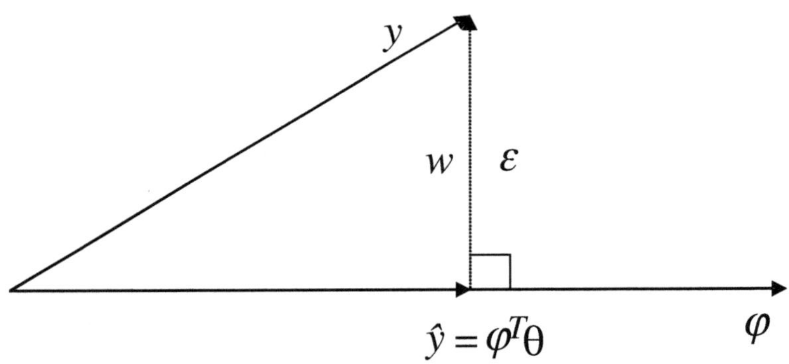

Figure 4.13 When w is orthogonal to φ, parameter estimation using the projection of y on φ gives the true vector θ in the LIP case

4.3.2 Estimation of Expected Values Through Time Averages

The solution to many problems in engineering is given in terms of expected values. The difficulty is that the probability functions needed to determine those expected values are very hard to obtain or they cannot be determined in practice. A possible solution for this problem is obtained if the expected values may be estimated through time averages of realizations (measurements) of random functions or random sequences (which are stochastic processes). These time averages may be quite meaningless unless certain conditions are met.

4.3.2.1 Random Functions or Stochastic Processes

A real random function (stochastic process) is a function $z(t,\xi)$ of two arguments, mapping a set of outcomes Ω of a probability space – the elements of which are denoted here by ξ – to a space H of functions or sequences if t is discrete [47]. Figure 4.14 illustrates this mapping. Then for a given value $\xi = \xi'$, $z(t,\xi')$ is a function of time, *e.g.*, a sequence if t is discrete. This function is called a realization of $z(t,\xi)$. In practice a realization is the resulting measurement made on a variable in a plant, out of all possible measurements that could have occurred for different ξ, which represents the random nature of the plant variable. On the other hand, for a particular $t = t'$, $z(t',\xi)$ is random variable and $z(t',\xi')$ is a real number [47].

In industrial plants usually only one realization $z(t,\xi')$ of a random function $z(t,\xi)$ is available through measurement of a variable in a given time interval. Thus certain statistics used for plant analysis, modeling and control must be obtained from single measurements, in particular, correlations between a set of plant variables, expected values and variances of plant variables, *etc.*

It turns out that in the stochastic process $z(t,\xi)$ argument ξ is almost always omitted in the technical literature, and is only implicitly considered by stating that $z(t)$ is a random sequence or random function (random process).

A real random variable (which might be better denoted as a function rather than a variable) is a function $\eta(\xi)$ which maps a set of outcomes Ω of a probability space to the real line $\Re$ and satisfies certain conditions that insure that the distribution function $F(x)$ may be obtained [47].

A random function (or random sequence) $z(t,\xi)$ is wide sense stationary if [47]:

1. $E\{z(t,\xi)\} = \bar{z}$ is constant, *i.e.*, does not depend on time t;
2. $E\{z(t,\xi)z(t-\tau,\xi)\} = R_{zz}(\tau)$, *i.e.*, its autocorrelation depends only on the displacement τ and not on t.

Similar definitions apply to joint wide sense stationarity [47]. For example, let the discrete time model of a plant be given by

$$y(i,\xi) = \varphi^T(i,\xi)\theta + w(i,\xi), \tag{4.116}$$

where $w(i,\xi)$ is a white noise random sequence, and

$$\varphi(i,\xi) = [\varphi_1(i,\xi)\ldots\varphi_m(i,\xi)]^T. \tag{4.117}$$

According to the above definition, each $\varphi_1(i,\xi)$ is a random sequence, $\varphi_1(i',\xi)$ is a real random variable, $\varphi_1(i,\xi')$ is one of the sequences that results (*e.g.*, through

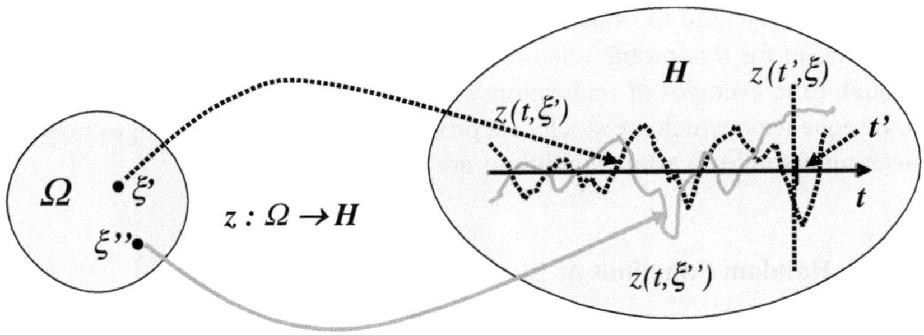

Figure 4.14 A random function (process) is a function $z(t,\xi)$ of two arguments mapping a set of outcomes Ω of a probability space to a space H of functions or sequences. For a particular $t = t'$, $z(t',\xi)$ is random variable and for a particular $\xi = \xi'$, $z(t,\xi')$ is a realization, *i.e.*, a known function. If t is discrete a random function is a random sequence, *i.e.*, a sequence of, in general, correlated random variables

measurement), $\varphi_1(i',\xi')$is a real number, and $\varphi(i',\xi) = [\varphi_1(i',\xi)\ldots\varphi_m(i',\xi)]^T$ is a random vector. Then

$$\varphi(i',\xi') = [\varphi_1(i',\xi')\ldots\varphi_m(i',\xi')]^T \tag{4.118}$$

is a vector of real numbers for a particular realization (measurement) $\xi = \xi'$ at a particular time $i = i'$ of (4.117).

Let

$$\hat{y}(i,\xi) = \varphi^T(i,\xi)\hat{\theta} \tag{4.119}$$

and

$$\varepsilon(i,\xi) = y(i,\xi) - \hat{y}(i,\xi). \tag{4.120}$$

Let $X(\xi)$ be a random matrix (*i.e.*, its elements are random variables) formed by stacking $\varphi^T(i,\xi)$, for given times $1,2,...,k,...,N$. Then

$$X(\xi) = \begin{bmatrix} \varphi_1(1,\xi) & \cdots & \varphi_m(1,\xi) \\ \vdots & & \vdots \\ \varphi_1(k,\xi) & \cdots & \varphi_m(k,\xi) \\ \vdots & & \vdots \\ \varphi_1(N,\xi) & \cdots & \varphi_m(N,\xi) \end{bmatrix} = \begin{bmatrix} \varphi^T(1,\xi) \\ \vdots \\ \varphi^T(1,\xi) \\ \vdots \\ \varphi^T(N,\xi) \end{bmatrix}. \tag{4.121}$$

Similarly, let

$$Y(\xi) = \left[y(1,\xi)\cdots y(k,\xi)\cdots y(N,\xi)\right]^T, \tag{4.122}$$

$$\hat{Y}(\xi) = \left[\hat{y}(1,\xi)\cdots \hat{y}(k,\xi)\cdots \hat{y}(N,\xi)\right]^T, \tag{4.123}$$

$$\varepsilon(\xi) = \left[\varepsilon(1,\xi)\cdots \varepsilon(k,\xi)\cdots \varepsilon(N,\xi)\right]^T, \tag{4.124}$$

$$W(\xi) = \left[w(1,\xi)\cdots w(k,\xi)\cdots w(N,\xi)\right]^T, \tag{4.125}$$

be random vectors also formed by stacking.

Vector $W(\xi)$ represents a window of a stationary white noise random sequence

$$\{\ldots,w(-r,\xi),\ldots,w(0,\xi),w(1,\xi),\ldots,w(N,\xi),w(N+1,\xi),\ldots\} \tag{4.126}$$

where $E\{w(k,\xi)\}=0$, $E\{w(k,\xi)w(j,\xi)\}=0$ for $k\neq j$, and $E\{w(k,\xi)w(k,\xi)\}=\sigma_w^2$.

Then,

$$Y(\xi)=X(\xi)\theta+W(\xi) \tag{4.127}$$

is the assumed model of the plant, and

$$\hat{Y}(\xi)=X(\xi)\hat{\theta} \tag{4.128}$$

is the model output. The error vector is

$$\varepsilon(\xi)=Y(\xi)-\hat{Y}(\xi). \tag{4.129}$$

Then, from (4.128) and (4.129)

$$Y(\xi)=X(\xi)\hat{\theta}+\varepsilon(\xi). \tag{4.130}$$

For a given set of measurements, $X(\xi')$ is a matrix of real numbers and $Y(\xi')$, $\hat{Y}(\xi')$,$\varepsilon(\xi')$ are vectors of real numbers.

4.3.2.2 Time Averages

Let

$$A_T(\xi)=\frac{1}{T}\sum_{i=n_1+1}^{n_1+T} z(i,\xi) \quad A_T(\xi)=\frac{1}{T}\int_t^{t+T} z(t,\xi)dt \tag{4.131}$$

be the time average of a random sequence or of a random function. Since the sum is with respect to i or the integration is with respect to t, $A_T(\xi)$ is only a function of ξ, and therefore has the form of a random variable, which it really happens to be [47].

It may be immediately seen that $A_T(\xi)$ is an unbiased estimator of $E\{z(t,\xi)\}$ if $z(t,\xi)$ is wide sense stationary, because

$$E\{A_T(\xi)\}=\frac{1}{T}\int_t^{t+T} E\{z(t,\xi)\}dt=\frac{1}{T}\int_t^{t+T}\bar{z}dt=\bar{z}, \tag{4.132}$$

where $E\{z(t,\xi)\}=\bar{z}$ and similarly for random sequences.

This fact is illustrated in Figure 4.15, where $p(A_T)$ is the probability density function of A_T.

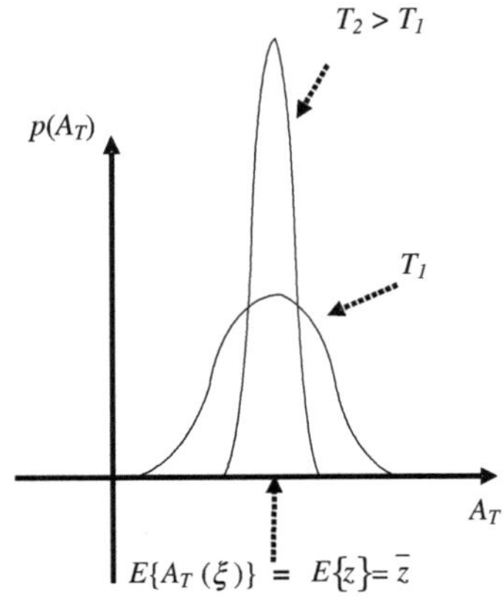

Figure 4.15 Unbiasedness and ergodicity of the time average estimator $A_T(\xi)$

But not only unbiasedness is a desired property for an estimator such as $A_T(\xi)$. It should also have a small variance so that with only one set of measurements in a time window of length T a result close to $\bar{z}$ is obtained. This can be achieved if $z(t,\xi)$, besides being wide sense stationary is also ergodic [47]. This means that as length T goes to infinity the variance $\sigma^2_{A_T}$ of A_T tends to zero. In practice this means that a small variance $\sigma^2_{A_T}$ may be obtained by increasing the length T of the time window of the measurements. Figures 4.15 and 4.16 illustrate the ergodic property.

4.3.2.3 Parameter Estimation

In practice $\hat{\theta}$ in (4.110) is estimated by using the time average estimators

$$E\{\varphi\varphi^T\} \approx \frac{1}{N}\sum_{i=1}^{N} \varphi(i,\xi)\varphi(i,\xi)^T \quad E\{\varphi y\} \approx \frac{1}{N}\sum_{i=1}^{N} \varphi(i,\xi)y, \tag{4.133}$$

then, from (4.110), the estimator using time averages is

$$\hat{\hat{\theta}}(\xi) = [\sum_{i=1}^{N} \varphi(i,\xi)\varphi(i,\xi)^T]^{-1} \sum_{i=1}^{N} \varphi(i,\xi)y. \tag{4.134}$$

It is immediately apparent that $\hat{\hat{\theta}}$ has the character of a random vector. Besides, by (4.121) and (4.122), (4.134) may be written

$$\hat{\hat{\theta}}(\xi) = [X(\xi)^T X(\xi)]^{-1} X(\xi)^T Y(\xi). \tag{4.135}$$

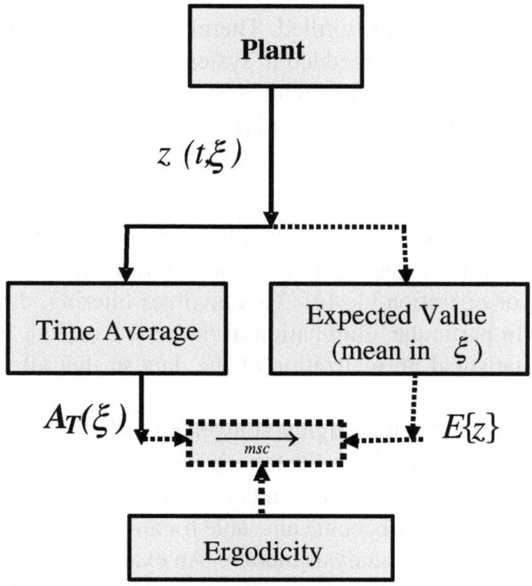

Figure 4.16 The ergodic property causes convergence in the mean square of estimator $A_T(\xi)$ to $E\{z\}$ (*i.e.*, the variance of their difference tends to 0 as T tends to infinity)

For the case treated here it may be shown that the random estimation vector is unbiased and that its variance tends to zero as the length T of the time window tends to infinity [44].

4.4 Soft Sensors in Industrial Concentrators

This first part of this section deals with actions that must be taken in order to ensure the availability of soft sensors. The second part gives some examples of soft sensor design and testing for industrial plants using actual plant data. Two cases are considered in some detail: (i) model based – a soft sensor for particle size and an operational work index soft sensor; and (ii) feature extraction based – a soft sensor for grindability estimation. More cases of soft sensors in mineral processing plants may be found in the references given in Tables 4.1 to 4.4 in the Introduction.

4.4.1 Soft Sensor Management Systems

From an industrial point of view a soft sensor must be reliable and robust, otherwise its main purpose, serving as an alternative for obtaining a measurement when a real sensor is not available, is not fulfilled. Therefore, in an industrial application the soft sensor models must be imbedded in systems that ensure robustness of the soft sensor performance [1]. As an example, a soft sensor management systems called soft sensor shell may be seen in [42] and [66].

In order to perform its task the functions of a soft sensor shell may be grouped in modules.

1. Signal analysis module. Data collected through measurements in the plant must be conditioned prior to being used in the determination of soft sensor models and soft sensor operation [1, 44]. This involves filtering, data reconciliation in general and, in particular, elimination of outliers, *e.g.* [31]. It is also suitable to carry out a statistical normalization of the data so that all variables have zero mean and unit variance.
2. Soft sensor set module. It is highly convenient to have not just one soft sensor model, but rather a set of candidate soft sensors models. For instance, this is necessary to keep the the primary measurement available in case one or more secondary measurements become unusable for any reason. This condition will be determined by the signal analysis module. An example is given below in Section 4.4.2 for the case of a particle size soft sensor. More examples are found in [11, 16, 17].
3. Model selection module. If at time t_F the signal analysis module detects that the actual sensor is unusable, then it must be substituted by one of the candidate soft sensors of the soft sensor set module. A decision must then be made as to which of the soft sensors candidates is best suited to replace the actual sensor. This on-going decision is based on the quality of the primary variable estimation by each soft sensors in the soft sensor set. This implies: (i) checking if the secondary measurements for the candidate soft sensor models of the soft sensor set are available and suitable; (ii) Estimating the mean square error between the outputs of these soft sensors and the primary measurement while the sensor is available before time t_F when the actual sensor becomes unavailable. For example, using a moving average in a time window $[t - T_1, t]$ where t is the present time (alternatively using a forgetting factor [44]); (iii) using other model validation indexes [1] that may be determined on-line in the period $[t - T_1, t]$.
4. Monitoring module. As long as the actual sensor is available the performance of each soft sensor in the soft sensor set must be monitored. If, *e.g.*, for any of these soft sensors the mean square error exceeds a given estimated threshold or fails any other test during a given time window $[t - T_2, t]$, its parameters should be updated through the parameter estimation module until, *e.g.*, the mean square error between the soft sensor output and the sensor measurement falls below the given threshold during a given time window.
5. Parameter estimation module. When it becomes necessary to update the parameters of the soft sensors of the soft sensor set, the secondary measurements should be made changed so that the parameters are properly estimated as a consequence of persistent excitation [44]. Measured disturbances may contain important system information so they should be included in the candidate set of variables used in the soft sensor design. But measured disturbances just happen and cannot be

made to change at will. Therefore, if they have an effect on some of the soft sensor components, and if measured disturbances are not sufficiently variable in a parameter estimation interval, the model may not adequately reflect their influence on dynamic conditions. Parameter updating may be performed using a time window or a recursive parameter estimation method using fixed or variable forgetting factors [44]. In any case it should be considered that the variance of the estimated parameters, which affects the soft sensor output, depends on the window length, or on the value of the forgetting factor in recursive estimation. As in many estimation problems a compromise must be made between having a small estimator variance and speed in achieving the parameter estimation. In general, subjecting the plant to excitation conditions may unduly alter the production objectives. But since proper excitation is needed for an adequate determination of model parameters, a compromise should be reached between length and magnitude of the excitation and the possibility of losing the information provided by the soft sensors when the actual sensors become unavailable.

6. Parameter selection module. This module uses past data to predict the parameter. At the moment of replacement this module decides whether to use frozen parameters values, their mean values, or their prediction according to the quality of the parameter prediction or expected value estimation, verified while the sensor is operating [19]. See also Section 4.2.4.
7. Sensor fault detection module. This function includes the detection of a fault in any sensor providing required secondary measurements for the model components, as well as in the primary measurement sensor. Fault detection and identification methods may be found in [1] and [67].

4.4.2 *Soft Sensors for Grinding Circuits*

4.4.2.1 Particle Size Soft Sensors

A particle size soft sensor has been designed to replace a real sensor which measures the proportion of particles retained by 65# in the hydrocyclone overflow of an industrial grinding circuit of a large copper concentrator. The layout of the grinding circuit is shown in Figure 4.17. A conveyor belt conveys the fresh feed ore to the sump where it is mixed with the sump water addition and the hydrocyclone reject flow. The sump discharge flow is pumped to the hydrocyclone feed. The hydrocyclone overflow is conveyed to a flotation plant. Measured circuit variables are given in Table 4.9.

The approach to robustness used here is through selection of the best available model out of a set of candidate models for the soft sensor.

Collected data.

Three sets of data were used:

Table 4.7 Measured grinding circuit variables

Symbol	Description	Units
G_s	Fresh feed ore flow	t/h
J_{BM}	Ball mill power draw	kW
Q_{1w}	Sump water addition flow	m^3/h
S_p	Solids proportion in hydrocyclone feed pulp S_p Solids proportion in the hydrocyclone feed pulp, computed using the measurement of the pulp density	g/cm^3
f_{65}	Mass proportion of particles retained at 210 μm mesh (+65#) in the hydrocyclone overflow	%65#

- Data Set 1. Data collected during four days of month 1 was used to determine the structure of the models. From this data set the period during which the plant appeared to be sufficiently excited was used.
- Data Set 2. Data collected during four days of month 2.
- Data Set 3. Data collected during days three additional days of month 2. It was used for tests including parameter updating of soft sensor models designed with Data Set 1.

Model determination.

The model structures were determined by means of the stepwise regression method, which sequentially selects model components (regressors, components, bases) from a set of candidate components [11, 48]. See also Section 4.2.4. Following the approach by Gonzalez *et al.* [10], the set of candidate components not only includes plant measurements (ore and water flows, densities, particle size, power) but also functions of these measurements determined using a phenomenological model of the grinding circuit according to the approach given by Casali *et al.* [11] and Gonzalez *et al.* [10].

From a phenomenological model more than 100 possible composite components (bases) were found for building LIP models. These composite regressors are nonlinear functions of the measured variables and, as pointed out in Section 4.2, make it possible to extend the validity of the models, for a given parameter set, to an operating region that may be much broader than if only direct plant measurements are used. In this way, a nonlinear model but which is linear in the parameters is used, *i.e.*, a NARX model, and linear regression may be employed.

In order to ensure the availability of the estimated +65#, soft sensor models were designed for the cases in which some measurements affecting the soft sensor components became unavailable. Therefore, when determining models for which a measurement was purposely omitted, components which are function of these omitted measurements are also omitted.

The composite regressors chosen from the candidate set by stepwise regression are

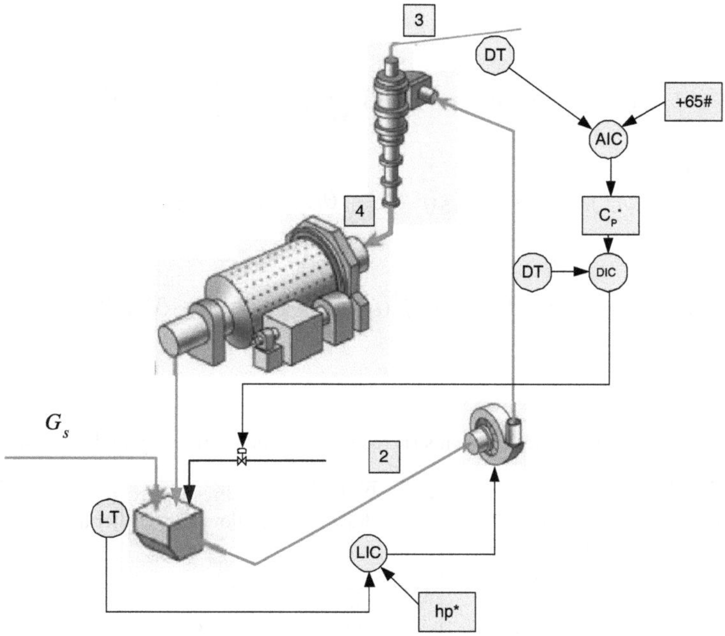

Figure 4.17 Grinding circuit for which the particle size soft sensors are designed. Fresh feed ore flow G_s is conveyed to the sump where it joins the sump water flow Q_{1w} and the hydrocyclones underflow [5]. The sump discharge flow [2] is pumped to the hydrocyclones feed. At the hydrocyclones overflow [3] a particle size sensor (AT) measures the proportion of particles retained by 65 mesh. Hydrocylones feed density is controlled to set-point $Cp*$ by a PID controller manipulating the sump water feed through a local flow control. The sump level is similarly controlled using the pump speed as manipulated variable

$$\begin{aligned} S_p(t-2),\ S_p^{-3}(t-1),\ J_{BM}(t-1)S_p^3(t), \\ J_{BM}(t-3)S_p(t-3),\ \frac{G_s(t-1)}{G_s(t-1)+Q_{1w}(t)}, \end{aligned} \tag{4.136}$$

(see Table 4.7).

Soft sensor set.

The best determined models are listed below. Table 4.8 shows the measurements used and omitted in the design of each soft sensor model. It can be seen that the solids proportion S_p is included in four out of the five models. Therefore the hydrocyclone feed density measurement is very important, since it determines S_p.

Model 1

$$\begin{aligned} f_{65}(t) = {} & \theta_{10} f_{65}(t-1) + \theta_{11} J_{BM}(t-1) S_p^3(t) \\ & + \theta_{12} J_{BM}(t-3) S_p(t-3) + \theta_{13}. \end{aligned} \tag{4.137}$$

Model 2

$$f_{65}(t) = \theta_{20} f_{65}(t-1) + \theta_{21} S_p(t-2) + \theta_{22}. \tag{4.138}$$

Model 3

$$f_{65}(t) = \theta_{30} f_{65}(t-1) + \theta_{31} \frac{G_s(t-1)}{G_s(t-1) + Q_{1w}(t)} + \theta_{32}. \tag{4.139}$$

Model 4

$$f_{65}(t) = \theta_{40} f_{65}(t-1) + \theta_{41} S_p^{-3}(t-1) + \theta_{42}. \tag{4.140}$$

Model 5

$$f_{65}(t) = \theta_{50} f_{65}(t-1) + \theta_{51} S_p^{-3}(t-1) + \theta_{52} Q_{1w}(t-3) + \theta_{53}. \tag{4.141}$$

Table 4.8 Availability of measurements (I) required for each soft sensor and not required (O)

Soft sensor model	S_p: % solids in HC feed from density measurement	J_{BM}: Ball mill power draw	G_s: Fresh feed ore flow	Q_{1w}: Sump water addition flow
1	I	I	O	O
2	I	O	O	O
3	O	O	I	I
4	I	O	O	O
5	I	O	O	I

Upon detecting a failure at time t_F of a measurement affecting the input to the current soft sensor, the soft sensor shell should give an alarm and possibly replace the current soft sensor by a soft sensor which does not use the failed measurement. For example, if Q_{1w} becomes unavailable, models 1, 2 and 4 may be used, since they do not depend on this measurement.

As examples, Figures 4.18, 4.19 and 4.20 show the performances of soft sensors with models 2 and 5 for two periods, with their parameters determined on different periods. In particular, in the case of Figure 4.18 and 4.20 there are many long periods when the actual sensor is down. Nevertheless, an indication that the soft sensor is performing well is that when the real sensor comes back into operation its measurement tends to be close to the soft sensor output. The other models also have satisfactory performances.

4.4.2.2 Global Operational Work Index Soft Sensor

A global operational work index was designed for a grinding plant whose layout is shown in Figure 4.21 [24]. It consists of a rod mill followed by three sections in parallel, each one formed by a ball mill in inverse circuit with a hydrocyclone battery. In this case an empirical relation is combined with an ARX model. The involved variables are:

- G_f is the feed ore flow to the rod mill, with size f_{80};
- G_k is the product flow from the rod mill, with size k_{80};
- G_p is the plant product ore flow, with size p_{80};
- P_{eR} is the power draw of the rod mill;
- P_{eBj} is the power draw of the ball mill in section j $(j = 1,2,3)$;
- $P_T = P_{eR} + \sum_{j=1}^{3} P_{eBj}$ is the total power draw of the mills.

The variables f_{80}, k_{80} and p_{80} are the sizes for which 80% of the ore has a smaller size, at the points shown in Figure 4.21.

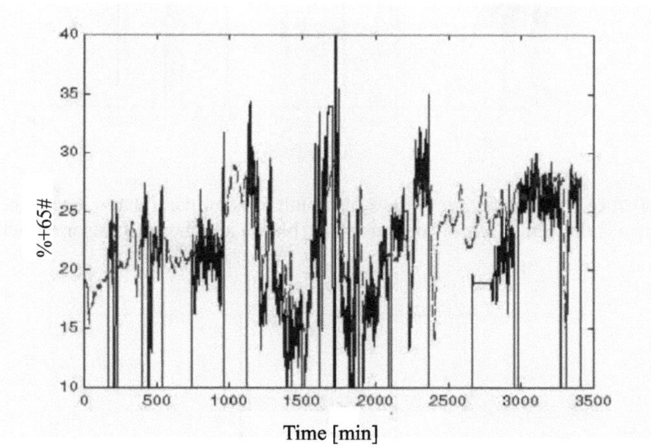

Figure 4.18 Test of soft sensor 2 in 2.4 days of month 1 using parameters determined on a different day of month 1. Shown is the actual sensor measurement (black) and the soft sensor output (segmented)

A global operational work index W_{iog} was defined for the grinding plant of Figure 4.21 – following the definitions of operational work indexes for rod mills and ball mills – as

$$W_{iog} = \frac{P_t/G_f}{11.02F_{TG}({p_{80}}^{-1/2} - {f_{80}}^{-1/2})}, \tag{4.142}$$

where $F_{TG} = 0.857$ [24].

From (4.142), there are four variables determining W_{iog}. On the one hand, the on-line measurements of two of them are almost always available; power P_T is measured by highly reliable wattmeters, and ore flow G_f measures the throughput of the plant so it is maintained to provide a highly reliable measurement. On the other hand, neither f_{80} nor p_{80} are measured. However, for this plant a relation has been found linking p_{80} with measured g_{65} giving +65# particle size percentage (percentage weight of particles larger than 210 μm). This relation is

$$p_{80} = 1.78 + 10.07g_{65}. \tag{4.143}$$

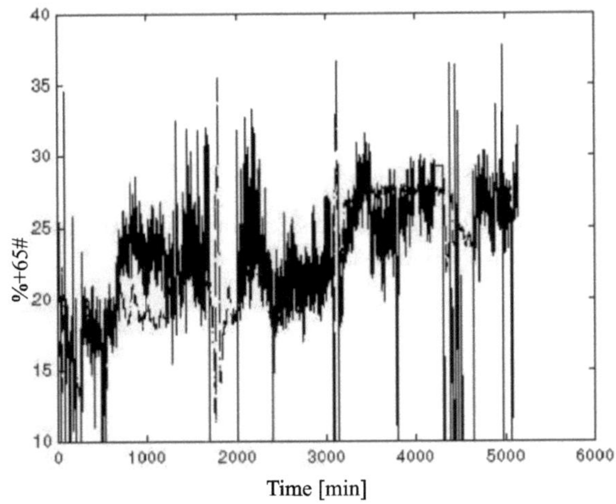

Figure 4.19 Test of soft sensor 2 in 3 days of month 2 using parameters determined on a day of month 1. Shown is the actual sensor measurement (black) and the soft sensor output (segmented)

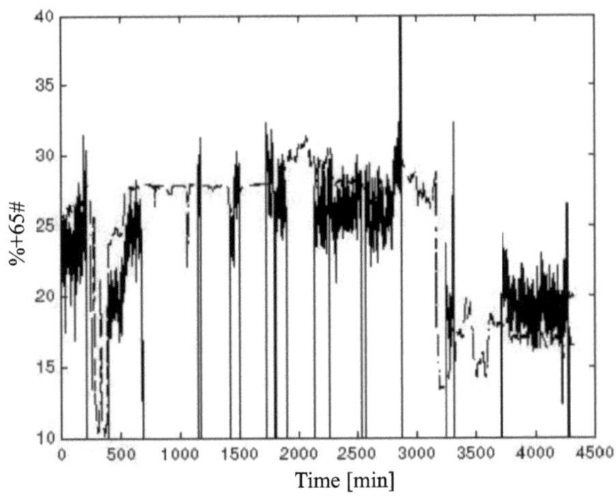

Figure 4.20 Test of soft sensor 5 in 3 days of month 2 using parameters determined on a day of month 2. Shown is the actual sensor measurement (black) and the soft sensor output (segmented)

But particle size sensors tend not to be highly available, so a back-up soft sensor system for g_{65} should be installed. Alternatively, a soft sensor for p_{80} may be designed that does not use g_{65} as a secondary measurement (input). But no measurements are available that may be used to estimate the feed f_{80}.

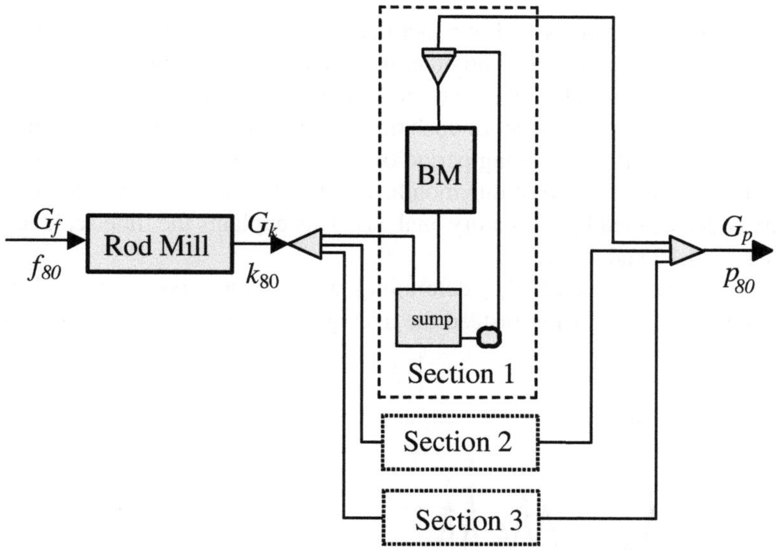

Figure 4.21 Grinding plant for which a work index soft sensor was developed

4.4.2.3 Development of Soft Sensors for f_{80} and p_{80}

The requirements of these two soft sensors were determined by finding the sensitivities S_{Wf} and S_{Wp} of W_{iog} with respect to f_{80} and p_{80}. These sensitivities, defined by the relative change of W_{iog} with respect to the relative changes of f_{80} and p_{80}, using (4.142) are found to be

$$S_{Wf} = \frac{\frac{dW_{iog}}{W_{iog}}}{\frac{df_{80}}{f_{80}}} = -0.5 \frac{1}{\sqrt{r_{80}} - 1} \tag{4.144}$$

and

$$S_{Wp} = \frac{\frac{dW_{iog}}{W_{iog}}}{\frac{dp_{80}}{p_{80}}} = 0.5 \frac{\sqrt{r_{80}}}{\sqrt{r_{80}} - 1}, \tag{4.145}$$

where $r_{80} = f_{80}/p_{80}$ gives the size reduction performed by the grinding plant [24].

Using measurements collected from the plant these sensitivities were found to be $S_{Wf} = 0.1084$, $S_{Wp} = 0.608$, $S_{Wp}/S_{Wf} = 5.61$.

Over a period of 46 months, the average value of f_{80} was 9774 μm with a standard deviation of 379 μm, while the average of p_{80} was 310 μm with a standard deviation of 45 μm.

If f_{80} changes by three times its standard deviation around its average, the variations of W_{iog} was found to be quite small; 1.26 %. Therefore the average value of f_{80} may be used. From a soft sensor point of view the soft sensor for f_{80} is simply its estimated average. This agrees with the fact that the optimal estimate of a random variable is its expected value, when no other information is available.

The operational work index W_{iog} may be calculated with a fast sampling rate, since it depends on power P_T (a continuous time measurement) and on p_{80} through g_{65} measuring +65# with a sampling period of 1 min. However, the plant data was available with a sampling time of one month, which is the period in which the ore work index is determined by laboratory analysis. This explains the time scale used in Figures 4.22, 4.23 and 4.24.

Figure 4.22 shows the actual W_{iog} determined using f_{80} and p_{80} found by laboratory measurements, and using instead the average value of f_{80} $\pm$ three standard deviations.

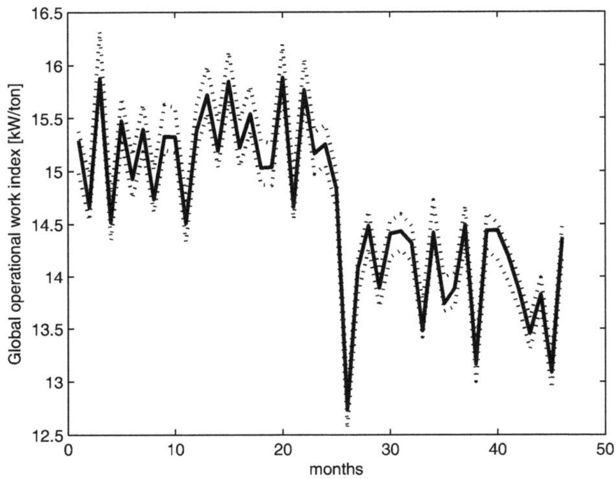

Figure 4.22 Actual W_{iog} determined using f_{80} and p_{80} found by laboratory measurements is given by the solid line. The W_{iog} determined using the average value of f_{80} $\pm$ three standard deviations instead is given by the two dotted lines

The situation is different in the case of p_{80}, since if p_{80} changes by three standard deviations around its average, the change in W_{iog} is 29.4%. This means that a soft sensor is required for p_{80}. Only the two highly reliable measurements have been used as secondary measurements: the feed ore flow G_s to the plant and the total power P_T drawn by the mills. Then the following soft sensor was determined:

$$\hat{p}_{80} = \theta_1 P_T(t) + \theta_2 P_T(t-1) + \theta_3 G_s(t) + \theta_4 l(t), \tag{4.146}$$

using both stepwise regression and all possible subsets from a set of candidate bases (components), which included $P_T(t)$, $G_s(t)$, their delays up to $t-1$, and $l(t)$ to accommodate a constant term.

Figure 4.23 shows a test of the prediction capabilities of soft sensor (4.146). The parameters were updated from month 1 to month 30 and the prediction was made from month 31 to 46 .

The prediction of W_{iog} using average f_{80} and p_{80} soft sensor output of Figure 4.23 is given in Figure 4.24.

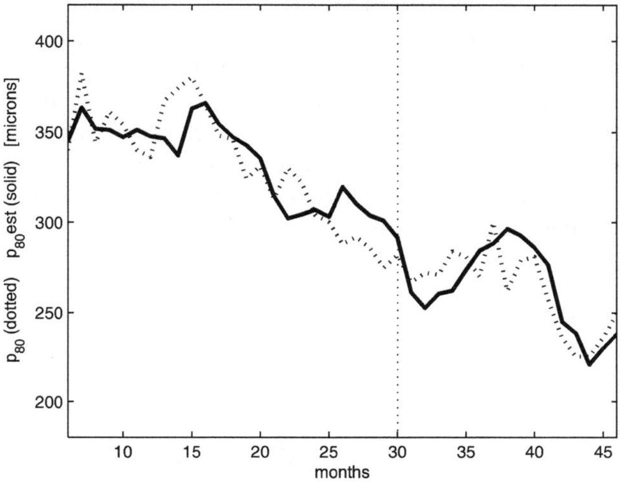

Figure 4.23 Soft sensor for p_{80} (dotted) and p_{80} predicted (black). The soft sensor model is updated from months 1 to 30 and the prediction is from months 31 to 46

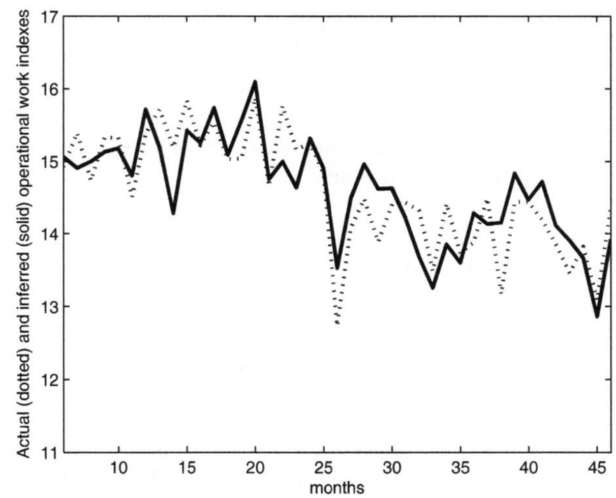

Figure 4.24 Global operational work index W_{iog} computed using mean f_{80} and p_{80} inferred by a soft sensor (solid) and Work Index found by actual f_{80} and p_{80} though laboratory analysis (dotted). The soft sensor model is updated from months $t-6$ to t and the prediction is from $t+1$ to $t+6$

4.4.2.4 Grindability Index Soft Sensor Using Feature Extraction

A grindability index soft sensor has been developed for ore contents in the mill of the SAG mill circuit is shown in Figure 4.25 [22]. No model was developed to estimate this grindability index. Instead this index was estimated using features extracted from three of the measured variables in the circuit: (i) mill power draw P_d; (ii) return ore flow F_r; and (iii) fresh feed ore flow F_f (Figure 4.25). The water feed flow F_w is controlled for a fixed proportion of the fresh feed flow using a ratio controller. The fresh feed ore flow is the manipulated variable for a PID control of the total hold-up (ore plus water hold-up) in the mill. Therefore, the feed ore flow F_f depends on the type of ore. For example, if the ore becomes harder the control reduces the feed ore flow – and conversely if the ore becomes softer – in order to attain and keep the total hold-up close to its set-point. A similar situation happens if the feed size distribution changes from finer to coarser.

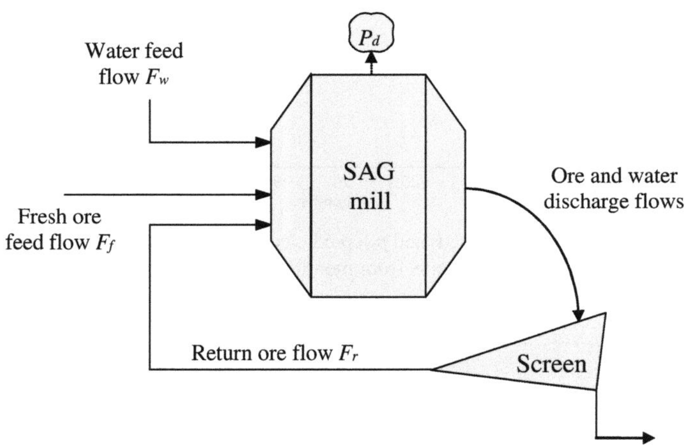

Figure 4.25 SAG circuit for which a grindability index soft sensor was developed based on the variance of wavelet transforms of measured variables

The feature used in this case is the variance of the continuous wavelet transform (CWT). The CWT is a transformation

$$W_y(a,b) = \frac{1}{\sqrt{a}} \int_{-\infty}^{\infty} y(t)\Psi(\frac{t-b}{a})dt \tag{4.147}$$

of a function $y(t)$ that produces a function $W_y(a,b)$. In (4.147) $\Psi(t)$ is the wavelet, a is the wavelet scale and b is the wavelet displacement. The Haar wavelet shown in Figure (4.26) has been used here [22].

As seen in (4.147) for a any given scale a, function $y(t)$ is multiplied by all possible shifts of the wavelet, and the product is integrated.

The CWT of function $y(t)$ may be compared with its Fourier transform, and particularly the Fourier sine or cosine transforms. One difference between these trans-

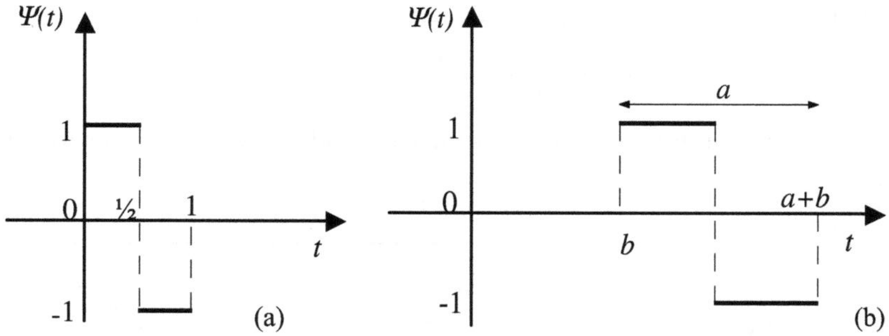

Figure 4.26 Basic Haar wavelet $\Psi(t)$ with unit scale and zero displacement (a) and with scale a and displacement b (b)

forms is that while in (4.147) the function $y(t)$ is multiplied by a function having finite support – *i.e.*, it vanishes outside a finite interval – in the Fourier sine transform of $y(t)$ is multiplied by a sine wave which is different from zero in intervals $[t,t+T](T>0)$ for all t. In addition, while the Fourier transform is a function of only one variable – frequency – the CWT is a function of two variables: scale and displacement.

If $y(t)$ is assumed to be a wide sense stationary random process (at least within a given time window), the expected value of $W_y(a,b)$ is zero, and its variance $Var\{W_y(a,b)\}$ does not depend on the displacement b [22], *i.e.*,

$$E\{W_y(a,b)\}=0,\quad Var\{W_y(a,b)\}=E\{W_y^2(a,b)\}=V_y(a). \tag{4.148}$$

4.4.2.5 Templates

When the grindability index $\Gamma=\Gamma_i$ is known, let $V_1(a,\Gamma_i)$ be the variance of the CWT for the feed ore flow $F_f(t,\Gamma_i)$, $V_2(a,\Gamma_i)$ for the power draw $P_d(t,\Gamma_i)$, and $V_3(a,\Gamma_i)$ for the return ore flow $F_r(t,\Gamma_i)$.

Figure 4.27 shows $V_1(a,\Gamma_i)$ for hard $(i=1)$, medium $i=2$ and soft ores $i=3$. These CWT variances for known grindability indexes serve as templates for estimating the current grindability of the ore in the SAG mill. They may be determined off-line using historical plant data [22]. Since the expected value of the measured variable $y(t)$ is not taken into account by (4.147) because of (4.148), the expected value has been added to the CWT variance, so $V_j(a,\Gamma_i)$ shall be considered to include this addition [22].

Also shown in Figure 4.27 is the CWT of the sample variance template $S_1(a,\Gamma)$ determined on-line using the current measurement of feed ore flow $F_f(t,\Gamma_i)$ – including its mean – in a given time window when the current grindability index is unknown and must be estimated. Then the current grindability index is the one for which the sample template $S_1(a,\Gamma)$ is closest to, *e.g.*, as determined by the minimum of

$$\min_{\Gamma_i} \sqrt{\int_0^T [S_1(a,\Gamma) - V_1(a,\Gamma_i)]^2 da}. \tag{4.149}$$

For example, in Figure 4.27 the estimated grindability corresponds to medium ore.

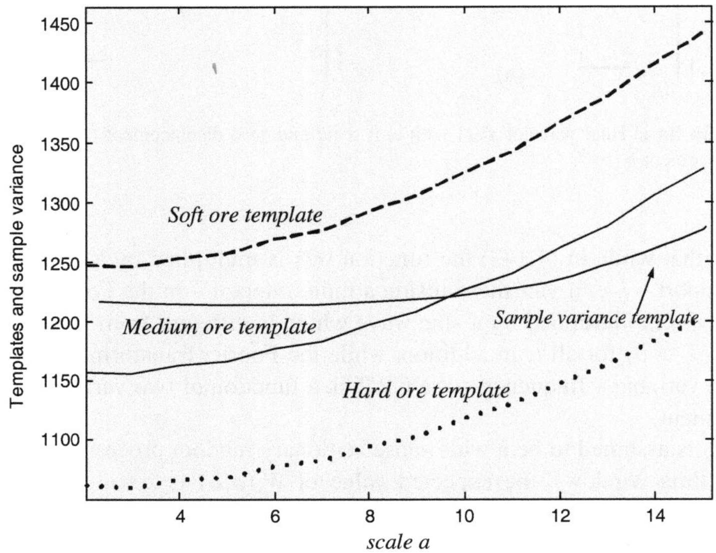

Figure 4.27 Templates of CWT variance corresponding to feed ore $F_f(t,\Gamma_i)$ for hard ore and coarse particle size, for medium ore and normal particle size, and for soft ore and fine particle size, under normal operating conditions. Also shown is sample variance template $S_1(a,\Gamma)$ for $F_f(t,\Gamma)$ corresponding to a given measurement when the ore grindability in the mill is medium, but unknown

Again, since probability functions are not known, expected values and variances are computed using time averaged estimators (see Sections 4.2.2 and 4.3.2).

In order to proceed with the process of identifying Γ all templates are converted into points in a Euclidean space of dimension M through discretizing the range of scale a into M discrete values $\{a_1, a_2, a_3, , a_M\}$ for each template; Figure 4.28 gives a pictorial view of this conversion.

Further processing includes (i) Concatenation [22] of the three templates corresponding to feed ore flow $F_f(t,\Gamma_i)$, to power draw $P_d(t,\Gamma_i)$, and to return ore flow $F_r(t,\Gamma_i)$ and of the sample templates; (ii) PCA to reduce dimensions while retaining relevant information; and (iii) projection to Fisher space for further dimension reduction and facilitating discrimination between the three possible grindability indexes [22]. Concatenation is a juxtaposition of the templates, so it increases the Euclidean space dimension from $M = 14$ to $3M = 42$, but this dimension is brought down to 8 using PCA, and may be further reduced to 2 by projection into the Fisher space [57]. Figure 4.28 has an equivalence in the PCA space and in the Fisher space.

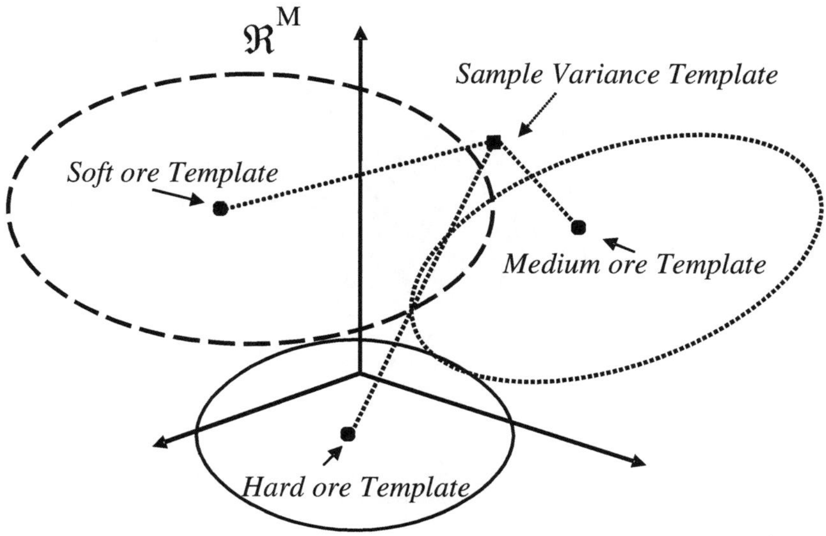

Figure 4.28 Pictorial representation of templates $V(\Gamma_i)$ in Euclidean space R^M for operating conditions corresponding to grindability indexes for hard, medium and soft ores. Also shown is a sample variance template $S(\Gamma)$ for an unknown operating condition of the plant. Ellipses represent sets of sample templates obtained from repeated on-line measurements for the three different kinds of ores. The points representing the soft, medium and hard ore templates are calculated by averaging similar sample templates for each case when the type of ore is known. The estimated grindability index would be medium in this case

Results were obtained through simulation using a dynamic SAG mill circuit model (able to handle two different kinds of ore) adjusted to data from a large grinding plant with a layout shown in Figure 4.25 [22].

Comparisons of Figures 4.29 and 4.30 show the advantages obtained through concatenation of the three measured variables. This advantage is reflected in Tables 4.9 and Table 4.10, which show that the improvement in the correct estimation of the grindability index brought about by concatenation is quite remarkable.

Table 4.9 Percentages of hits (diagonal) and misses (off-diagonal) of actual grindability index when only the template for return ore flow F_r is used

Actual index	Identified index		
	0.94	1.00	1.06
0.94	64.2	34.2	1.6
1.00	31.4	34.0	34.6
1.06	3.4	29.4	67.2

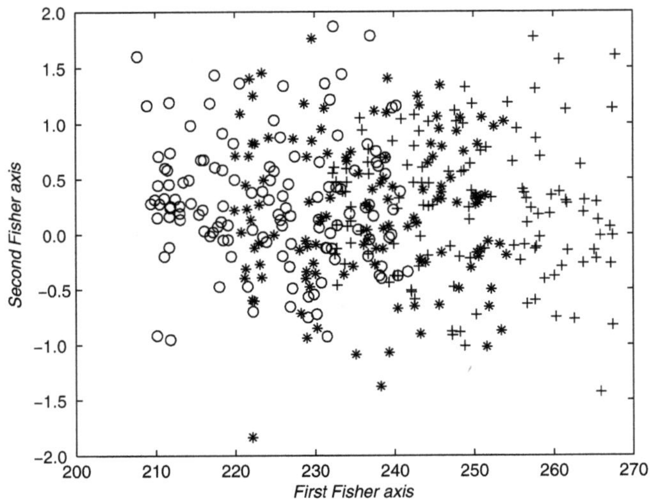

Figure 4.29 Sample variances in the Fisher space obtained with several measurements of F_f for the three grindabilities corresponding to hard ore (o), medium ore (*) and soft ore (+)

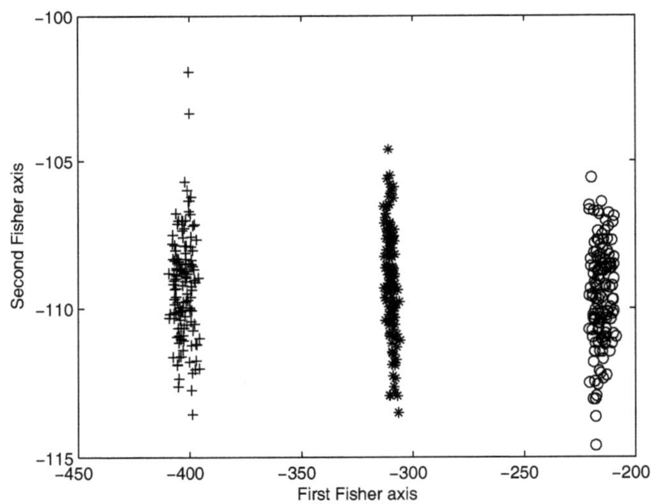

Figure 4.30 Sample variances in Fisher space (obtained for several measurements) for the three grindabilities corresponding to hard ore (o), medium ore (*) and soft ore (+) using concatenation of the CWT variances of F_f, P_d and F_r

Figure 4.31 shows the estimation of the grindability index when the ore changes from medium to hard at $t = 10(\mathrm{h})$. This is done by determining to which ore template the distance between the sample template and the three ore templates is minimum (see Figure 4.28). For $t < 10$ the smallest distance of the sample template is clearly to the medium ore template (solid line). Hence the grindability is correctly estimated

Table 4.10 Percentages of hits (diagonal) and misses (off-diagonal) of actual grindability index using concatenated templates combining F_s, P_d, and F_r templates

Actual index	Identified index		
	0.94	1.00	1.06
0.94	100	0	0
1.00	0	100	100
1.06	0	0	0

as that of medium ore. As t increases beyond 10h, this distance increases, while the distance of the sample template to hard ore template becomes minimum, correctly implying that the grindability is that of hard ore. Further details and other tests may be found in [22].

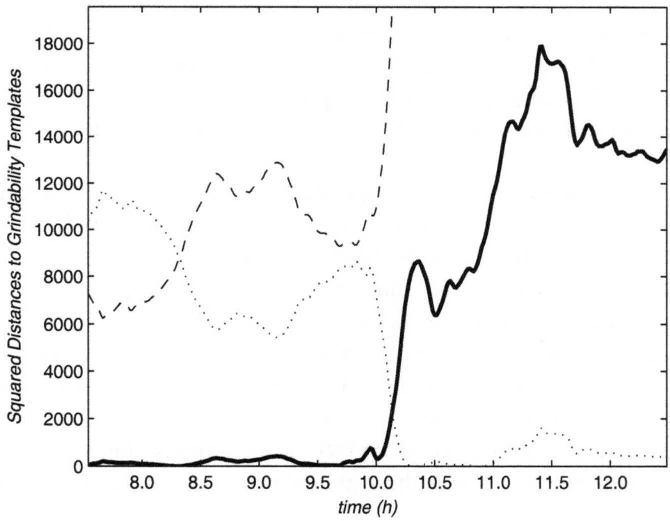

Figure 4.31 Squared distances in the Fisher space between a sample template and the templates for hard, medium and soft ores, for a step change of grindability index from medium to hard at $t = 10$(h). The dotted line is the squared distance from the sample template to the hard ore template, the solid line and the segmented line are the squared distances from the sample template to templates for medium and soft ores, respectively. (From [22])

4.5 Final Remarks

Soft sensors have become widely used in many industrial plants to replace sensors either as a back up for installed sensors that cease to be available, or as alternative sensors when no real sensors are installed, if appropriate conditions are met.

The growing area of soft sensor R&D and applications has also encompassed the mineral processing industry, where the relatively difficult modeling problems are inherited by the problem of finding suitable soft sensor models. Soft sensors have been developed for ore particle size, slurry density, ore grade, ore grindability, mill contents, operational work index and weight in mineral processing plants. But sensors or soft sensors are still missing for on-line measurement or estimation of important variables in mineral processes.

Different classes of models that have been used include LIP models, which may be linear or nonlinear in plant measurements. Among these models, nonlinear models of the NARX class have the advantage of allowing the introduction of process phenomenology, while retaining the relatively simple and fast updating properties of the linear models. An added advantage of LIP models is the geometrical insight involved. Also employed are neural network models, fuzzy models, support vector machine based models, and soft sensors designed using characteristics of measures variables such as mean, variances of the variables themselves or of transformations of them.

The issue of soft sensor robustness, specifically low sensitivity to disturbances caused by failing secondary measurements and to changes in operating conditions and plant characteristics, has been approached. On the one hand, by using models that cover relatively large regions of plant characteristics and operating conditions, such as models based on neural networks, fuzzy sets, clustering, and models of the NARX class. On the other hand, robustness to secondary measurement unavailability has been obtained by introducing a set of candidate soft sensors – instead of having only one soft sensor – based on models whose components have forcibly been made independent of different secondary measurements. Soft sensor shell systems have been designed to manage these and other aspects of soft sensor operation in industrial plants.

New data processing and sensor technologies in general, and new equipment in the mineral processing industry may be expected to promote the development of new sensors and soft sensors. On the one hand, nanotechnologies, mechatronics and developing wireless technologies will allow networks of very high speed miniature processors, allowing in turn networks of miniature sensors and soft sensors profusely located throughout the plant equipment. New sensors for as yet unmeasured variables, developed using these new technologies, will surely appear. On the other hand, it is also reasonable to expect that the development of new mineral processing equipment based on new technologies may require new sensors and soft sensors. All these developments should have an important impact on improving plant control and maintenance and, as a result, on plant technical and economic performance.

Acknowledgements The author of this chapter is grateful for the contributions of various colleagues and students during his research and development work on soft sensors. Among them, Professors Aldo Casali and Gianna Vallebuona of the Mining Engineering Department of the University of Chile have provided their knowledge and experience making possible a symbiosis between their field of mineral processing and the author's field of systems and control engineering. Moreover, through the work on their engineering theses chaired by this author, a number of students have also contributed to the R&D on soft sensors, most recently Roberto Paut and David Miranda. Also to be thanked in this aspect are Chilean funding agencies, among them FONDEF, FONDECYT, CORFO, and CODELCO special fund for R&D, in addition to the resources provided by the Departments of Electrical Engineering and of Mining Engineering of the University of Chile.

References

[1] Fortuna F, Graziani S, Rizzo A, Xibilia MG (2007) Soft sensors for monitoring and control of industrial processes. Springer

[2] Control Eng. Eur. (2001) Process Control's latest tool: Soft Sensors. Control Engineering Europe, 42–45, June, 2001

[3] González GD (1999) Soft-sensors for processing plants. Proceedings Second Intelligent Processing and Manufacturing of Materials International Conference, Honolulu, Hawaii, 10–15 July, 1999. Vol. 1, 59–70

[4] Basso M, Bencivenni F, Giarre L, Groppi S,Zappa G (2002) Experience with NARX model identification of an industrial power plant gas turbine. Proceedings 41st IEEE Conference on Decision and Control, Las Vegas, Nevada, USA, Dec. 2002

[5] Romero F (2007) Automation, Information & Communication Technologies: Challenges and Opportunities for the Mineral Industry. Collahuasi's Case. Plenary session paper in Application of Computers and Operations Research in the Mineral Industry –APCOM 2007, Santiago, Chile. fromero@anglochile.cl

[6] Valery W, Jankovic A (2004) Multi-stage crushing versus AG/SAG mill and HPGR, Metso Minerals, Crushing and Grinding 2004

[7] Xstrata Plc (2006) Welcome to Isa mill. http://www.isamill.com

[8] Xstrata Plc (2006a) Welcome to Jameson cell. http://www.jamesoncell.com

[9] González G, Mendez H, de Mayo F (1985) A dynamic compensation for particle size distribution estimators. ISA Transactions 25(1): 47–51

[10] González GD, Odgers R, Barrera R, Casali A, Torres F, Castelli L, Gimnez P (1995) Soft-sensor design considering composite measurements and the effect of sampling periods. Proceedings Copper 95, International Conference, Santiago, Chile, II, 213–224

[11] Casali A, González GD, Torres F, Vallebuona G, Castelli L, Giménez P (1998) Particle size distribution soft-sensor for a grinding circuit. Powder Technology 99:15–20

[12] Du Y-G, del Villar R, Thibault J (1997) Neural net-based softsensor for dynamic particle size estimation in grinding circuits. International Journal of Mineral Processing 52(2):121–135

[13] Sbarbaro D, Ascencio P, Espinoza P, Mujica F, Cortes G (2008) Adaptive soft-sensors for on-line particle size estimation in wet grinding circuits. Control Engineering Practice 16(2):171–178

[14] Sbarbaro D, Barriga J, Valenzuela H, Cortes G, Mujica L, Perez N (2001) A comparison of neural networks architectures for particle size distribution estimation in wet grinding circuits. ISA 2001 Conference and Exhibition, Houston TX, USA

[15] Sun Z, Wang H, Zhang Z (2008) Soft sensing of overflow particle size distributions in hydrocyclones using a combined method. Tsinghua Science And Technology, ISSN 1007-0214 08/19 Volume 13, Number 1, pp 47–53

[16] Casali A, González G, Torres F, Cerda I, Castelli L, Gimnez P (1995) Pulp density soft-sensor for a grinding circuit, Proceedings XXV APCOM, Brisbane, Australia, 371–376
[17] Casali A, Vallebuona G, Bustos M, González G, Giménez P (1998a) A soft-sensor for solids concentration in hydrocyclone overflow. Minerals Engineering 11(4):375–383
[18] Chuk OD, Ciribeni V, Gutierrez LV (2005) Froth collapse in column flotation: a prevention method using froth density estimation and fuzzy expert systems. Minerals Engineering 18:495–504
[19] González GD, Aguilera MA, Castelli L (1993) Development of a density soft-sensor for a mineral grinding plant. Prepr. 12th IFAC World Congress, Sydney, Australia, 5, 355–358
[20] Amestica R, González G, Menacho J, Barría J (1993) On-line estimation of fine and coarse ore, water, grinding rate and discharge rates in semiautogenous grinding mills. Proceedings XVIII International Mineral Processing Congress, Sydney, Australia, 1, 109–115
[21] Apelt TA, Asprey SP, Thornhill NF (2002) Inferential measurement of SAG parameters II: state estimation. Minerals Engineering 15(2):1043–1053
[22] González GD, Miranda D, Casali A, Vallebuona G (2008) Detection and identification of ore grindability in a semiautogenous grinding circuit model using wavelet transform variances of measured variables. International Journal of Mineral Processing 89(1-4):53–59
[23] Casali A, González, GD, Vallebuona G, Perez C, Vargas R (2001) Grindability soft-sensors based on lithological composition and on-line measurements. Minerals Engineering 14(7):689–700
[24] González GD,Casali A, Perez C, Vallebuona G (2000) Operational work index for grinding plants. Proceedings XXI International Mineral Processing Congress, Rome, July, 2000, Vol. C4:1–7
[25] Perez C, Casali A, González G, Vallebuona G, Vargas R (1999) Lithological composition sensor based on digital image feature extraction, genetic selection of features and neural classification. ICIIS'99, Proceedings IEEE International Conference on Information
[26] Espinoza PA, González GD, Casali A, Ardiles C (1995) Design of soft-sensors using cluster techniques. Proceedings International Mineral Processing Congress, Oct. 22–27, San Francisco, CA., USA., 1, vol. 1, Chapter 45, 260–265
[27] Pan XW, Metzner G, Selby N, Visser C, Letwaba T, Coleman K (2003) Development of weightometer soft sensor. Application of Computers and Operations Research in the Minerals Industries (APCOM), South African Institute of Mining and Metallurgy, 2003, pp. 261–266
[28] Herbst JA, Pate WT (1989) Experiences in the use of model based expert control systems in autogenous and semi autogenous grinding circuits. SAG Milling Conference, Vancouver, B.C., Canada, 669–686
[29] Herbst JA, Pate WT (1996) On-line estimation of charge volumes in semiautogenous and autogenous grinding mills. In:SAG 1996, Second International Conference on Autogenous and Semiautogenous Grinding Technology, vol. 2. Vancouver, BC, Canada, pp. 817–827
[30] González GD, Redard JP, Barrera R, Fernandez M (1994) Issues in soft-sensor applications in industrial plants. Proceedings IEEE International Symposium on Industrial Electronics, 25–27 May 1994, 380–385
[31] González GD, Orchard M, Cerda JL, Casali A, Vallebuona G (2003) Local models for soft-sensors in a rougher flotation bank. Minerals Engineering 16(5):441–453
[32] Hyötyniemi H, Raimo Ylinen, Jorma Miettunen (2000) AI in practice: case study on a flotation plant. Helsinki University of Technology, Control Engineering Laboratory, Konemiehentie 2, FIN-02150 Espoo, Finland. http://www.control.hut.fi/Hyotyniemi/publications/00_step_4.pdf
[33] Crisafulli S, Pierce RD, Dumont GA, Ingegneri MS, Seldon JE, Baade CB (1996) Estimating sugar cane fibber rate using Kalman filtering techniques. Proc. 13th IFAC Triennial World Congress, San Francisco, USA, 361–366
[34] Dunia R, Qin SJ, Edgar TF, McAvoy TJ (1996) Identification of faulty sensor using principal component analysis. AIChE Journal 42(10):2797–2812

[35] González GD, Odgers R (1996) Issues in the design of control loops using soft-sensors. Proceedings 13th IFAC Triennial World Congress, San Francisco, USA. A, 499-504. CD
[36] González G, González M, Cartes JC (1992) Control problems due to replacement of sensors by soft-sensors. Advances in Instrumentation and Control, Proceedings ISA/92 International Conference and Exhibition, Houston, USA, Vol 47, part 2, 1193–1200
[37] Wang X, Luo R, Shao H (1996) Designing a soft sensor for a distillation column with the fuzzy distributed radial basis function neural network, Proceedings of the 35th IEEE Decision and Control, 1996, Vol. 2, 1714–1719
[38] Wendt C, Weyer E, Mareels I (1999) Soft sensor for flow measurement: overshot regulator. Proceedings IEEE Information, Decision and Control, 1999, 471–475
[39] Herbst JA, Pate WT (1999) Object components for comminution softsensor design. Powder Technology, 105(1–3):424–429
[40] Zhang H, , Wang X, Zhang C, Lv G (2006) New soft sensor method based on SVM. 2006 IEEE International Conference on Granular Computing, 546 – 549
[41] King MJ (2004) How to lose money with inferential properties: improving process control. Hydrocarbon Processing 83(10):47–52
[42] Barrera R, Gonzalez G, Casali A, Vallebuona G (1996) SENVIR: a soft-sensor system for industrial applications (in Spanish). Project FONDEF MI-17, Dept. of Electrical Engineering, University of Chile
[43] Casali A, González GD, Agusto H, Vallebuona G (2002) Dynamic simulator of a rougher flotation circuit for a copper sulphide ore. Minerals Engineering 15(4):253–262
[44] Ljung L (1987) System identification: theory for the user. P.T.R. Prentice Hall, Information and System Sciences Series, New Jersey
[45] Sjöberg J, Zhang Q, Ljng, Benveniste L, Delyon A, Glorennec B, Hjalmarsson P-Y, Juditsky H (1995) Nonlinear black-box modeling in system identification: a unified overview. Automatica 31(12):1691–1724
[46] Bakshi BR, Utojo U (1998) A common framework for the unification of neural thermometric and statistical modeling methods. Analytical Chimica Acta 384:227–247
[47] Papoulis A, Pillai SU (2002) Probability, Random Variables, and Stochastic Processes, fourth edition. McGraw-Hill.
[48] Billings SA, Voon WS F (1986) A prediction error and step-wise regression estimation algorithm for non-linear systems International Journal of Control 44(3):803–822
[49] Ljung L (2001) System identification toolbox, The Mathworks, Inc., Natick, MA, USA. www.mathworks.com
[50] Amestica R, González G, Menacho J, Barría J (1996) A mechanistic state equation model for a semiautogenous grinding mill. International Journal of Mineral Processing 44-45:349–360
[51] Mathworks, findcluster. http://www.mathworks.com/access/helpdesk/help/ toolbox/fuzzy/index.html?/access/helpdesk/help/toolbox/fuzzy/findcluster.html
[52] Haber R, Unbehauen H 1990 Structure identification of nonlinear dynamic systems – a survey on input-output approaches. Automatica 26(4):651–677
[53] Mathworks, stepwisefit. http://www.mathworks.com/access/helpdesk/help /toolbox/stats/index.html?/access/helpdesk/help/toolbox/stats/stepwisefit.html
[54] Holtsberg A (1994) lsselect. http://read.pudn.com/downloads102/sourcecode/ math/419128/stixbox/lsselect.m__.htm
[55] Amestica R, González G, Barria J, Mange L, Menacho J, Castro O (1993) A SAG mill circuit dynamic simulator based on a simplified mechanistic model. Proceedings XVIII International Mineral Processing Congress, Sydney, Australia, 1, 117–130
[56] Myers RH (1990) Classical and modern regression with applications, 2nd edition. Duxbury Press, Belmont, California
[57] Mardia KV, Kent JT, Bibby JM (1979) Multivariate analysis. Academic Press
[58] Jackson JE (1991) A user's guide to principal components. John Wiley, New York
[59] Dunia R, Qin SJ, Edgar TF, McAvoy TJ (1996) Sensor fault identification and reconstruction using principal components. Proceedings 13th Triennial IFAC World Congress, San Francisco, USA, N, 259–264

[60] Smola A, Schlkopf B (2003) A tutorial on support vector machines. RSISE Australian National University, Canberra, Australia. http://eprints.pascal-network.org/archive/00002057/01/SmoSch03b.pdf
[61] National Semiconductor (1980) Application Note 236, http://www.national.com/an/AN/AN-236.pdf
[62] Tham, MT, Montague GA, Morris AJ, Lant PA (1991) Soft sensors for process estimation and inferential control. Journal of Process Control 1:3–14
[63] Frazier MW, (1999) An introduction to wavelets through linear algebra. Springer
[64] Naylor AW, Sell GR (2000) Linear operator theory in engineering and science. Springer Applied Mathematical Sciences, Vol. 40
[65] Davis HF (1963) Fourier series and orthogonal functions. Allyn and Bacon, Boston, USA
[66] Tapia H (2000) Software and hardware platform for industrial application of soft sensors (in Spanish). Electrical Engineering thesis, Department of Electrical Engineering, University of Chile. http://200.89.70.219/web2/tramp2.exe/do_keyword_search/A2g6mni6.000
[67] Chen, J, Patton RJ (1999) Robust Model-Based Fault Diagnosis for Dynamic Systems. Kluwer Academic Publishing Group, Dordrecht, The Netherlands

Chapter 5
Dynamic Simulation and Model-based Control System Design for Comminution Circuits

Daniel Sbarbaro

Abstract Design and validation of control strategies requires a knowledge of both control theory and the process to be controlled. These activities can be time consuming if a process is complex, involving many interacting equipments with nonlinear characteristics. Simulators have proven to be a powerful tool in the design and validation of control structures for this type of complex process. The aim of this chapter is to describe a dynamic simulator for comminution plants implemented in a commercial simulation platform, and to illustrate its use in the design of control strategies for different stages in the comminution process.

5.1 Introduction

Comminution circuits, as described in Chapter 1, are among the main processes in a concentrator plant, being also the most energy-demanding ones. Due to their relevance in plant operation, there have been many studies regarding their optimization. Key tools in the quest for optimal comminution plant design and operation have been modeling and simulation [1–3].

Modeling of these processes has been carried out to satisfy different objectives such as plant design and optimization, operational optimization, control system design and verification, as well as operator training. Nowadays, models play an important role in the planning and operation of mineral processing plants, as described in [4].

Basically, each model representing a process has an associated set of variables, classified as follows:

- *Manipulated variables* can be modified in order to change the state of the model.

Daniel Sbarbaro
Department of Electrical Engineering, Universidad de Concepción, Concepción, Chile, e-mail: daniel.sbarbaro@udec.cl

- *Output variables* are associated with the objective of analysis. These are usually measurable.
- *Disturbances* are input variables which cannot be modified at will. These generally represent the effect of the environment on the process. These variables can be measured in some cases.
- *Parameters* are variables associated with the characteristics of the system. Generally, these do not change during study.
- *States* are internal variables which uniquely define the behavior of the model.

A mathematical model consists of a series of mathematical relationships between these variables. These relationships can be established either through empirical knowledge or by the use of first principles, such as energy and mass balance equations. As mineral processes involve very complex phenomena, it is common to use a combination of empirical and phenomenological models. These mathematical relationships describe how characteristics associated with a given material flow are transformed along the process.

The equations describing each process can be encapsulated in modules in order to simplify programming and debugging. In addition, modules makes it easier for the user to configure a simulation scenario, since the addition and deletion of models does not change the simulation strategy. A simulator can then be defined as a set of interconnected modules representing a process. Graphical interfaces of modern simulation tools also make it possible to give the simulation diagram a look much closer to actual engineering documents such as Process and Instrumentation (P&I) diagrams.

If the mathematical relationships are independent of time, then the models are static. These models serve mainly for determining operating points (through the steady-state effect of constant manipulated variables within a time window), disturbances, and plant parameter changes. Although these steady-state conditions may be arrived at using dynamic simulators, it is faster to use static simulators, which are simpler than dynamic ones [4]. Static models have been successfully used in the design of new plants and in the optimization of existing ones through the definition of new operational parameters.

Dynamic models consider all time-dependent variables. In this way, it is possible to simulate transient conditions generated by changes in manipulated variables, disturbances, different kinds of faults (*e.g.*, process, instrumentation, control system faults), plant parameter changes, *etc.* These types of models are very useful for the design and verification of control systems, for on-line optimization and operator training.

5.2 Dynamic Simulation of Comminution Circuits Based on MATLAB®/Simulink®

As described in the previous section, in order to model a comminution circuit for control system design, it is necessary to have a flexible modular simulation environment. This environment must have the flexibility to configure and blend processes and control equipment models. There are many simulation tools available on the market that satisfy these requirements. One of these is MATLAB®/Simulink®[1], as described in [5]. This tool has many desirable features, such as a modular approach, an open model structure, ease in changing circuit configurations, a powerful graphical interface, advanced nonlinear system solvers and control design toolboxes. Several authors have recognized these advantages and they have proposed simulators for comminution operations [5, 6]. This section describes the efforts carried out at Universidad de Concepci6n during the last eight years in the building of a flexible dynamical simulator for control system design. The simulator includes a library of equipment and interconnection models which can be blended to configure complex plants. The library also considers sensor modules that extract information from the interconnections. This library has been under constant development and refinement since its conception. Figure 5.1 depicts the modules available to date.

The process modules are interconnected by a flow of material with the following attributes:

- solid mass flowrate f;
- water volumetric flowrate (for wet circuits only) q;
- particle size distribution $\bar{\mathbf{f}}$;
- hardness γ.

To change these attributes, there are special mixing modules. For example MIXWP adds water to a specific flow.

Each model represents the transformation of flow attributes as they advance through the process equipment. In the following section, main assumptions and equations considered in the simulator are described.

The notation of this chapter follows these simple rules: lower-case Latin letters are scalar quantities representing flows, lower-case Latin bold letters are vectors representing size distributions, overline lower-case Latin bold letters are vectors representing size distributions in terms of mass fraction, and bold Latin capital letters represent matrices associated with equipments. For instance $\mathbf{f}$ represents ore feed flow at n-size intervals and $f = \sum_{i=1}^{n} f_i$ represents total ore feed flow. Thus, the discrete size distribution associated with $\mathbf{f}$ can be written as $\bar{\mathbf{f}} = \mathbf{f}/f$.

[1] MATLAB® and Simulink® are registered trademarks of The MathWorks, Inc., 3 Apple Hill Drive, Natick, MA 01760-2098, USA, http://www.mathworks.com

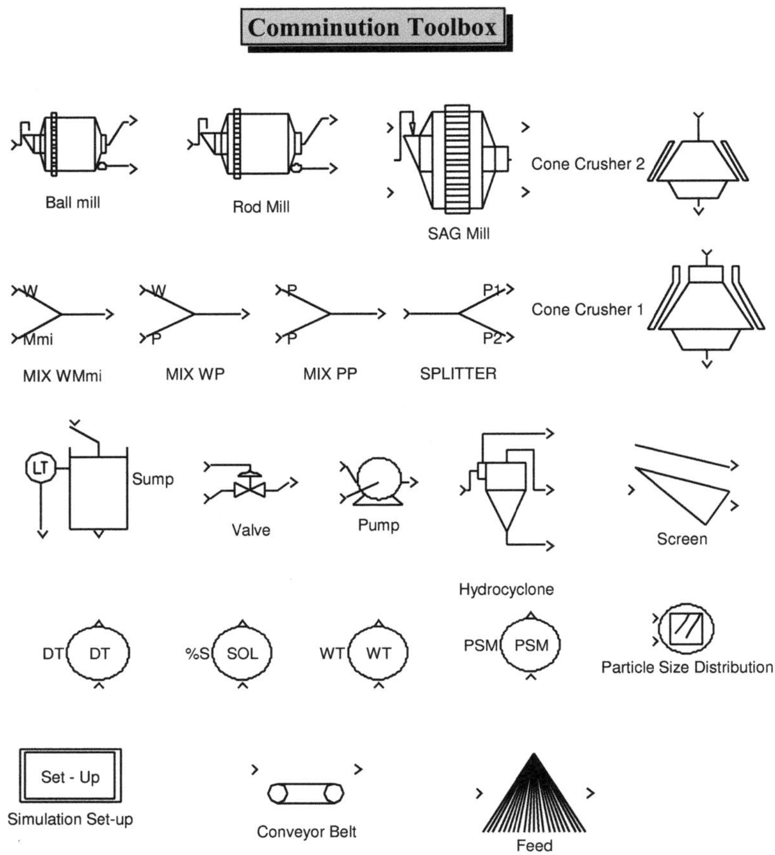

Figure 5.1 Modules developed to date

5.3 Dynamic Modeling of Crushing Circuits

The main elements of a crushing circuit are crushers, screens, conveyor belts, bins, and stock piles. From a dynamic point of view, screens are considered to have an instantaneous response; *i.e.*, there is no accumulation of mass. In the current version, there are no models for bins and stock piles.

5.3.1 Dynamic Model of a Cone Crusher

The modeling of a crusher must consider the different factors affecting the size reduction process and the flow of material through the crushing chamber. Semi-empirical models have been used for simulating crushing plants, with excellent results for operational purposes [7, 8]. On the other hand, there have been few studies on models based on first principles; *i.e.*, mechanical laws [9]. These models, which are usually more complex, describe in detail the factors affecting the reduction process. Their use is oriented to the design and optimization of crusher structure. Semi-empirical models provide a reasonable compromise between representability and simplicity. For this reason, the model used in the simulator is based on Whiten's perfect mixing model [7].

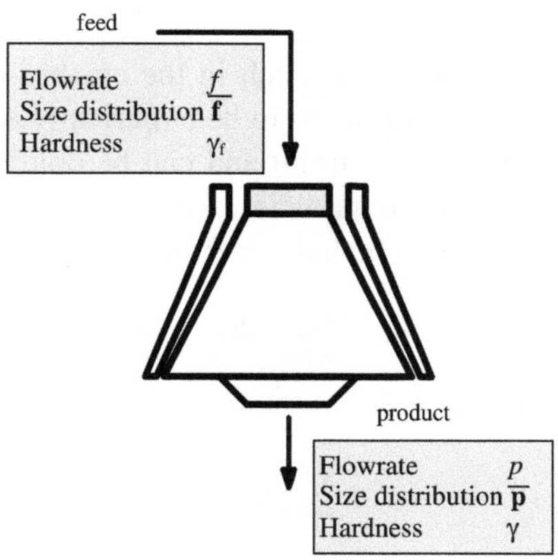

Figure 5.2 A cone crusher and its main variables

The variables associated to the crusher model are depicted in Figure 5.2. Mass balance of the contents is given by the following equation [7]:

$$\frac{d\mathbf{m}(t)}{dt} = \mathbf{f}(t) - \mathbf{p}(t) - \gamma(t)(\mathbf{S} - \mathbf{BS})\mathbf{m}(t), \tag{5.1}$$

where $\gamma(t)$ is a variable representing ore hardness, $\mathbf{f}(t)$ and $\mathbf{p}(t)$ are vectors having as elements $f_i(t)$ and $p_i(t)$, which are the mass flowrate in the ith size fraction of the feed and the product, respectively. The mass in the ith size fraction of the contents is noted as $\mathbf{m}$. The matrix $\mathbf{S}$ is a diagonal matrix representing the specific breakage rate of size i, $\mathbf{B}$ is a lower diagonal matrix, where b_{ij} represents the fraction of particles of size fraction j appearing in the size fraction i after breakage. Product mass flow is assumed to be proportional to the mass contents; *i.e.*,

$$\mathbf{p} = \mathbf{Dm}, \tag{5.2}$$

where $\mathbf{D}$ is a diagonal matrix representing each element in the specific discharge rate of size j. The steady-state solution of (5.1) can be found by setting the first term to zero and expressing $\mathbf{p}$ in terms of $\mathbf{f}$ as follows:

$$\mathbf{p} = [\mathbf{I} - \mathbf{C}][\mathbf{I} - \mathbf{CB}]^{-1}\mathbf{f}, \tag{5.3}$$

where the vector $\mathbf{C}$ is the classification matrix defined as $\mathbf{S}(\mathbf{S}+\mathbf{D})^{-1}$.

The matrix $\mathbf{B}$ can be broken down into two parts as follows:

$$\mathbf{B} = \alpha\mathbf{B}_1 + (1-\alpha)\mathbf{B}_2. \tag{5.4}$$

The matrix $\mathbf{B}_1$ is associated with primary breakage and $\mathbf{B}_2$ is associated with the production of fines. The term α must be found empirically as a function of the operational variables. Here, the following expression is considered:

$$\alpha = a_o + a_1 f + a_2\bar{s} + \frac{a_3}{c_{ss}}, \tag{5.5}$$

where f is the feedrate, $\bar{s}$ is the % +1 inch in the crusher feed, c_{ss} is the crusher closed side setting, and a_i are parameters to be empirically adjusted.

The model c describes crusher current and can be obtained by considering the vector $\mathbf{Rm}$ which contains no fine particles. The following empirical relationship has been proposed in [7], but also includes a term γ as a function of hardness:

$$c = b_0 + b_1 r + b_2 r^2 + b_3\gamma, \tag{5.6}$$

where the variable r is defined as

$$r = \sum \frac{z_i}{s_i + s_{i+1}}, \tag{5.7}$$

where z_i is the ith element of $\mathbf{Rm}$, and s_i are the upper and lower limits of the ith fraction.

Mass retained inside the crusher chamber can be obtained as $m(t) = \sum_{i=1}^{n} m_i(t)$. Volume is given by $\frac{m(t)}{\rho(t)}$, and bulk density $\rho(t)$ is calculated as a function of product size distribution [9]. Thus, level can be calculated as a function of crusher geometry and volume of retained mass by solving the equation $\frac{m(t)}{\rho(t)} = V(h(t))$.

To obtain the value of hardness inside the crusher chamber, a simple mixing model is considered:

$$\frac{d\gamma(t)}{dt} = \frac{f(t)}{m(t)}(\gamma_f(t) - \gamma(t)), \tag{5.8}$$

where γ_f is the ore hardness in the feed. According to the mixing model principle, ore hardness in the crushing chamber, γ, is also product ore hardness.

5.3.2 Screens

Screens are modeled as static systems; *i.e.*, no dynamic is associated to this equipment.

As it is illustrated in Figure 5.3, undersize $\mathbf{p}_u$ and oversize $\mathbf{p}_o$ are given by the following relationships:

$$\mathbf{p}_u = (\mathbf{I} - \mathbf{C})\mathbf{f},$$
$$\mathbf{p}_o = \mathbf{C}\mathbf{f}. \tag{5.9}$$

There many forms for the efficiency curve [3]. In this work, the empirical efficiency curve proposed by Hatch and Mular [8] is used:

$$E(x_i) = c_i + \frac{(1 - c_i)}{1 + exp(\frac{x_{50}^3 - x_i^3}{k})}, \tag{5.10}$$

where $E(x_i)$ is the weight fraction of a narrow size range x_i of feed that reports to oversize, c_i is a term accounting for bypassing small particles to oversize and x_{50} is constant for a given set of operating conditions. In other words, it is modeled as a polynomial function depending on the screen opening h, the screen feedrate of solids and the percentage of +1 inch material in the screen feed. Ore hardness of undersize and oversize is assumed to have the same value as feed hardness.

5.3.3 *Conveyor Belts*

Fixed speed conveyor belts are modeled as a pure time delay; *i.e.*, the product is

$$\mathbf{p}(t) = \mathbf{f}(t - d) \tag{5.11}$$

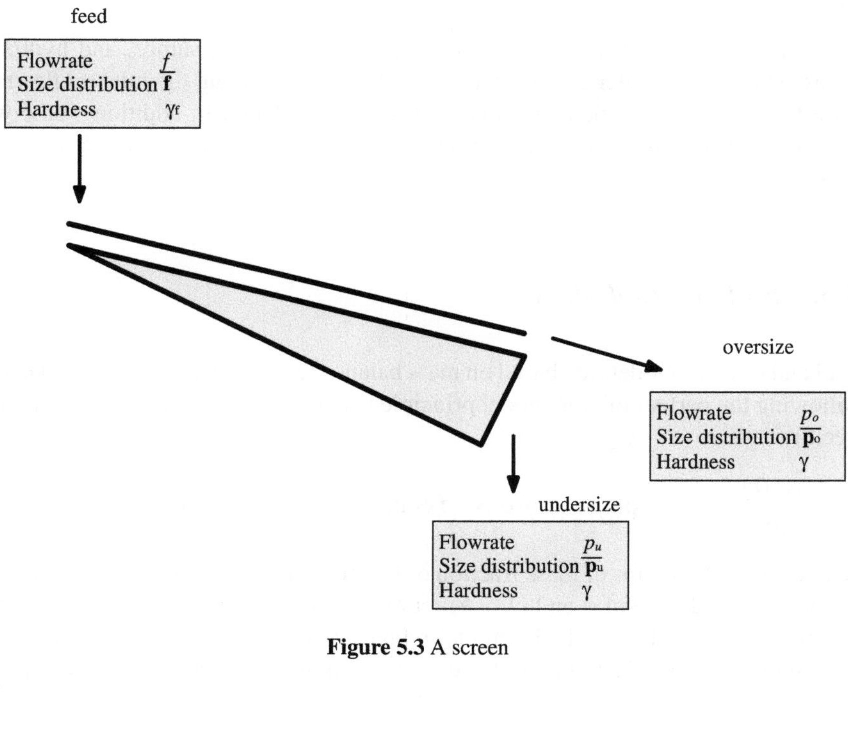

Figure 5.3 A screen

where d is the time delay. This parameter can be estimated by using the length of the conveyor belt l and its speed v, $d = l/v$. As depicted in Figure 5.4 both the feed size distribution and the associated hardness are delayed.

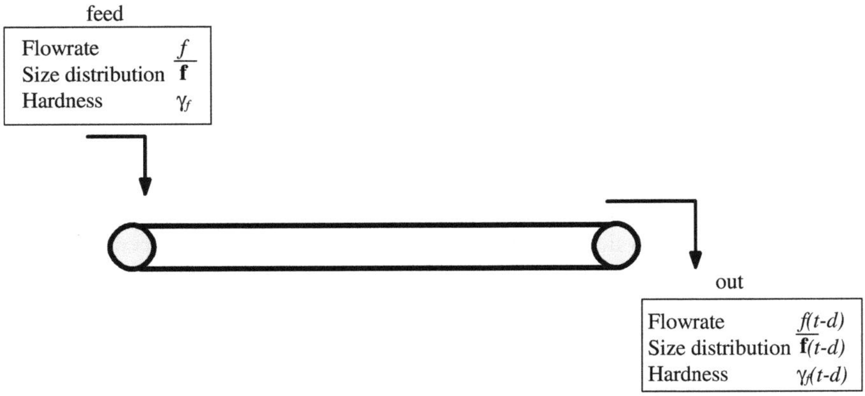

Figure 5.4 A conveyor belt

5.4 Dynamic Modeling of Wet Grinding Circuits

The main elements of a grinding circuit are grinding mills, sumps, and hydrocyclons. The flow of material considers the water flowrate along the mineral flowrate with the associated particle size distribution and hardness. In addition, some wet grinding circuits involving semi-autogenous grinding (SAG) mills will consider cone crushers.

5.4.1 Ball and Rod Mills

Ball and rod mill models are based on mass balance for material in each size interval, following the perfect mixing model principle [2]. These equations can be written in vector form as

$$\frac{d\mathbf{m}(t)}{dt} = \mathbf{f}(t) - \mathbf{p}(t) - \gamma(t)(\phi[\mathbf{S} - \mathbf{BS}]\mathbf{m}(t) + (1-\phi)[\mathbf{S} - \mathbf{BS}]\mathbf{m}(0)), \qquad (5.12)$$

where $\mathbf{f}(t)$ is the vector of mass fraction of the mill charge, $\mathbf{p}(t)$ is the mass fraction of the mill product, and ϕ and $(1-\phi)$ represent the fraction of first and zero order kinetics assumed. If $\phi = 1$, the model represents a ball mill and if $0 < \phi < 1$ it represents a rod mill [10]. Figure 5.5 depicts the main input and output variables.

The mass fraction inside the mill **m** can also be expressed as a function of the mass distribution as

$$\mathbf{m}(t) = h\overline{\mathbf{m}(t)}, \tag{5.13}$$

where h is the total hold-up in the mill. Thus Equation 5.12 can be written as

$$\frac{dh\overline{\mathbf{m}(t)}}{dt} = \mathbf{f}(t) - \mathbf{p}(t) - \gamma(t)(\phi[\mathbf{S} - \mathbf{BS}]h\overline{\mathbf{m}}(t) + (1-\phi)[\mathbf{S} - \mathbf{BS}]h\overline{\mathbf{m}}(0)), \tag{5.14}$$

where matrices **S** and **B** are breakage matrices.

The total mass retained in the mill, m, can be written as

$$m = Vc, \tag{5.15}$$

where V is the volume of slurry in the mill and c is the solids concentration; *i.e.*, the mass of solids per unit volume of slurry. For overflow mills, the volume of slurry present in the mill is generally constant over a wide range of operating conditions [11]. The variation of concentration is described by

$$\frac{dVc(t)}{dt} = q_f(t)c_f(t) - q(t)c(t). \tag{5.16}$$

Since V is constant $q = q_f$, Equation 5.16 can be written as

$$\frac{dc(t)}{dt} = \frac{q_f(t)}{V}(c_f(t) - c(t)). \tag{5.17}$$

The ore hardness is assumed to be transported through the mill following the perfect mixing model:

$$\frac{d\gamma(t)}{dt} = \frac{q_f(t)}{V}(\gamma_f(t) - \gamma(t)). \tag{5.18}$$

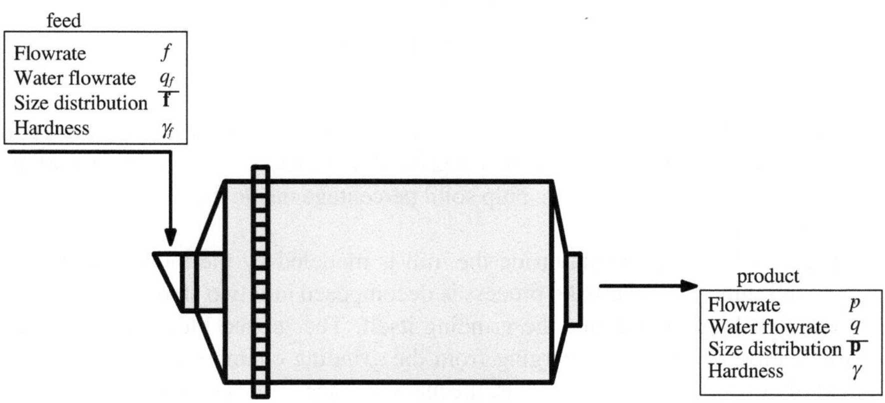

Figure 5.5 A ball mill

In order to model the residence time more accurately, a set of mixing tanks in series can be considered. Thus, the total residence time can be obtained by connecting several models in series. In other words, the discharge of one mill will be fed to the next mill. The volume hold-up of each model V can be used to adjust the residence time of each sub-model.

5.4.2 SAG Mill

The modeling of SAG mills has been addressed by different researchers [6, 12–14]. In this simulator, the model developed by Amestica [13] was implemented. The main input and output variables are defined in Figure 5.6.

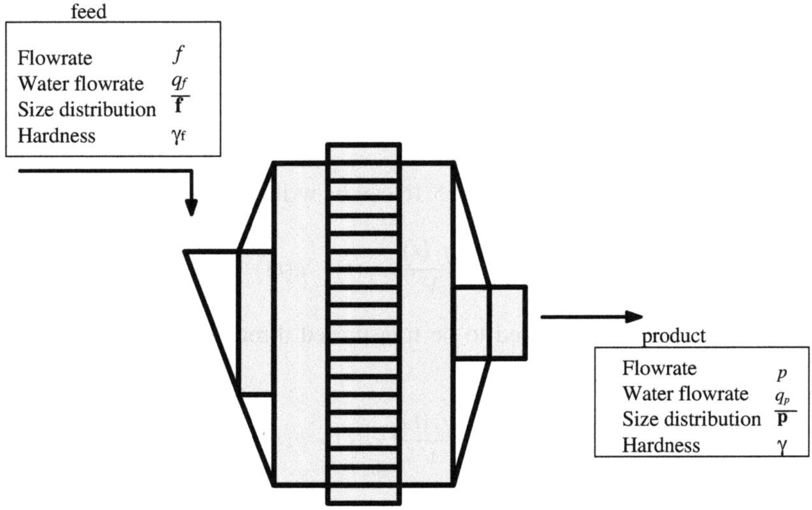

Figure 5.6 A SAG mill

This model is based on a dynamic balance equation for ore and water mill contents, as well as several sub-models for grinding, classification, mass transport and power draw in terms of hold-up, pulp solid percentage inside the mill, and volumetric filling fraction.

The grinding phenomena inside the mill is modeled by the block diagram depicted in Figure 5.7, where the process is decomposed into two steps.

The first block represents the grinding itself. The second block represents the classification of ore flow emerging from the grinding chamber and considers the effect of grate and pulp lifters. Thus, the mass balance of ore content in the grinding chamber can be determined by the following equation:

$$\frac{d\mathbf{m}(t)}{dt} = \mathbf{f}(t) + \mathbf{C}\mathbf{p}^*(t) - \mathbf{p}^*(t) - \gamma(t)\mathbf{D}\mathbf{m}(t), \tag{5.19}$$

where $\mathbf{f}$ is the fresh ore feed, $\mathbf{C}$ is the classification matrix and $\mathbf{D}$ is a lower triangular matrix defined as

$$\mathbf{D} = \frac{M_p}{H_m}\mathbf{R}^{-1}\mathbf{K}^E\mathbf{R}. \tag{5.20}$$

The variable H_m is the ore content in the grinding chamber, M_p is the mill power draw, matrix $\mathbf{K}^E$ is a diagonal matrix with K_i^E being the specific grinding rate of the production of fines below size interval i, and $\mathbf{R}$ is a unitary lower triangular matrix. Assuming perfect mixture, the ore discharged from the grinding chamber is given by

$$\mathbf{p}^* = \frac{p^*}{H_m}\mathbf{m}. \tag{5.21}$$

The classification matrix $\mathbf{C}$ is a diagonal matrix, its elements being a function of the mill pulp solid percentage, defined as $p_s = H_f/(H_f + W)$.

The power draw model is given by the Austin empirical model:

$$M_p = kD^{2.5}L(1 - AJ)f(\phi)\frac{H_m + W + W_b}{V_m}, \tag{5.22}$$

where D and L are the diameter and length of the SAG mill, V_m is the internal mill volume, J is the volumetric mill fraction occupied by the total hold-up, and W_b is the mass hold-up of steel balls. Function $f(\phi)$ depends on the critical speed fraction ϕ [15]. The parameters k and A are experimental parameters.

The water balance is given in terms of inflow water q_f and discharge flow from the mill, which is given by the following empirical expression:

$$q_p = (\alpha_0 + \frac{\alpha_1}{H_m^4})W, \tag{5.23}$$

where α_0 and α_1 are experimentally adjusted.

Thus, the final expression for water balance is

$$\frac{dW(t)}{dt} = q_f(t) - (\alpha_o + \frac{\alpha_1}{H_m^4(t)})W(t). \tag{5.24}$$

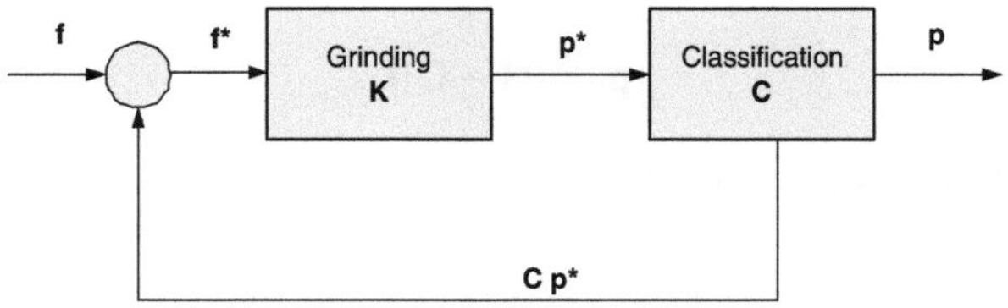

Figure 5.7 Block diagram of a SAG mill

Ore hardness is assumed to be transported through the mill as

$$\frac{d\gamma(t)}{dt} = \frac{f(t)}{H_m(t)}(\gamma_f(t) - \gamma(t)). \tag{5.25}$$

5.4.3 Hydrocyclone

A hydrocyclone is modeled as a system with negligible dynamics, since its residence time is much smaller than mills and sumps. The model equations used in the simulator correspond to the Lynch and Rao empirical model [11]. The input and output variables are defined as in Figure 5.8.

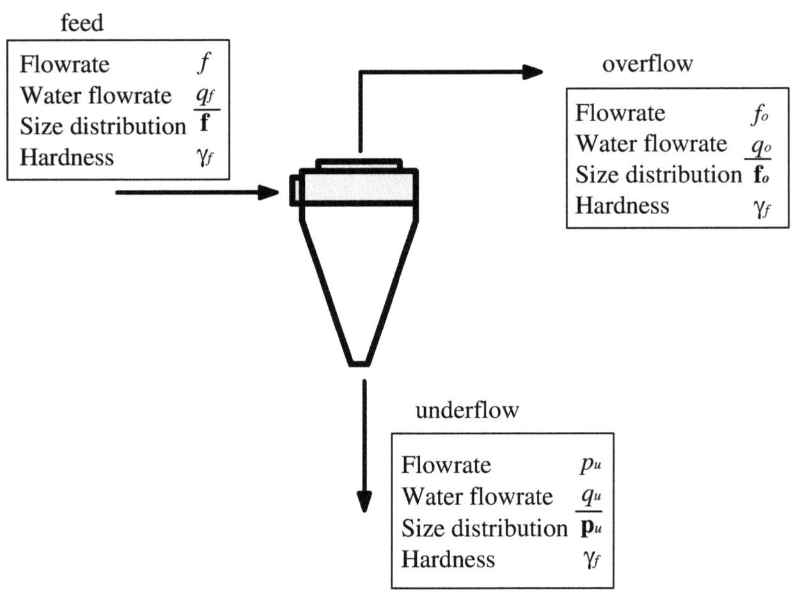

Figure 5.8 A hydrocyclone

The model describes water flowrates in the overflow and cut size as a function of cyclone parameters and operational conditions:

$$\mathbf{p}_u = (\mathbf{I} - \mathbf{C})\mathbf{f}. \tag{5.26}$$

The efficiency curve is given by

$$E(x_i) = \frac{e^{\alpha \bar{x}_i}}{e^{\alpha \bar{x}_i} + e^{\alpha} - 2}, \tag{5.27}$$

where $\bar{x}_i = x_i/d_{50}$, α is a parameter, and d_{50} is parameterized in terms of operational variables and parameters of the hydrocyclone:

$$log_{10}d_{50}(c) = k_1 v_f - k_2 Spig + k_3 In + k_4 c - k_5 q_s + k_6, \tag{5.28}$$

where v_f is the vortex diameter, $Spig$ is the spigot diameter, In is the inlet diameter, c is the feed percentage of solids, and q_s is the slurry volumetric flowrate.

The water balance is

$$q_u = q_f - r_f q_f, \tag{5.29}$$

where r_f is the percentage of water in the underflow, which has been correlated with inlet volumetric flowrate w_f and spigot diameter $Spig$ as follows:

$$r_f = k_7 \frac{Spig}{w_f} - k_8 \frac{1}{w_f} + k_9. \tag{5.30}$$

All of the parameters k_i, $i = 1,...,9$ can be adjusted so as to model a specific hydrocyclone. The ore hardness of the feed is transferred to the over and underflows.

5.4.4 Sump

The sump can be modeled as a perfect mixed tank, due to its relatively small size and the turbulence produced by the feed flow. Thus, a model for a perfect mixing tank is considered [11]. The main input and output variables are depicted in Figure 5.9.

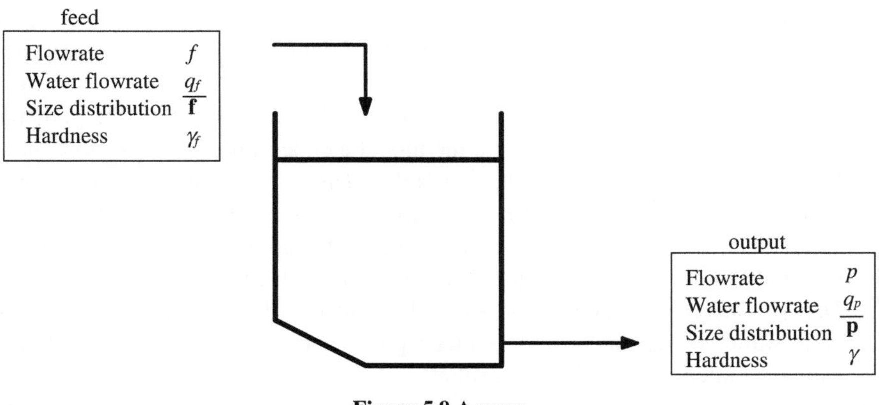

Figure 5.9 A sump

Under the assumption that no change in particle size occurs in the sump, the following mass balance equation provides a description of mass size distribution in the sump:

$$\frac{d\mathbf{m}(t)}{dt} = \mathbf{f}(t) - p(t)\overline{\mathbf{m}}(t). \tag{5.31}$$

$p(t)$ represents the mass rate of solids discharged from the sump. Sump level variations can be found by combining a volumetric balance of slurry and solid balance:

$$\frac{dc(t)}{dt} = \frac{1}{V}(f(t) - q_s(t)c_s(t) - (p(t) - q_{sp}(t)c(t))), \tag{5.32}$$

where $q_s(t)$ is the volumetric feed flowrate defined as $q_s(t) = q(t) + (f(t)/\rho_s)$ with ρ_s being density of the ore. Volumetric flowrate $q_{sp}(t)$ is fixed by the pump. The level of the slurry can then be calculated as a function of the volume.

Ore hardness is assumed to be transported trough the sump as

$$\frac{d\gamma(t)}{dt} = \frac{q(t)}{V(t)}(\gamma_f(t) - \gamma(t)). \tag{5.33}$$

5.5 Model-based Control Systems Design

This section illustrates the use of a simulator in the design of model-based control strategies for different processes. According to classical design procedure, the major steps for an advanced model-based controller using a simulator can be summarized as follows:

1. *Define objectives.* Identify control objectives with respect to economics and process behavior (including safety).
2. *Variables identification.* Identify manipulated variables, disturbances and controlled variables based on control objectives, process knowledge and experience. Identify possible restrictions.
3. *Experimental testing.* Perform plant tests to obtain suitable data for dynamic modeling. If possible, measure disturbance variables as well. Collect data for further analysis.
4. *Model identification.* Identify a linear empirical model based on the gathered data. If the fit is not good enough, the model can be improved by seeking nonlinear transformations on selected variables. This can be guided by using phenomenological knowledge. After fitting the model, it is always useful to perform some kind of validation; *i.e.*, to analyze the static and dynamic characteristic of the model and check whether they are consistent with conditions at the plant.
5. *Design.* Design the final controller based on the identified model. In some cases, an identified model can be used to tune a pre-defined controller such as a PID, or can be an integral part of the controller.
6. *Closed loop simulation.* Perform off-line testing and tuning of the proposed controller by testing it in a closed loop against the nonlinear simulated process model.

After finishing the design stage, an implementation and commissioning stage follows, with final tuning adjustments made on-line. In the following examples, the simulator will be used to carry out most of the steps described above.

5.5.1 Models for Control Design

Process models are the core of many control strategies. They can be used for tuning a strategy or as an integral part of the strategy itself.

Linear models are widely used to represent the process dynamic around a given operating point. These models, as described in Chapter 4, have a given structure and a set of parameters that can be adjusted to represent a specific process. Chapter 4 has provided a thorough review of techniques for finding both structure and parameters for models. This section focuses on the model structure that is suitable for predictive control design.

A basic structure generally used for the design of discrete time controllers is an ARMAX structure (additional structures are commented upon in Chapter 7):

$$A(z^{-1})y(t) = z^{-d}B(z^{-1})u(t-1) + C(z^{-1})w(t), \tag{5.34}$$

where $y(t)$ and $u(t)$ are the model output and input, respectively, and $w(t)$ is a white noise signal. The polynomials $A(z^{-1})$, $B(z^{-1})$ and $C(z^{-1})$ are defined as follows:

$$\begin{aligned} A(z^{-1}) &= 1 + a_1 z^{-1} + \cdots + a_{n_a} z^{-n_a}; \\ B(z^{-1}) &= b_0 + b_1 z^{-1} + \cdots + b_{n_b} z^{-n_b} \quad n_b \leq n_a; \\ C(z^{-1}) &= 1 + c_1 z^{-1} + \cdots + c_{n_c} z^{-1n_c} \quad n_c \leq n_a. \end{aligned} \tag{5.35}$$

Parameters for these polynomials can be estimated using the techniques described in Chapter 4. However, in the model predictive control framework, $C(z^{-1})$ is considered as a design parameter as it has a direct effect on closed loop sensitivity [17].

In order to account for the effect of unmeasurable disturbances normally found in practice (such as step-like disturbances), the so-called controlled auto-regressive integrated moving average (CARIMA) structure has been proposed:

$$A(z^{-1})y(t) = z^{-d}B(z^{-1})u(t-1) + \frac{C(z^{-1})}{\Delta(z^{-1})}w(t), \tag{5.36}$$

where $\Delta(z^{-1}) = 1 - z^{-1}$. The above equation can also be written as

$$A(z^{-1})\Delta(z^{-1})y(t) = z^{-d}B(z^{-1})\Delta(z^{-1})u(t-1) + C(z^{-1})w(t) \tag{5.37}$$

and in the time domain:

$$\sum_{i=1}^{n_a} a_i \Delta y(t-i) = \sum_{i=1}^{n_b} b_i \Delta u(t-i-d-1) + \sum_{i=1}^{n_c} c_i w(t), \tag{5.38}$$

where $\Delta y(t) = y(t) - y(t-1)$ and $\Delta u(t) = u(t) - u(t-1)$. These models can be easily extended to systems with more inputs and outputs. For instance, if the system has a control input $u(t)$ and a measurable disturbance $p(t)$ acting on the process input, then a suitable model could be

$$A(z^{-1})\Delta(z^{-1})y(t) = z^{-d}B(z^{-1})\Delta(z^{-1})u(t-1) + z^{-d}H(z^{-1})\Delta(z^{-1})p(t-1) \\ +C(z^{-1})w(t), \tag{5.39}$$

where $H(z^{-1})$ is a polynomial of degreen_h, $n_h \leq n_a$.

These models can be used for making predictions of system output, given future control inputs and disturbance signals. In order to obtain a prediction, it is necessary to use Equation 5.38 recursively. In other words, first compute $y(t+1|t)$, then substitute the result back into (5.38) at $k+2$ to obtain $y(t+2|t)$. A thorough derivation of the optimal prediction can be found in [16]. In this section, a simplified derivation considering $C(z^{-1}) = 1$ will be described. In order to obtain $y(t+1|t)$, Equation 5.38 can be written as

$$\Delta y(t) = -\sum_{i=i}^{n_a} a_i \Delta y(t-i) + \sum_{i=1}^{n_b} b_i \Delta u(t-i-d) + \sum_{i=1}^{n_c} c_i \Delta p(t-i-d) + w(t). \tag{5.40}$$

Thus, the next step prediction with information up to time t will be

$$\Delta y(t+1|t) = -\sum_{i=i}^{n_a} a_i \Delta y(t-i+1) + \sum_{i=1}^{n_b} b_i \Delta u(t-i-d+1) \\ + \sum_{i=1}^{n_d} d_i \Delta p(t-i-d+1) + w(t+1) \tag{5.41}$$

As the noise term represents future values, the best prediction for $y(t+1|t)$ will be

$$y(t+1|t) = y(t) - \sum_{i=i}^{n_a} a_i \Delta y(t-i+1) + \sum_{i=1}^{n_b} b_i \Delta u(t-i-d+1) + \sum_{i=1}^{n_d} d_i \Delta p(t-i-d+1). \tag{5.42}$$

This procedure can easily be systematized by considering the following Diophantine equation:

$$1 = E_j(z^{-1})\Delta(z^{-1})A(z^{-1}) + z^{-j}F_j(z^{-1}), \tag{5.43}$$

where the polynomials E_j and F_j have degrees $j-1$ and n_a, respectively.

Multiplying (5.39) by $E_j(z^{-1})z^j$:

$$E_j(z^{-1})z^j A(z^{-1})\Delta(z^{-1})y(t) = z^{-d}E_j(z^{-1})z^j B(z^{-1})\Delta(z^{-1})u(t-1) \\ +E_j(z^{-1})z^j H(z^{-1})\Delta(z^{-1})p(t-1) + E(z^{-1})z^j w(t) \tag{5.44}$$

and using (5.43)

$$(1 - z^{-j}F_j(z^{-1}))y(t+j) = z^{-d}E_j(z^{-1})z^j B(z^{-1})\Delta u(t-1) \\ +E_j(z^{-1})z^j H(z^{-1})\Delta p(t-1) + E(z^{-1})w(t) \tag{5.45}$$

the prediction $y(t+j)$ is given by

$$\begin{aligned} y(t+j) = &F_j(z^{-1}))y(t)+E_j(z^{-1})B(z^{-1})\Delta u(t-d+j-1) \\ &+E_j(z^{-1})H(z^{-1})\Delta p(t-d+j-1)+E_j(z^{-1})w(t+j) \end{aligned} \tag{5.46}$$

As the last term associated with noise $E_j(z^{-1})w(t+j)$ only has terms in the future, the best prediction is

$$y(t+j|t) = F_j(z^{-1}))y(t)+G_j(z^{-1})\Delta u(t-d+j-1)+L_j(z^{-1})\Delta p(t-d+j-1), \tag{5.47}$$

where $G_j(z^{-1}) = E_j(z^{-1})B(z^{-1})$ and $L_j(z^{-1}) = E_j(z^{-1})H(z^{-1})$.

Thus, the set of N-ahead predictions starting from d steps in the future is:

$$\begin{aligned} y(t+d+1|t) &= F_{d+1}(z^{-1}))y(t)+G_{d+1}(z^{-1})\Delta u(t)+L_{d+1}(z^{-1})\Delta p(t) \\ y(t+d+2|t) &= F_{d+2}(z^{-1}))y(t)+G_{d+2}(z^{-1})\Delta u(t+1)+L_{d+1}(z^{-1})\Delta p(t+1) \\ &\vdots \\ y(t+d+N|t) &= F_{d+N}(z^{-1}))y(t)+G_{d+N}(z^{-1})\Delta u(t+N)+L_{d+N}(z^{-1})\Delta p(t+N). \end{aligned} \tag{5.48}$$

The coefficients of F_j, G_j and L_j can be efficiently calculated by recursive methods [16] or by matrix calculations [17].

These predictions can be expressed in matrix notation as follows:

$$\mathbf{y} = \mathbf{G}\mathbf{u}+\mathbf{F}(z^{-1})y(t)+\bar{\mathbf{G}}(z^{-1})\Delta u(t-1)+\mathbf{L}(z^{-1})\mathbf{p}, \tag{5.49}$$

where

$$\begin{aligned} \mathbf{y} &= [y(t+d|t),\ldots,y(t+d+N|t)]^T \\ \mathbf{u} &= [\Delta u(t),\ldots,\Delta u(t+N-1)]^T \\ \mathbf{p} &= [\Delta p(t),\ldots,\Delta p(t+N-1)]^T \end{aligned} \tag{5.50}$$

$$\mathbf{G} = \begin{bmatrix} g_0 & 0 & \ldots & 0 \\ g_1 & g_0 & \ldots & 0 \\ \vdots & \vdots & \ddots & \vdots \\ g_{N-1} & g_{N-2} & \ldots & g_0 \end{bmatrix} \tag{5.51}$$

$$\bar{\mathbf{G}}(z^{-1}) = \begin{bmatrix} (G_{d+1}(z^{-1})-g_0)z \\ (G_{d+2}(z^{-1})-g_0-g_1z^{-1})z^2 \\ \vdots \\ (G_{d+N}(z^{-1})-g_0-g_1z^{-1}-\ldots-g_{N-1}z^{N-1})z^N \end{bmatrix} \tag{5.52}$$

$$\mathbf{L}(z^{-1}) = \begin{bmatrix} L_{d+1}(z^{-1}) \\ L_{d+2}(z^{-1}) \\ \vdots \\ L_{d+N}(z^{-1}) \end{bmatrix}^T, \quad \mathbf{F}(z^{-1}) = \begin{bmatrix} F_{d+1}(z^{-1}) \\ F_{d+2}(z^{-1}) \\ \vdots \\ F_{d+N}(z^{-1}) \end{bmatrix} \tag{5.53}$$

Parameters g_i, $i = 0,\ldots,N-1$ are the coefficients of G_{N-1} and can be obtained by solving (5.43) for E_j and F_j.

In order to obtain a prediction, it is necessary to know the future value for the disturbance. A common practice is to assume that it will have a constant value equal to the last measured value [17]. Taking this observation into account, Equation 5.49 can be written as:

$$\mathbf{y} = \mathbf{G}\mathbf{u} + \mathbf{f}, \tag{5.54}$$

where $\mathbf{f}$ is a vector composed of both controlled and control variables known at time t.

5.5.2 Generalized Predictive Control

In this section, a brief summary of a model-based control algorithm is given. A thorough description can be found in [16] and [18]. A practical presentation, with only a few equations but deep insight, is given in [19]. Chapter 7 provides a survey of current industrial control products based on model-predictive approaches.

The generalized predictive control (GPC) algorithm applies a control sequence that minimizes a multistage cost function of the form:

$$\begin{aligned} J = {} & \textstyle\sum_{j=N_1}^{N_2} \left[y(t+j|t) - r(t+j)\right]^2 \\ & + \lambda \textstyle\sum_{j=1}^{N_u} \left[\Delta u(t+j-1)\right]^2 \end{aligned} \tag{5.55}$$

subject to $\Delta u(t+j-1) = 0$ for $j > N_u$. The variable $y(t+j|t)$ is an optimum j-step-ahead prediction of system output on data up to time t. N_1 and N_2 are the minimum and maximum costing horizons, N_u is the control horizon, λ is the weight of future control actions and $r(t+j)$ is the desired trajectory.

Using the predictive model (5.54) with

$$\begin{aligned} \mathbf{y} &= [y(t+N_1|t), \dots, y(t+N_2|t)]^T \\ \mathbf{u} &= [\Delta u(t), \dots, \Delta u(t+N_u-1)]^T \\ \mathbf{f} &= [f(t+N_1), \dots, f(t+N_2)]^T \end{aligned} \tag{5.56}$$

and

$$\mathbf{G} = \begin{bmatrix} g_0 & 0 & \dots & 0 \\ g_1 & g_0 & \dots & 0 \\ \vdots & \vdots & \ddots & \vdots \\ g_{N_2-N-1-1} & g_{N_2-N-1-2} & \cdots & g_0 \end{bmatrix} \tag{5.57}$$

Equation 5.55 can be expressed as follows:

$$\begin{aligned} J = {} & \left[\mathbf{G}\mathbf{u} + \mathbf{f} - \mathbf{r}\right]^T \left[\mathbf{G}\mathbf{u} + \mathbf{f} - \mathbf{r}\right] \\ & + \lambda \mathbf{u}^T \mathbf{u}, \end{aligned} \tag{5.58}$$

where $\mathbf{r} = [r(t+N_1), \dots, r(t+N_2)]^T$.

Supposing that no constraints exist on the control signal, the minimum value of J can be found through the gradient zero:

$$\mathbf{u} = \left[\mathbf{G}^T\mathbf{G} + \lambda\mathbf{I}\right]^{-1}\mathbf{G}^T(\mathbf{r} - \mathbf{f}). \tag{5.59}$$

In order to deal with constraints and profit from efficient optimization programs such as linear programming (LP) algorithms, the objective function is recast in the following form:

$$\begin{aligned} J(\Delta u) = & \textstyle\sum_{j=N_1}^{N_2} \sum_{i=1}^{n} |y_i(t+j|t) - r(t+j)| \\ & + \lambda \textstyle\sum_{j=1}^{N_u} \sum_{i=1}^{m} |\Delta u_i(t+j-1)| \end{aligned} \tag{5.60}$$

where N_1 and N_2 define the costing horizon, N_u defines the control horizon and n and m are the number of outputs and inputs, respectively. Absolute values of the output tracking error and absolute values of control increments are taken from their squares, as is usual in GPC. If a set of auxiliary variables $\mu_i \geq 0$ and $\beta_i \geq 0$ are defined so that

$$\begin{array}{rcl} -\mu_i \leq & y_i(k+j|t) - r(k+j) & \leq \mu_i \\ -\beta_i \leq & \Delta u_i(k+j-1) & \leq \beta_i \\ 0 & \leq \sum_{i=1}^{n\cdot(N_1-N_2)} \mu_i + \lambda \sum_{i=1}^{m\cdot Nu} \beta_i \leq \zeta, & \end{array} \tag{5.61}$$

then ζ is an upper bound of $J(\Delta u)$ and the problem is reduced to minimize the upper bound ζ [16].

When constraints on output variables $[\mathbf{y}_{min}, \mathbf{y}_{max}]$, manipulated variables $[\mathbf{u}_{min}, \mathbf{u}_{max}]$ and slew rate of the manipulated variables $[\tilde{\mathbf{u}}_{min}, \tilde{\mathbf{u}}_{max}]$ are taken into account, the problem can be formulated as a LP problem with

$$\min_{\mathbf{x}} \; \mathbf{c}^T\mathbf{x} \quad \text{subject to} \quad \mathbf{A}\mathbf{x} \leq \mathbf{b}, \quad \mathbf{x} \geq \mathbf{0} \tag{5.62}$$

with

$$\mathbf{x} = \begin{bmatrix} \tilde{\mathbf{u}} - \tilde{\mathbf{u}}_{min} \\ \mu \\ \beta \\ \zeta \end{bmatrix} \qquad \mathbf{c} = \begin{bmatrix} \mathbf{0} \\ \mathbf{0} \\ \mathbf{0} \\ 1 \end{bmatrix} \tag{5.63}$$

$$\mathbf{A} = \begin{bmatrix} \mathbf{G} & -\mathbf{I} & \mathbf{O} & 0 \\ -\mathbf{G} & \mathbf{O} & \mathbf{O} & 0 \\ \mathbf{G} & \mathbf{O} & \mathbf{O} & 0 \\ -\mathbf{G} & \mathbf{O} & \mathbf{O} & 0 \\ \mathbf{I} & \mathbf{O} & -\mathbf{I} & 0 \\ -\mathbf{I} & \mathbf{O} & -\mathbf{I} & 0 \\ \mathbf{I} & \mathbf{O} & \mathbf{O} & 0 \\ \mathbf{T} & \mathbf{O} & \mathbf{O} & 0 \\ -\mathbf{T} & \mathbf{O} & \mathbf{O} & 0 \\ \mathbf{0} & \mathbf{1^t} & \mathbf{1^t}\lambda & -1 \end{bmatrix} \tag{5.64}$$

$$\mathbf{b} = \begin{bmatrix} -\mathbf{G}\tilde{\mathbf{u}}_{min} - \mathbf{f} + \mathbf{r} \\ -\mathbf{G}\tilde{\mathbf{u}}_{min} + \mathbf{f} - \mathbf{r} \\ \mathbf{y}_{max} - \mathbf{G}\tilde{\mathbf{u}}_{min} - \mathbf{f} \\ -\mathbf{y}_{max} + \mathbf{G}\tilde{\mathbf{u}}_{min} + \mathbf{f} \\ -\tilde{\mathbf{u}}_{min} \\ \tilde{\mathbf{u}}_{min} \\ \tilde{\mathbf{u}}_{max} - \tilde{\mathbf{u}}_{min} \\ \mathbf{u}_{max} - \mathbf{T}\tilde{\mathbf{u}}_{min} - \mathbf{1} \cdot u(k-1) \\ -\mathbf{u}_{min} + \mathbf{T}\tilde{\mathbf{u}}_{min} + \mathbf{1} \cdot u(k-1) \\ 0 \end{bmatrix} \tag{5.65}$$

where matrix $\mathbf{T}$ is a lower triangular matrix whose entries are ones, $\mathbf{O}$ is a matrix whose entries are zeros, $\mathbf{I}$ is the identity matrix and $\mathbf{1}$ is a vector composed of ones.

5.6 Model-based Control Strategies for Grinding Circuits

Developing a control strategy for a grinding circuit is an extremely complex task. This is due to the large number of variables involved in the process, the high degree of interaction between them, highly nonlinear and time-varying dynamics, long time delays and strong non-measured disturbances such as ore hardness. Two examples illustrate the use of a simulator for designing and testing model-based control strategies for grinding circuits.

5.6.1 Regulatory Control of a Conventional Grinding Circuit

Control of conventional grinding circuits has been widely studied, through both numerical simulation and plant experiments. Proposals for grinding control strategies include, for instance, PID regulators [20, 21]; multivariable control systems [21–23]; adaptive control [24, 25] and optimal control [26]. Most of these strategies try to maintain tight level control over the sump box by manipulating pump speed. This approach results in a highly coupled system, as inlet flow disturbances of the sump box are propagated towards the battery of hydrocyclones, producing undesirable transient dynamics, which can affect the recovery of valuable minerals in the rougher flotation stage.

This section describes a constrained model-predictive control system formulated as a LP problem. This dynamically optimizes the operation of the grinding circuit, enabling the incorporation of some operational practices usually seen at grinding plants, such as maximization of production at a constant fineness [27]. Furthermore, under this formulation, inlet flow disturbances from the sump box towards the hydrocyclones are decoupled in a natural way, achieving a smoother grinding circuit operation. The solution to the optimization problem results in a control strategy similar to the stabilizing strategy at Amax's Buick Mine, as described by Herbst [20].

The basic phenomenological model for a classical grinding circuit as depicted in Figure 5.10 consists of a rod mill model, a ball mill model, a sump model and pump-cyclone models.

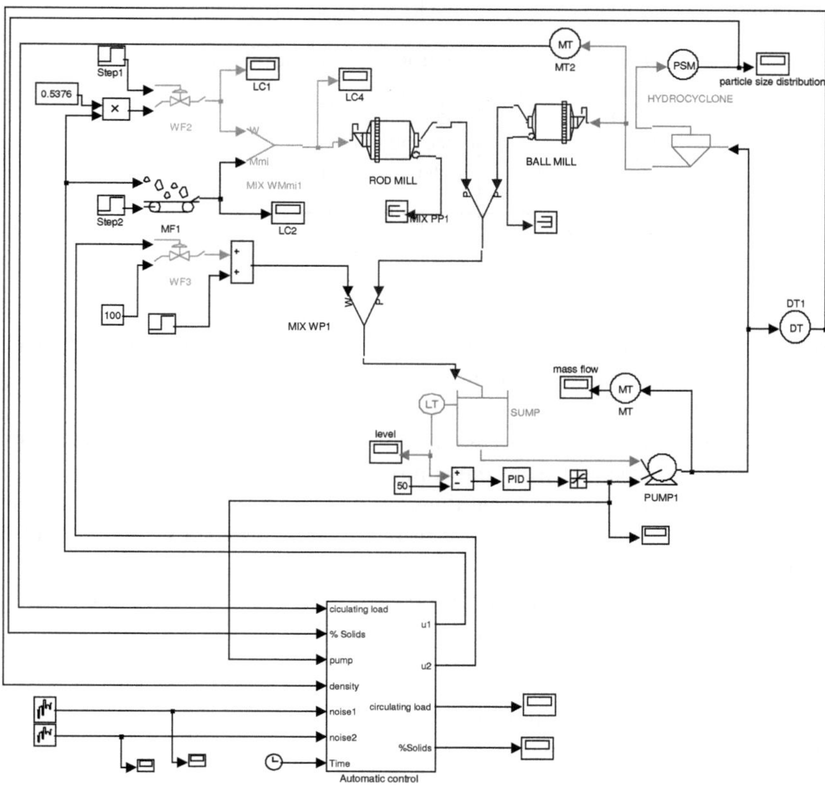

Figure 5.10 Block diagram of a conventional grinding circuit

The grinding process is represented by a multiple-input multiple-output (MIMO) system for which the following control inputs are considered:

1. water addition to the sump box (u_1);
2. fresh ore feedrate to the rod mill (u_2);
3. pump speed (u_3).

The outputs are:

1. product fineness (y_1);
2. circulating load (y_2);
3. sump level (y_3).

Water addition to the rod mill is a constant fraction of the fresh ore feedrate (m^3/h of water = ton/h of fresh ore/1.86) [24]. Sump box level is regulated with a

PI controller. Table 5.1 lists control and controlled variables and their corresponding controllers.

Table 5.1 Control and controlled variables and their corresponding controller

Control var.	Controlled var.	Controller
1. Pump speed	Sump box level	PI
2. Ore feedrate	Water addition to the rod mill	P
3. Water addition to the sump box	Product fineness	GPC
4. Ore feedrate	Circulating load	GPC

Process model parameters are identified by using a set of identification data and a least-squares method. Figure 5.11 shows controlled and control variables through several changes in set-points and ore hardness.

As a result of using a tight level control around the sump box, the speed pump propagates flow disturbances from the sump box towards the battery of hydrocyclones, producing undesired transient dynamics in the grinding circuit. These are emphasized when an increment in ore hardness occurs, as seen in Figure 5.11 at 410 min.

In this case, the grinding process is represented by a 3×2 MIMO system, in which the following control variables are considered:

1. water addition to the sump box (u_1);
2. fresh ore feedrate to the rod mill (u_2);
3. pump speed (u_3).

The outputs considered are:

1. product fineness (y_1);
2. sump box level (y_2).

Under this formulation, several operation strategies can be designed by changing weight parameters in the cost function to be optimized.

The control objective for a typical operating strategy used in grinding plants maximizes fresh ore feedrate to the rod mill and circulating load indirectly, considering constraints in product size, sump box level, fresh ore feedrate, speed pump, as well as the slew rate of control variables.

The control of circulating load is important, as high recirculation at a fixed product size means lower energy consumption [20]. However, circulating load is a non-measured variable most of the time. Therefore, sump box level and speed pump are taken as a good indication of the circulating load. A high level indicates maximum circulating load and a low level, minimum load [20]. Therefore, tonnage is increased to a point where sump box level constraints are met.

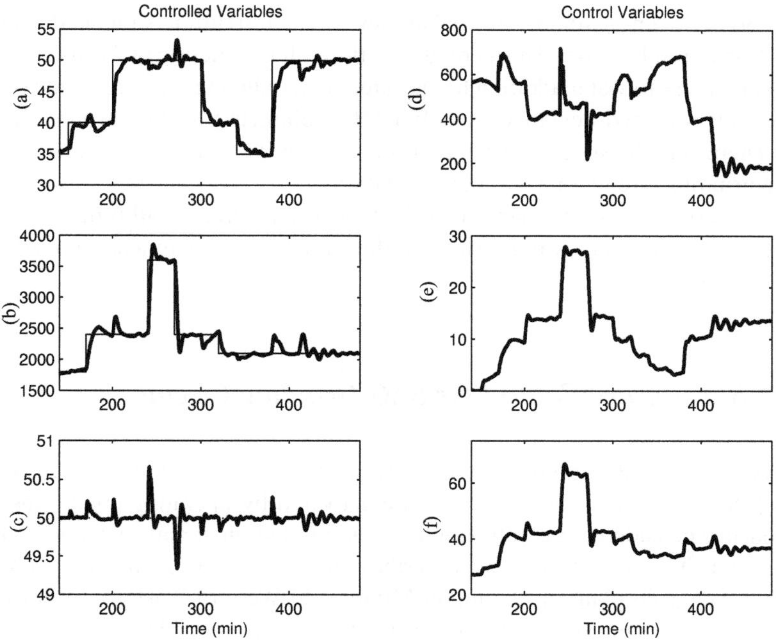

Figure 5.11 Controlled variables. (a) Product size (%). (b) Circulating load (ton/h). (c) Sump box level (%). Control variables. (d) Rod mill feedrate. (e) Water to sump box. (f) Speed pump

Under this formulation, the sump box level is used to decouple inlet disturbances from the sump box towards the hydrocyclones. The proposed control strategy maintains the speed pump at its maximum operating value, considering the number of hydrocylones and ball mill capacity. This fixes the volumetric flow Q towards the hydrocyclones, so that $d_{54}(c)$ is governed solely by the percentage of solids in the inlet to the hydrocyclones, which can be regulated through the fresh ore feedrate and with water addition to the sump box level, considering volumetric flow balance around the sump box. This formulation is particularly useful for rougher flotation stages which are extremely sensitive to particle size distribution .

The strategy is first tested without considering constraints in the fresh ore feedrate. Figure 5.13 shows the controlled and control variables, respectively. It can be seen from Figure 5.13 that the speed pump reaches its constraint level at 60% and remains there until 300 min, when an increase of the constraint occurs. During this time, the fresh ore feedrate is maximized and the the sump box level is maintained close to its maximum restriction at 80%. Therefore, circulating load is indirectly maximized, maintaining a desired product fineness, which means lower energy consumption. At 230 minutes, a change in the index function is made, increasing the weight parameters of the feedrate. Figure 5.13 shows that the fresh ore feedrate is increased until product fineness reaches its lower constraints and the sump box level remains close to its highest constraint.

At 362 min, an increment of ore hardness to the rod mill is simulated, causing a reduction to product size. This is regulated through a reduction of fresh ore feedrate and an increase to water addition to the sump box, reducing the percentage of solids to the battery of hydrocyclones, satisfying the volumetric balance around the sump box. Therefore, the speed pump remains at its upper constraints, achieving a more stable grinding circuit, one without oscillations.

As the control strategy maximizes the fresh ore feedrate at all times, it is necessary to constrain the maximum value of this variable, as it can lead to overloads in the rod mill.

5.6.2 *Supervisory Control for SAG Grinding Circuits*

The control of SAG grinding circuits has historically been done with expert systems [28], where a set of controllers are activated by operating conditions. Early approaches considered a combination of logic based on signal selectors and feedback controllers. Reference [29] describes an application of this approach at the Paddy's Flat gold plant of Dominion Mining Ltd Western Australia. With the advent of expert system shells, the logic was enhanced and coded as rules, which were easier to adapt to the different requirements of each installation. Some examples can be found in [30], where an application in Boliden Minerals Aitik concentrator is described. Additional examples can be found in two gold ore grinding circuits in South Africa, [31] and [32].

Recently, Thorton [33] has summarized the main challenges faced by control strategy design for SAG mills. He has proposed the use of both feedback and feedforward controllers to account for the effects of major plant dynamics and disturbances. Their implementation has led to a successful application (95% operator utilization) of conventional control techniques at two SAG milling circuits at Northparkes Copper Mines.

In recent years, model-based strategies have gained attention due to their successes in many process-based industries, such as the petrochemical industry. Early work on the application of model-based controllers to SAG mills can be found in [34] and [35]. Recent industrial applications of this approach have been carried by Ideas Simulation and Control [36].

The following sub-section is based on work carried out by Garrido [37]. It demonstrates the versatility of a simulator for studying process dynamic and advanced control strategies at SAG mills.

5.6.2.1 Open Loop Responses

As a first step, the simulator was configured for a simple SAG mill circuit, as depicted in Figure 5.14.

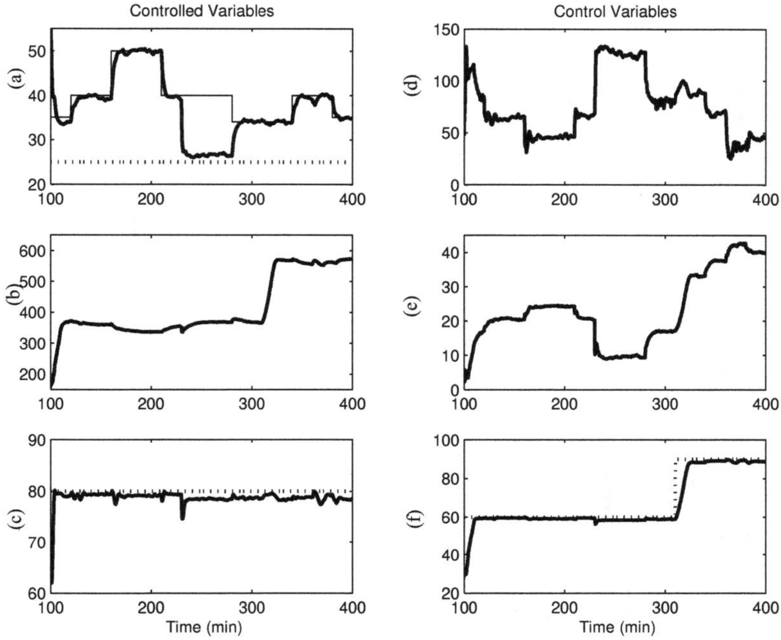

Figure 5.12 Controlled variables. (a) Product size (%). (b) Circulating load (ton/h). (c) Sump box level (%). Control variables. (d) Rod mill feedrate. (e) Water to sump box. (f) Speed pump

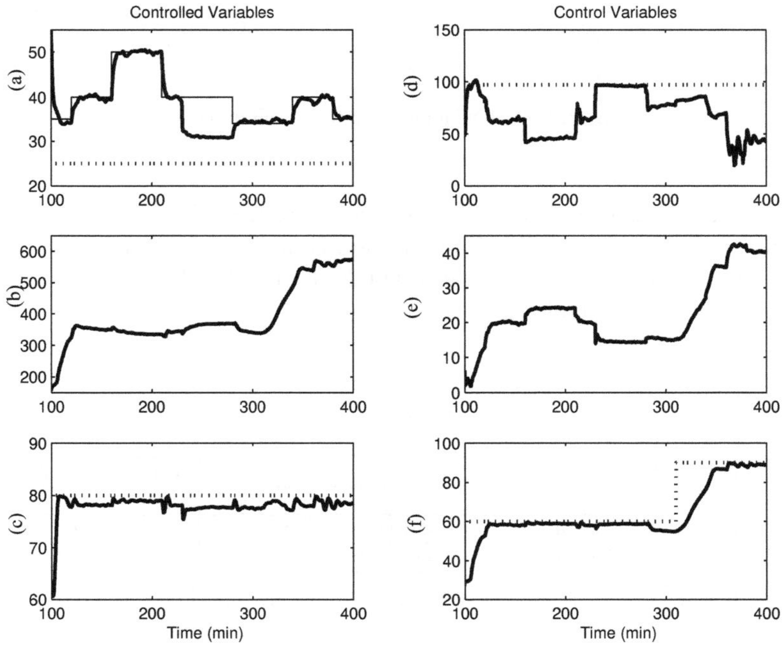

Figure 5.13 Controlled variables. (a) Product size(%). (b) Circulating load (ton/h). (c) Sump box level (%). Control variables. (d) Rod mill feedrate. (e) Water to sump box. (f) Speed pump

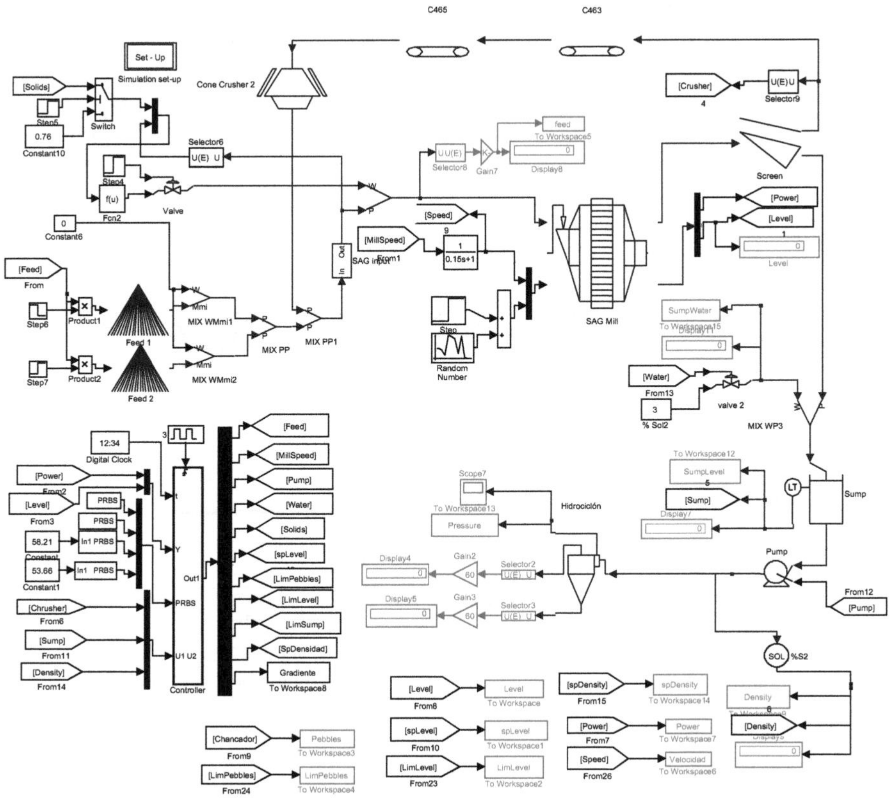

Figure 5.14 Simulation set-up for a SAG mill circuit

Simulation examples illustrate the effect of ore feed size distribution and ore hardness on mill filling level and power draw. Four feed size distributions are considered in order to demonstrate the effect of these variables. The different feed compositions are given in Table 5.2.

Table 5.2 Feed size distributions

Size distribution	Fines	Medium	Coarse
1. Normal	62.5	10.7	26.8
2. increased fines and medium and decreased coarse	64.02	13.34	21.64
3. increased fines and coarse and decreased medium	40.63	8.34	51.03
4. no medium size	65.1	0	34.9

In Table 5.2, fines are defined as being a size smaller than 50mm, medium as a size in the interval of [50, 1000]mm, and coarse for size bigger than 100mm. It is worth noting that these size distributions have been chosen solely to illustrate their effects. However, the distribution labeled normal corresponds to a real distribution obtained from a SAG mill. The simulation conditions were feed flowrate of 100 ton/h, water-mineral ratio controlled at a 76% of solids, speed at 9.5rpm and no changes to ore hardness. The simulation starts with distribution 1, at 300 minutes it changes to distribution 2, at 340 minutes to distribution 3 and finally, at 600 minutes to distribution 4.

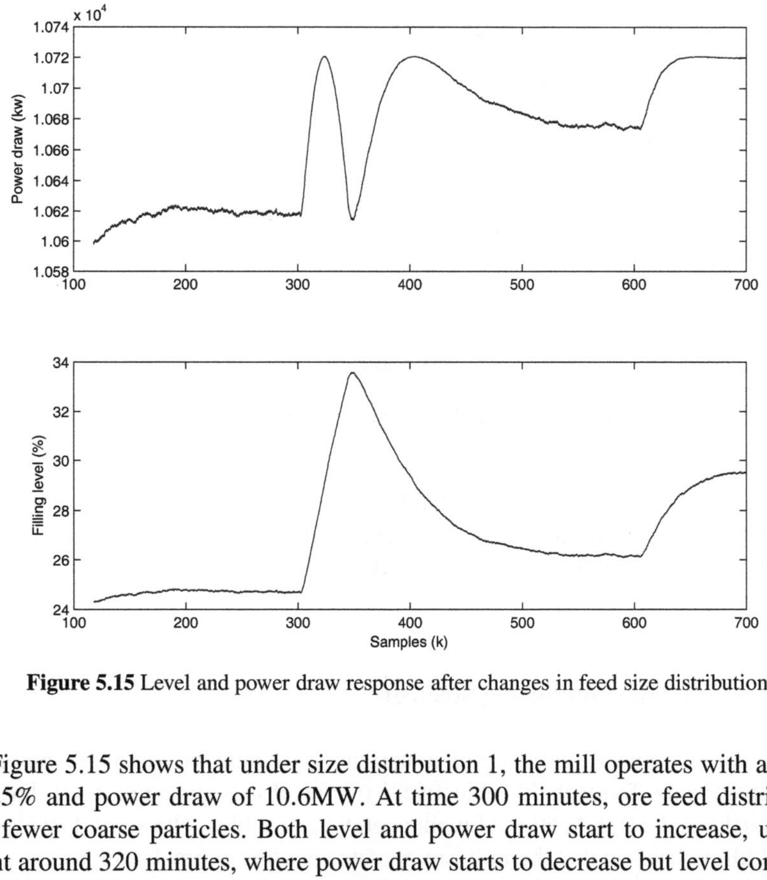

Figure 5.15 Level and power draw response after changes in feed size distribution

Figure 5.15 shows that under size distribution 1, the mill operates with a filling of 25% and power draw of 10.6MW. At time 300 minutes, ore feed distribution has fewer coarse particles. Both level and power draw start to increase, up to a point around 320 minutes, where power draw starts to decrease but level continues increasing. This is an undesirable operating region. At time 340 minutes, power draw starts to increase again, due to a distribution with coarser particles (distribution 3). After reaching a peak, power draw follows the same tendency as filling level. From time 600 minutes on, the feed does not have medium-sized particles. The mill increases filling level and power draw up to 29.7% and 10.72MW, respectively. These simulations show how sensitive SAG mill operation is to changes in feed particle size distribution. This variable represents one of the main unmeasurable disturbances affecting operation.

The next two simulations illustrate the effects of a 30% increase in hardness on two ore feedrates. Level and power draw for an ore feedrate of 1095 ton/h is depicted in Figure 5.16, where the change occurs at 200 minutes. Only after 50 minutes is there a variation in level and power draw, reaching a stable, steady-state.

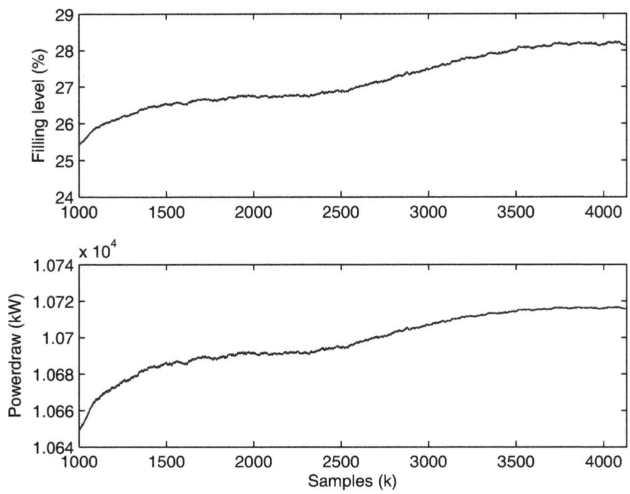

Figure 5.16 Level and power draw for a change in the ore hardness with a 1095 ton/h ore feedrate

However, if the ore flowrate is 1200 ton/h, the SAG mill enters a region where level and power draw do not follow the same tendencies. As level increases, power draw decreases, stabilizing at 30.5% and 10.71MW respectively, as seen in Figure 5.17.

5.6.2.2 A Two-level Optimizing Strategy

The goal of an optimizing strategy is to maximize feedrate without violating operational constraints. To reach this objective, a two-level strategy is proposed. The first level considers the stabilization of main operating variables. The second level sets a desired filling level so that the mill can operate near this optimal point.

The design of the control strategy considers a system composed of a SAG mill and a sump as a MIMO system with four inputs, five outputs and two unmeasurable disturbances as listed below.

The inputs are:

1. water addition to the sump box (u_1);
2. fresh ore feedrate (u_2);
3. pump speed (u_3);
4. mill speed (u_4).

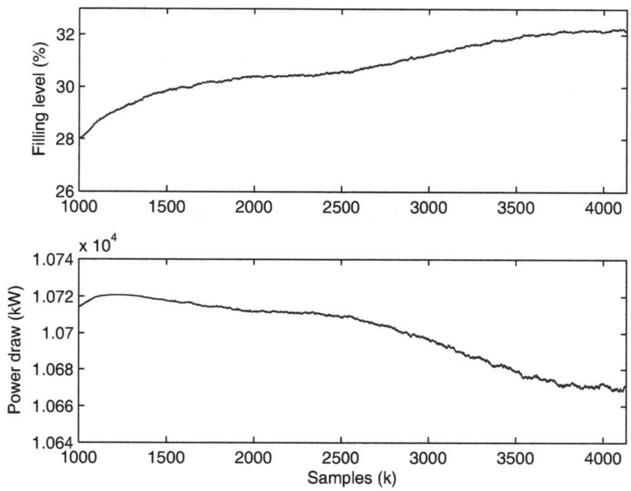

Figure 5.17 Level and power draw for a change in the ore hardness with a 1200 ton/h ore feedrate

Outputs are:

1. power draw (y_1);
2. pebbles flowrate (y_2);
3. sump level (y_3);
4. pulp density of sump output flowrate (y_4);
5. level filling (y_5);

and disturbances:

1. hardness (d_1);
2. feed particle size distribution (d_2).

In this application, all output variables are not required to reach a set-point. Some of them, for example the flow rate of pebbles, are only required to stay within a certain range.

In the first level, four control loops are implemented: inlet pressure in the hydrocyclones is controlled by the number of hydrocyclones in operation. (Therefore, this variable is considered as constant in this simulation.) Sump level is controlled by pump speed, considering a model in terms of pump speed, water flowrate and fresh ore feedrate. Density is controlled by the water feedrate and also considers ore feed rate as a measurable disturbance. Filling level is controlled by the ore feedrate and mill speed is subject to constraints in the pebble flowrate. The controller considers a model including mill speed, fresh ore feedrate and the pebble flowrate, which is modeled in terms of mill speed and fresh ore feedrate.

The stabilizing strategy was simulated by considering a constant particle size distribution and a ratio between water flowrate and ore feedrate of 76% as well as a change of 30% in ore hardness at time 250 minutes. Figure 5.18 shows the response of the strategy for different changes in the filling level set-point. Between 100 and 160 minutes, the set-point is smaller than the constraint and the controller is able to take the process at its value. However, at 160 minutes the set-point is larger than the constraint and the controller stabilizes the processes at the constraint. At 230 minutes, the constraint on the filling level is lifted and the controller is able to take the process to the reference point. The density is controlled at its set-point and is only perturbed when the controller activates the constraints, as seen in Figure 5.20. Finally, all supervised variables (pebble flowrate and sump level) are maintained in their normal operational ranges, as seen in Figure 5.19.

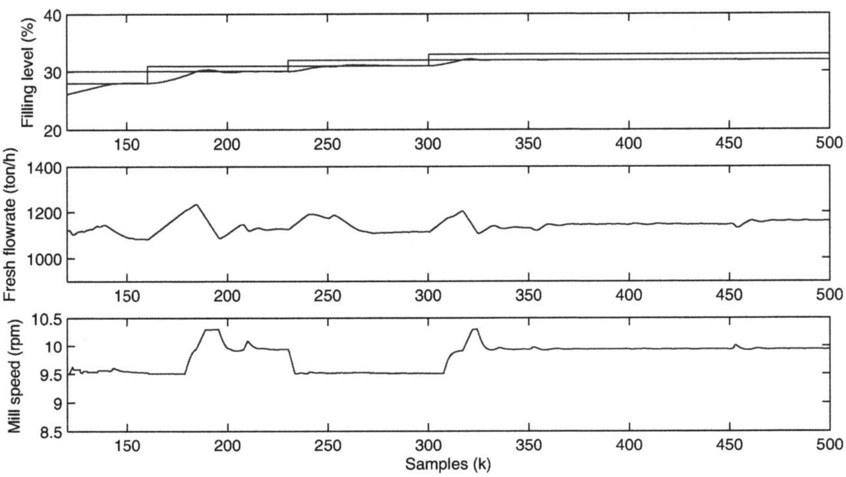

Figure 5.18 Stabilizing controller: filling level, fresh flowrate, mill speed

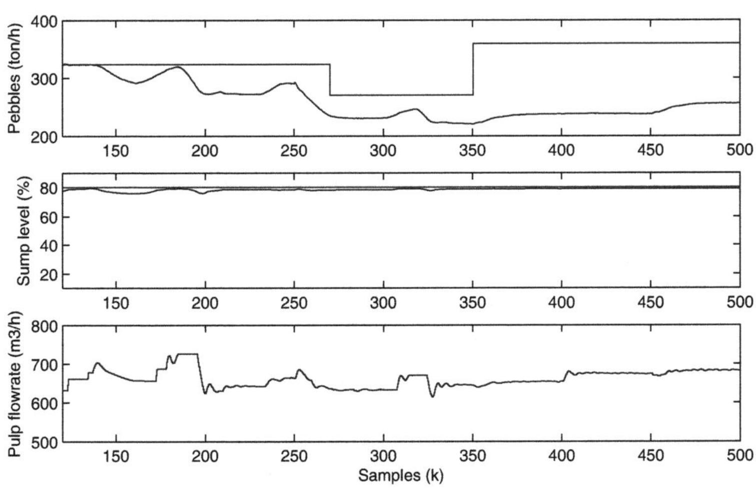

Figure 5.19 Stabilizing controller: pebbles flowrate, sump level, pulp flowrate

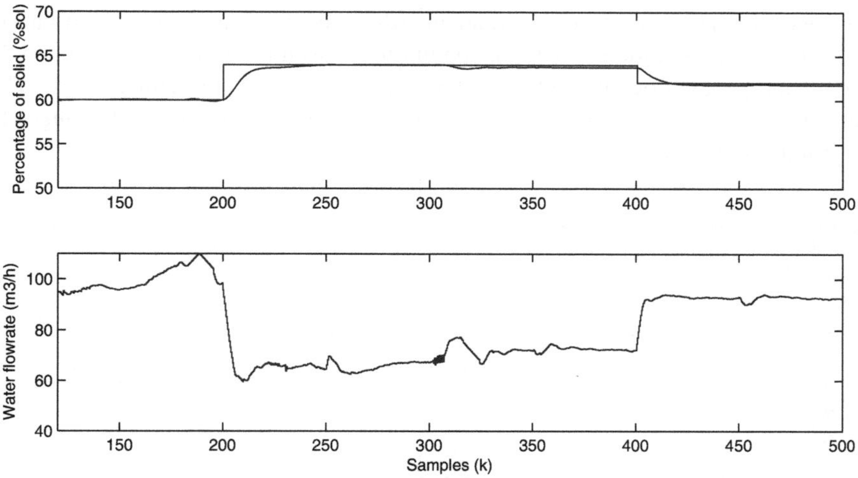

Figure 5.20 Stabilizing controller: percentage of solids and water flow rate

The supervisory level (second level) calculates a set-point for filling level so that the SAG mill always operates at a point near its maximum possible power draw. Power draw level is modeled as a quadratic function of mill filling level [37]. The strategy tries to maximize power draw while keeping all variables within defined constraints. If no constraint is active, then optimal operation will be obtained when the gradient of this quadratic function is zero.

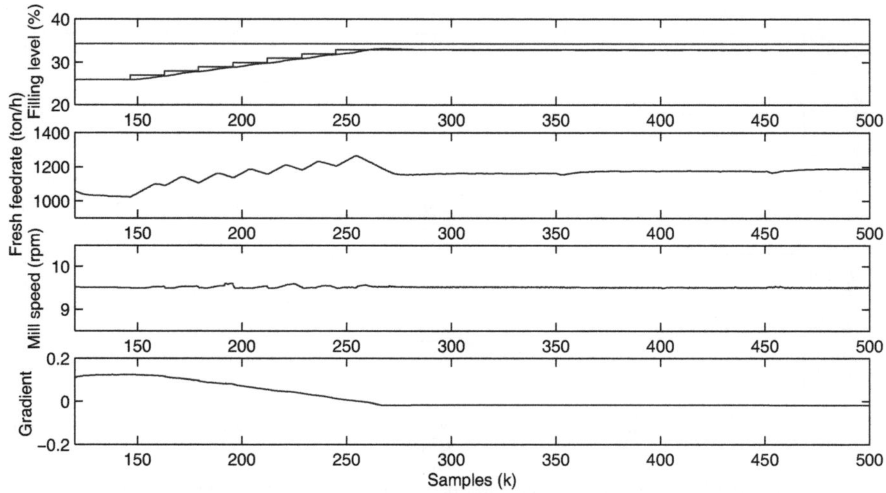

Figure 5.21 Optimizing controller: filling level, fresh flowrate, mill speed, gradient

Figure 5.21 shows how the supervisory level changes the set-point every 50 minutes so that ore feedrate is maximized. Gradient values near zero provide an indication of being at an optimal value. At time 250 minutes, ore hardness increases and the controller decreases the ore feedrate in order to compensate for this disturbance. This keeps mill operation at optimal conditions, as seen in Figure 5.21. Figure 5.22 shows that pebble flowrate and sump level stay within the pre-defined limits. The percentage-of-solids control loop is able to cope with interactions and changes to set-point, as seen in Figure 5.23.

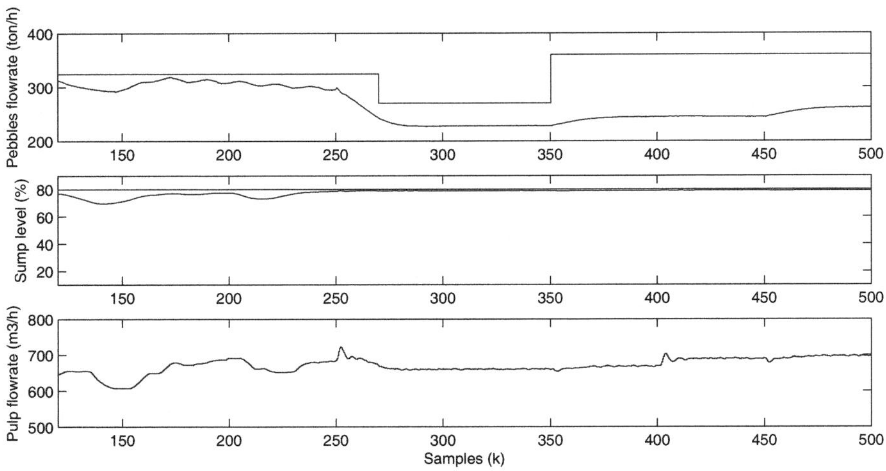

Figure 5.22 Optimizing controller: pebbles flowrate, sump level, pulp flowrate

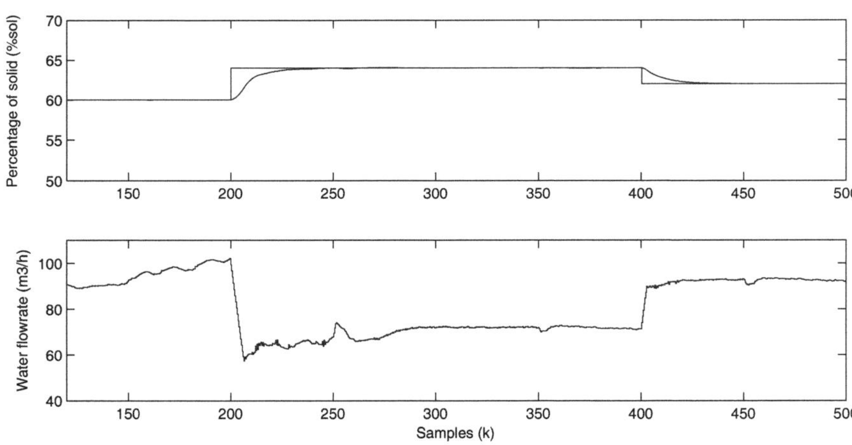

Figure 5.23 Optimizing controller: percentage of solids and water flowrate

5.7 Conclusions

This chapter has presented a dynamic simulator for comminution circuits. The implemented models are standard simplified models of the main equipments found in a comminution circuit, representing the main temporal features of the variables involved. A modular approach allows for future model upgrades, as well as the possibility of adding new ones. The use of this type of simulator simplifies the development and testing of complex control strategies. Two case studies have demonstrated the flexibility of the simulator and the potential of a simulation environment for blending different types of models. This tool not only enables the design of model-based controllers, but it can also be used for testing and debugging other types of control strategies such as those based on expert systems.

Real-time integration of this dynamic simulator with industrial control systems is also possible via object link embedding for process control. This standard specifies the communication of real-time plant data between different control devices and provides new opportunities for process and control engineers to improve plant knowledge and control strategies.

Acknowledgements This chapter would not have been possible without the diligent work carried out by my former students Johnatan Barriga, Fabian Urzua and German Palma. Special thanks to Miguel Maldonado and Cesar Garrido for their help in writing this chapter. I would also like to thank Andre Ribet for taking care of the figures.

References

[1] Mular AL, Barratt DJ, Halbe DN (2002) Mineral Processing Plant Design, Practice, and Control (2-volume set). Society for Mining Metallurgy & Exploration, New York
[2] Lynch AJ (1979) Mineral Crushing and Grinding Circuits: Their Simulation, Design, and Control (Developments in Mineral Processing Series, Vol.1). Elsevier Science Ltd. Amsterdam
[3] Napier-Munn TJ, Morrell S, Morrison RD, Kolovic T (1996) Mineral Comminution Circuits: Their Operation and Optimisation. JKMRC, University of Queensland, Brisbane
[4] Cipriano A, González G, Sbarbaro D (2007) Training in automation for optimising the profitability of technology investments. Final report P893A, Amira International
[5] Liu Y, Spencer S (2004) Dynamic simulation of grinding circuits. Minerals Engineering 17(11–12):1189–1198
[6] Salazar JL, Magne L, Acuna G, Cubillos F (2008) Dynamic modelling and simulation of semi-autogenous mills. Minerals Engineering 22(1):70–77
[7] Whiten WJ (1972) The simulation of chrushing plants with models developed using multple spline regression. Journal of the South African Institute of Mining and Metallurgy 72:257–264
[8] Hatch CC, Mular AL (1982) Simulation of the Brenda Mine Ltd. secondary crushing plant. AIME Transactions Mining Engineering 272:1354–1362
[9] Evertsson CM (2000) Cone crusher performance. PhD thesis. Department of Machine and vehicle design, Chalmers University of Technology, Goteborg, Sweden
[10] Grandy G, Fuerstenau DW (1970) Simulation of nonlinear grinding systems: rod mill grinding. Society of Mining Engineers AIME 247: 348–354

[11] Rajamani R, Herbst J (1991) Optimal control of a ball mill grinding circuit-I. grinding circuit modeling and dynamic simulation. Chemical Engineering Science 46(3):861–870
[12] Valery W, Morrell S (1995) The development of a dynamic model for autogenous and semi-autogenous grinding. Minerals Engineering 8(11):1285–1297
[13] Amestica R, Gonzalez GD, Menacho J, Barria J (1996)A mechanistic state equation model for semiautogenous mills. International Journal of Mineral Processing 44-45:349–360
[14] Magne L (1999) Mass transport in semiautogenous mills. PhD thesis Department of Metallurgical Engineering, Universidad de Concepción, Concepción, Chile
[15] Austin LG (1990) A mill power equation for SAG mills. Minerals and Metallurgical Processing 7(1):57–62
[16] Camacho EF, Bordons C (2000) Model Predictive Control. Springer Verlag
[17] Rossiter JA (2003) Model-based Predictive Control: a Practical Approach. CRC Press
[18] Maciejowski JM (2000) Predictive Control with Constraints. Prentice Hall
[19] Blevins TL, McMillan GK, Wojsznis WK, Brown MW (2003) Advanced control unleashed plant performance management for optimum benefit. ISA - The Instrumentation, Systems, and Automation Society
[20] Herbst J, Rajamani R (1980) Control of grinding circuits. In: Computer Methods for The 80's, 770–786
[21] Pomerleau A, Hodouin D, Gagnon E (2000) A survey of grinding circuit control methods: from decentralized PID controllers to multivariable predictive controllers. Powder Technology 108:103–115
[22] Hulbert DG, Koudstaal J, Braae M, Gossman GL (1980) Multivariable control of an industrial grinding circuit. In Proceeding of the 3rd IFAC Symposium on Automation in Mining, Minerals and Metal Processing, 311–322
[23] Duarte M, Sepulveda F, Castillo A, Contreras A, Lazcano V, Gimenez P, Castelli L (1999) A comparative experimental study of five multivariable control strategies applied to a grinding plant. Powder Technology 104:1–28
[24] Desbiens A, Pomerleau A, Najim K (1994) Adaptive predictive control of a grinding circuit. International Journal of Mineral Processing 41:17–31
[25] Najim K, Hodouin D, Desbiens A (1995) Adaptive control: state of the art and an application to a grinding process. Powder Technology 82:59–68
[26] Lestage R, Pomerleau A, Hodouin D (2002) Constrained real-time optimization of a grinding circuit using steady-state linear programming supervisory control. Powder Technology 124:254–263
[27] Maldonado C (2001) Constrained predictive control strategies for conventional grinding circuits. (in Spanish). Engineering thesis, Department of Electrical Engineering, Universidad de Concepción, Concepción, Chile
[28] Hales LB, Vanderbeek JL, Herbst JA (1988) Supervisory control of a semi-autogenous grinding circuit. International Journal of Mineral Processing 22(1–4):297–312
[29] Elber L, Thomson D, Johnson G (1994) Development of a control system for a SAG grinding circuit. Minerals Engineering 7(2/3):153–167
[30] Borell M, Bäckström P, Sijderberg L (1996) Supervisory control of autogenous grinding circuits. International Journal of Mineral Processing 44–45:337–348
[31] Bouche Ch, Brandt C, Broussaud A, van Drunick W (2005) Advanced control of gold ore grinding plants in South Africa. Minerals Engineering 18:866–876
[32] van Drunick WI, Penny B (2005) Expert mill control at AngloGold Ashanti. The Journal of The South African Institute of Mining and Metallurgy 105:498–506
[33] Thornton J, Pethybridge T, Rivett T, Dunn R (2006) SAG mill control at Northparkes Mines (Not So Hard After All). http://www.mipac.com.au
[34] Valenzuela J, Bourassa M, Najim K, Del Villar R (1994) Dynamic matrix control of an autogenous grinding circuit. Mineral Engineering 7(1):105–114
[35] Shang H (2006) Improving SAG mill performance using nonlinear MPC. Mineral Process, Modelling, Simulation and Control, editors T. Yalcin and H. Shang, Laurentian University, Sudbury, Canada June 6-7, 2006, pages 89–96

[36] Ideas simulation and control (2007) Brainwave SAG mill, product sheet rev. 6 august 2007
[37] Garrido C (2003) Model based control strategies for SAG mills. (in Spanish). Engineering thesis, Department of Electrical Engineering, Universidad de Concepción, Concepción, Chile

Chapter 6
Automatic Control of Flotation Columns

René del Villar, André Desbiens, Miguel Maldonado, Jocelyn Bouchard

Abstract Since their first commercial application for mineral separation in the early 1980s, flotation columns have become a standard piece of equipment in mineral concentrators particularly for cleaning operations. This chapter presents and discusses the most recent advances in instrumentation and automatic control of flotation columns. It also examines how current industrial practice could benefit from recent academic developments in these areas. A particular emphasis is placed on the development of specific sensors for the continuous monitoring of process operations and their regulation.

6.1 Introduction

Flotation is a commonly used method for separating valuable minerals (metal containing) from useless mineral (gangue). It consists in injecting air bubbles in a vessel where a finely ground mixture of minerals (ore) is also introduced as a slurry. Particles are maintained in suspension by means of an agitator in the so-called mechanical cells, and by a rising flow of bubbles in column-cells. If valuable mineral particles are adequately conditioned (hydrophobicized), they should attach to the rising bubbles forming bubble–solid aggregates and move up as result of the difference in specific gravity between pulp and mineralized bubbles. These aggregates finally

René del Villar
Department of Mining, Metallurgical and Materials Engineering, Université Laval, Québec, Canada, e-mail: rene.delvillar@gmn.ulaval.ca

André Desbiens, Miguel Maldonado
Department of Electrical and Computer Engineering, Université Laval, Québec, Canada, e-mail: desbiens@gel.ulaval.ca, e-mail: miguel.maldonado.1@ulaval.ca

Jocelyn Bouchard
XStrata Process Support, Falconbridge, Ontario, Canada, e-mail: jbouchard@xstrataprocess support.ca

reach the top of the vessel as a stable froth, where they are removed, usually constituting the valuable product (concentrate). The non-floatable particles (hydrophilic), which do not attach to the bubbles, settle and leave the vessel through the bottom port as non-valuable product called tailings. This separation scheme is called direct flotation. In some rare cases, the tailings stream may constitute the valuable product, whereas the concentrate is the discarded product (inverse flotation). In some flotation devices (*i.e.*, columns) the froth is sprayed with clean water, which, if adequate conditions prevail, moves downwards (bias flow) cleaning the froth from undesired entrained particles.

Despite the highly sophisticated devices (instrumentation and data acquisition) often found in mineral processing plants, the huge amount of data usually accumulated in their historians is only partially used. Their utilization for automatic control purposes is rather limited, especially when compared with other types of process plants such as chemical or petrochemical industries

For instance, typical regulatory control strategies implemented in industrial columns are basically limited to: (a) controlling the froth depth by manipulating the tailing flow rate (to indirectly influence the concentrate recovery); (b) manipulating the gas rate to attain the desired recovery. Since the bias rate is not currently measured, the cleaning action of the froth zone is, at most, manually controlled by manipulating the wash water rate. Besides gas rate, none of the gas dispersion properties, nor the bias rate are used for automatic control purposes.

However, some control strategies based on gas dispersion properties as secondary variables, such as gas hold-up and gasrate, have been proposed by academic researchers. Recent studies have shown that another gas dispersion property, called "bubble surface area flux (S_b)", a combination of gas rate and some average bubble size, is strongly related to flotation performance. Thus, an extended control approach considering S_b is worth considering. This means controlling bubble size and gas rate to get a desired S_b value. All these matters will be discussed in Sections 6.4 and 6.5.

The metallurgical performance of a flotation column is determined by the valuable-mineral concentrate grade and recovery, often called primary variables, as they determine the revenue the plant is able to get from the sale of its concentrate. Whereas the grade can be measured on-line using an X-ray on-stream analyzer (OSA), the recovery must be estimated from steady-state mass balance, which usually introduces some sort of estimation error, therefore strongly limiting its use for control purposes. The long sampling times and cost of these OSA devices, usually multiplexed, also favour the use of a hierarchical control. In such a scheme, secondary variables, ideally strongly correlated with the primary ones and measured or at least precisely estimated, are controlled. In the specific case of the flotation column, the interface position (froth depth), bias rate and gas hold-up or bubble surface area flux are convenient secondary variables. Ideally, the controller must adequately deal with the interactions between the variables and take into account various operational constraints. The set points of the secondary variable controllers are optimally calculated by a supervisory control strategy which optimizes an economic criterion. This cost function is based on the relationships between the secondary and primary variables, the operating costs and the market value of the sellable product (*e.g.* the Net Smelter Return).

It is evident that to develop some form of real-time optimization (RTO) for a given process, some previous steps must first be completed, namely to dispose of: (a) physical or virtual sensors for measuring the critical variables of the process; (b) efficient control strategies able to adequately face the inherent complexity of multi-variable processes with interactions; (c) mathematical tools to continuously modify the secondary variable set points to maximize the economic profit of the considered process. Many industrial and/or academic research groups have been working on these subjects. The LOOP group (Université Laval) has been especially interested in the application of all these aspects to the optimization of flotation column operation. Their experience and that of other research groups (particularly the McGill Mineral processing group on sensors) will be described in this chapter. The chapter is organized as follows. First, Section 6.2, Process Description, presents a brief description of the column flotation process and its main operating variables. Then, Section 6.3, Sensor Development and Applications discusses the available methods for measuring the main variables specific to flotation columns. Section 6.4, Automatic Control discusses current approaches for automatic control. Finally, Section 6.5, Future Trends and Industrial Applications, reviews possible research orientations of flotation column sensors and control research and opportunities for industrial applications.

6.2 Process Description

The flotation column being considered in this chapter was conceived by P. Boutin and R. Tremblay in the early 1960s [1, 2], and was first commercialized by Column Flotation of Canada (CFC) in the 1970s [3]. The first successful industrial application was at Les Mines Gaspé (North-East Québec, Canada) in 1981 [4]. Since then, this unit has acquired great popularity in the mineral separation business owing to its capability for producing clean concentrates at good recoveries. Although the original CFC column had a squared section (36 in or 72 in), present columns are of a circular section, typically 3 m in diameter. Large capacity plants though, have chosen rectangular shapes, internally baffled to approach plug flow conditions (*e.g.*, 20 m by 4 m baffled at 1 m by 1 m). A schematic representation of a flotation column is given in Figure 6.1. A conditioned pulp stream is fed at 2–3 m from the top of the column, whereas a flow of gas (usually air) is injected at the bottom through a bubbling system or sparger. A unique feature of the flotation column is the addition of a fine spray of water above the overflowing concentrate stream; its objective will be explained later. Under normal operating conditions, the column exhibits two distinct regions, each having a different objective and air content (volume fraction). The lower zone, between the feed port (roughly) and the air bubbling system at the bottom, typically contains less than 20% of air. The zone between the feed port and the column overflow exhibits much greater air content (above 70%); thus it is named the froth zone.

As mentioned before, water is usually sprayed on the overflowing froth. Provided adequate conditions are met (water balance in the upper zone and gas rate value), part of this water finds its way down to the bottom, washing down from the froth zone any liberated hydrophilic particles that could have been hydraulically entrained there by the ascending bubbles. In this form, the overflow concentrate will be freed from undesired gangue particles, thus improving its final grade. As a result of this washing action the froth zone is also called the cleaning zone.

The *sine qua non* condition for an effective flotation separation is the different behavior with respect to water and air, of the various mineral components of the processed ore. Some particles, typically the valuable mineral, will attach themselves to air bubbles (hydrophobic) whereas others, typically the undesired minerals or gangue, will remain in the water phase (hydrophilic). Almost all minerals are naturally hydrophilic. To render some minerals hydrophobic, while keeping hydrophilic the remaining ones, various chemical reagents are generally required: collectors, activators, depressants and pH modifiers. These chemicals are normally added in the preliminary stage of pulp conditioning. However, chemical reagents may be directly added to the flotation device. The reagent addition has obviously an important impact on the metallurgical performance as it will determine the ability of the valuable particles to be "collected" by the bubbles and thus transferred to the concentrate. Air bubble properties also play an important role in the collection process; for instance smaller bubbles exhibit a greater specific area (area/volume) and thus have a greater particle-carrying capacity. Bubble diameter also influences the ascending speed of bubbles, whereas its volume affects the buoyancy characteristics of the particle–bubble aggregates. The reagent used to modify the bubble properties, acting only upon the air–water interface, is the so-called frother, typically a water-soluble alcohol. Preliminary laboratory testing is usually carried out to determine the proper dosage of all these reagents.

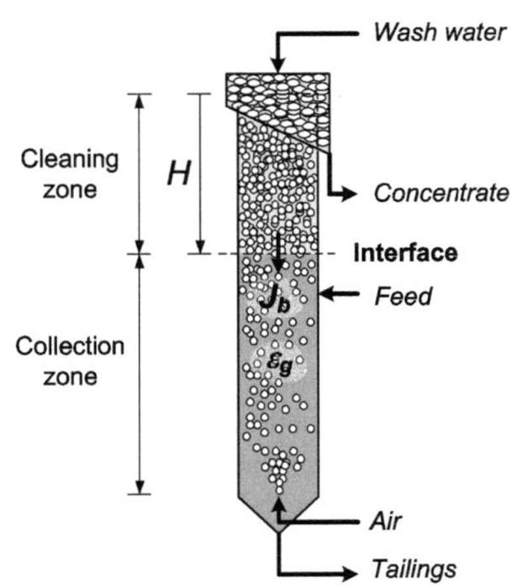

Figure 6.1 Schematic representation of a flotation column

Flotation column operation is characterized by specific variables, namely froth depth, bias rate, gas rate, gas hold-up, bubble size and bubble surface area flux. The last three variables, known as gas dispersion properties, partly define the hydrodynamic conditions which govern the metallurgical performance, and as such, current research has centered on gas dispersion as being the key hydrodynamic characteristic of flotation. All these variables are defined hereafter and their measurement is explained in more detail in Section 6.3.

6.2.1 Froth Depth

As its name indicates, this variable (H_f) corresponds to the distance between the top of the column (overflow) and the position of the pulp–froth interface. It determines the relative height of cleaning and collection zones, and consequently, the mean pulp residence time in the collection zone of the column. As such, it strongly contributes to the value of the collection zone recovery, although other variables such as gas flow rate, bubble diameter, *etc.* also have an influence. As a result of its integrator behavior, it must be closely monitored and controlled for stable column operation. This variable is currently measured by devices described in Section 6.3 and its regulatory control loop is normally implemented in most mineral separation plants (concentrators).

6.2.2 Bias Rate

Perhaps the most characteristic feature of flotation columns is the addition of a fine spray of water, called wash water, at the top of the device, on the surface of the overflowing concentrate stream. In addition to stabilizing the froth and facilitating the removal of the bubble–particle aggregates from the concentrate launder, this stream washes down to the pulp zone, hydrophilic particles that might have been entrained with these aggregates. This washing action can only take place if a proper water mass balance is assured in the lower part of the column, *i.e.*, the collection zone, so that part of this wash water can make its way down the froth zone.

This net downward water stream is called the bias rate [5]. For scaling purposes, it is commonly expressed as a superficial bias rate (J_b), *i.e.*, the bias flow rate (cm^3/s) divided by the column cross-sectional area (cm^2). The bias rate is said to be positive if water moves down the column, and negative otherwise. When this latter case occurs, the concentrate becomes stained with entrained hydrophilic particles and the grade decreases, which is a non-desirable situation. While some authors claim that a slight, but positive, bias rate ensures an adequate cleaning action [5], others have established that small values do not contribute to reducing the entrainment of gangue [6]. On the other hand, high bias rates (above 0.4 cm/s) are detrimental since they might generate increased froth mixing and thus, drop-back of collected particles [7]. During column operation, bias rate can become negative as a result of an excessive gas flow rate, entraining most of the wash water over the top of the column, diminishing the cleaning action of the froth zone. On the other hand, increasing bias rate would affect the residence time of ascending bubbles and that of settling particles in the collection zone, thus altering the recovery. Its action may also reflect on the froth zone recovery. Unfortunately, all these observations are mostly qualitative since bias rate measurements, particularly in three-phase systems, are not very reliable, as shown later on.

6.2.3 Gas Hold-up

This represents the volumetric fraction of gas in the considered zone (froth or collection). It is estimated through Equation 6.1, where the term volume might refer to the whole zone (overall gas hold-up) or part of it (local gas hold-up):

$$\varepsilon_g(\%) = 100 \cdot \frac{\text{Volume of bubbles}}{\text{Total volume}} \tag{6.1}$$

Since gas is often considered as a flotation "reagent", and thus influencing the concentrate recovery, plant metallurgists have been interested in monitoring the gas dispersion within the column using the air hold-up. Its value strongly depends on prevailing values of gas rate, frother concentration, bias rate and solid concentration.

6.2.4 Bubble Size

This is defined as the diameter of an equivalent spherical bubble (D_b). Note that bubbles in different sizes give rise to a bubble size distribution (BSD) as shown in Figure 6.2(b) here represented by a histogram. The collection of mineral particles by bubbles greatly depends on the amount of bubble surface available as they pass through the column. Therefore, both their surface and their ascending speed are important. Consequently, a more adequate way of monitoring the influence of gas on the flotation process is to use the amount of bubble surface area per unit time of a cross-sectional area of the column instead of the gas hold-up. This variable, called bubble surface area flux, or S_b, can be evaluated from the number of bubbles per unit time (n_b), the surface of one bubble (A_b) of diameter (D_b), and the cross-sectional area of the column (A_c) as shown in the left part of Equation 6.2 [5].

$$S_b = \frac{n_b A_b}{A_c} = \frac{6J_g}{D_b}. \tag{6.2}$$

Figure 6.3 illustrates the bubble surface area flux concept. Assuming a suitable mean bubble diameter, it can be estimated by the right part of Equation 6.2, where

J_g is the gas superficial velocity (gas volumetric flow rate divided by column cross-sectional area, m/s). Typically, the Sauter mean diameter (D_{32}) is used.

However, industrial bubbling devices produce a rather broad BSD, making the use of a single mean diameter inaccurate, since the same value could be obtained from quite different size distributions. This point will be discussed in Section 6.3. Recent studies made by Gorain *et al.* [8] have shown that flotation performance, expressed in terms of the flotation kinetic constant, is strongly related to bubble surface area flux. It was demonstrated that neither the bubble size, nor the gas hold-up, nor the gas superficial velocity, independently considered, could adequately correlate to the kinetic constant. On the other hand, it was shown that the bubble surface area flux exhibited a linear relation with the kinetic constant, at various impeller speeds. Similar results were later reported by other researchers [9, 10]. Although S_b seems a potential control variable to achieve a given metallurgical performance, its present definition disregards important information related to the shape of the BSD, such as multi-modal, narrowness and tail properties. Therefore, formal control strategies should take into account the S_b value in conjunction with the shape of the BSD. This issue will be discussed in Section 6.5. Controlling the bubble diameter implies manipulating some key variables, such as the gas superficial velocity, the frother concentration, and in some cases the bubble generator (either the pore size for laboratory porous spargers, or the output orifice size in the case of some commer-

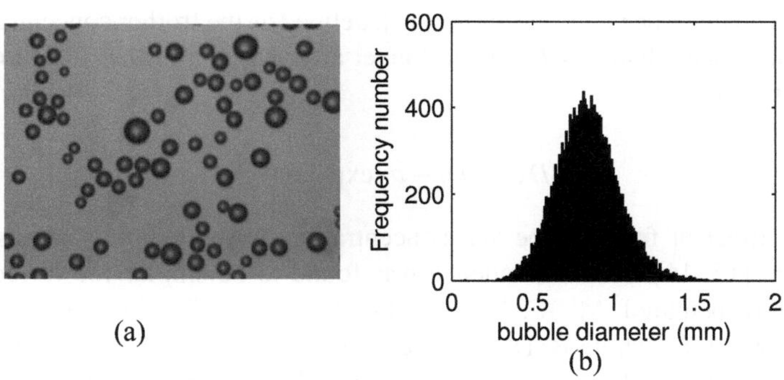

Figure 6.2 (a) Picture of laboratory column bubbles. (b) Histogram representation of bubble size

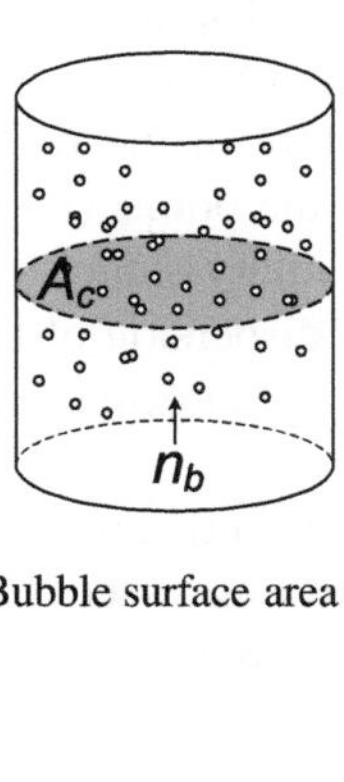

Figure 6.3 Bubble surface area flux concept

cial bubblers). The gas superficial velocity can easily be measured using commercially available gas flow meters. However, the frother concentration measurement presents a major challenge, since no sensor exists for a continuous measurement. Some alternatives are possible though, as will be explained later. Nesset *et al.* [11] proposed empirical models relating D_{10} and D_{32} to both superficial gas velocity and frother concentration for forced air mechanical cells. Both models are only valid under steady-state conditions where a large number of bubbles are counted, typically 10000.

$$D_{32} = D_0 + C \cdot J_g^n. \tag{6.3}$$

In this relationship D_0, C and n are parameters to be adjusted. This is a modified version of that proposed by Dobby and Finch [12] for porous spargers in columns, where the term D_0 was not included. In this way, the S_b can be estimated for a given frother concentration as follows:

$$S_b = \frac{6 \cdot J_g}{D_0 + C \cdot J_g^n}. \tag{6.4}$$

To model the effect of frother concentration on bubble size, an empirical model was proposed by Finch *et al.* [13]. An exponentially decreasing function plus a constant was proposed for modeling the Sauter mean diameter D_{32} in terms of the normalized frother concentration (Equation 6.5). In this equation, D_I, a, b are adjustable parameters, C is the frother concentration and CCC is a parameter called "critical coalescence concentration" [14], defined as the frother concentration where the Sauter mean diameter D_{32} is no longer affected by a further increase in frother concentration.

$$D_{32} = D_I + a \cdot \exp\left\{ - \frac{b \cdot C}{CCC} \right\}. \tag{6.5}$$

The effect of frother type and concentration on the bubble size has also been explored [15, 16]. This dependence was found to be similar for various types of frothers (n-pentanol, MIBC, DF-250, F-150, F-140). The impact of frother concentration on the whole BSD is only described qualitatively by looking at the histograms generated. It is observed that an initially bi-modal BSD at very low frother concentration becomes narrower and a uni-modal shape is obtained as the frother concentration increases. In these works, no information regarding the shape of the BSD was provided, as histograms are only being used for visualization purposes. To gain some insight into the nature of the BSD, a plot of D_{32} against D_{10} was proposed by Nesset *et al.* [11]. If the points (D_{32}, D_{10}) belonged to a very narrow distribution, they would sit on a unitary-slope straight line, the scatter from such a line would indicate the degree of dispersion of the size distribution. However this type of analysis does not provide any information about the actual size distribution shape.

An adaptive observer has been recently proposed to on-line estimate frother concentration [17]. It consists of identifying a set of linear dynamic models representing the effect of gas rate on the collection zone gas hold-up at different known frother concentrations (calibration step). An adaptive law is then obtained using Bayes' rule to evaluate the conditional probability of each model representing the observed system behavior. In this way, a frother concentration estimate is calculated as a weighted sum of model probabilities and their associated (known) frother concentrations. This method has been successfully tested in a two-phased system working with Dowfroth and MIBC frothers. Extension to three-phase systems (solid–liquid and gas) must consider the effect of solids on gas hold-up. Another concern is related to the well-known fact that some collectors or even salted water may partly act in some degree as frothers. Unknown changes in collector dosages or salt content in process water after calibration will change the frother estimate given by the proposed method. Therefore, the algorithm will estimate an "apparent" frother concentration, not necessarily resulting from the effect of frother itself.

6.3 Sensor Development and Applications

As mentioned in the beginning of Section 6.2, the first successful commercial application of the flotation column was in Canada where it was invented. Not surprisingly, the bulk of the research was initially done in Canada or more precisely at McGill University, where a great deal of work was accomplished in the 1980s on column flotation modeling, simulation and scale-up. In the second part of that decade, McGill research work focused on the development of specific sensors for adequately monitoring column operation. Various graduate students who worked there during these years, returned to their original countries or other cities in Canada, and the initial research work was thus spread world-wide. In the 1970s, a research group (GRAIIM) was formed at Université Laval to work on computer applications for the mineral industry (modeling, simulation, data reconciliation by material balances, *etc.*) This same group later became what is now known as LOOP, interested in observation and optimization techniques, particularly automatic control. In the 1990s, these two groups (Laval and McGill) began a close collaboration regarding sensor development and their use for automatic control of flotation columns.

For more than two decades, researchers from both McGill University and Université Laval have been working on the development of specific sensors for the measurement of flotation column operating variables: (a) the pulp-froth interface position (hereafter simply refer to as interface position); (b) the bias rate; (c) the gas hold-up, (d) the bubble diameter and its distribution (BSD); and (e) the bubble surface area flux. The sensors developed for the first three variables are based on the use of pulp electric conductivity, whereas the bubble distribution sensor (McGill Bubble Viewer) applies an image analysis technique to pictures of bubbles. The Laval research group has concentrated its efforts on the development of the first two sensors, whereas the McGill group has focused on the last three.

This section and Section 6.4 will present some of their achievements in sensor development and process control, as well as some others resulting from research work elsewhere in the world.

6.3.1 Interface Position Sensor

A combination of a float and an ultrasonic sensor to measure the froth depth is now commonly used in industrial flotation plants because of the simplicity of its concept and its non-invasive implementation (the sensor's electronics are not in contact with the pulp). Even though its accuracy is limited by the assumption of uniform pulp and froth densities and by the absence of the accumulation of solids on the float gage, its use seems satisfactory to industrial flotation column practitioners. Pressure gages have also been used to measure the interface position [5]. Theoretically, a single pressure transducer should suffice, as long as the average value of collection zone (ρ_c) and froth zone (ρ_f) specific gravity are known, as shown in Equation 6.6, where H_P indicates the position of the pressure transducer measured from the column lip, P the pressure reading, and g the gravitational constant:

$$H = \frac{\rho_c g H_P - P}{g(\rho_c - \rho_f)}. \tag{6.6}$$

Unfortunately, ρ_c and ρ_f depend on time-varying operating conditions and consequently froth depth values calculated from initial pressure transducer calibration may not represent later operating conditions.

The use of multiple pressure transducers partially solves this problem since it provides the extra necessary information to estimate the average value of both densities at prevailing operating conditions. Normally three pressure transducers are used, one located in the uppermost part of the froth zone, and the other two in the collection zone, one close to the interface and the other well below. If one assumes that the collection and froth zone specific gravities are uniform up to or from the interface, often an unrealistic situation [18], the resulting three-equation system can be solved for H as shown in Equation 6.7, where P_i and H_i are the pressure measurement and the position (distance) of transducer i, measured from the top of the column:

$$H = H_3 \frac{H_1(P_1 - P_3) - P_1(H_1 - H_2)}{H_3(P_1 - P_2) - P_3(H_1 - H_2)}. \tag{6.7}$$

Despite the improvement in froth depth estimation provided by this method, some problems still remain unsolved. Assuming a uniform specific gravity in each zone might lead to incorrect results, particularly in the case of the froth density, which is very dependent on operating conditions [18]. Moreover, the solution of the system of equations requires that the interface be located above the intermediate transducer otherwise the system of equations becomes underdetermined. To overcome these problems, some researchers have looked at methods based on intensive properties of the pulp involved. Moys *et al.* [19] first reported a method for measuring froth depth based on conductivity and temperature profiles across the interface. Since the former has been studied in depth, it will be discussed in the next few paragraphs. To measure the conductivity profile across the interface, Gomez *et al.* [20] proposed a probe consisting of a series of modules mounted around a stainless steel pipe. Each module is made of a non-conductive PVC cylinder machined to support a stainless steel ring electrode. The conductivity value of each consecutive pair of electrodes (conductivity cell) is sequentially measured. Once all conductivity cells have been scanned, the conductance values are processed by an algorithm that searches for the largest change in the conductance profile ("largest slope"); then the interface position is assigned to the middle position of the cell generating the largest slope. The accuracy of this "largest slope" algorithm is strongly related to the separation between electrodes. It has been reported though, that the largest conductivity change gives not one but two possible cells containing the interface [21].

The accuracy problem of the largest slope method was tackled by Uribe-Salas *et al.* [22], by using a "fairly linear relationship" between conductance and the interface position, enabling data interpolation to increase the accuracy of interface detection. Although good results were obtained using this technique, its major drawback is the extensive experimentation required to obtain the linear relation between cells' conductance and actual interface position, which obviously has to be known. To describe the nonlinear conductivity profile across the interface (rotated S-shape), Pérez *et al.* [23] proposed the use of an artificial neural network (ANN). Although this method provides excellent estimates, it is not suited for practical application because it requires extensive experimentation for the ANN training, where measuring the interface position by an independent method (*e.g.*, visual observation of transparent laboratory columns or a properly calibrated probe in industrial columns) is a must. In order to avoid the determination of linear equations in each conductivity cell of the probe or the extensive experimentation associated with ANN training, Grégoire proposed a semi-analytical method in which the cell containing the interface is represented as a series of two electrical resistances, corresponding to its two components: froth and pulp [21]. If the conductivity of the adjacent cells (above and below) and the cell containing the interface are known, then these two conductivities (froth and pulp) can be estimated, and the exact interface position, X, within that cell can be calculated.

As previously mentioned, the detection of the cell containing the interface is fundamental to the success of this approach. This is done through an iterative procedure where the first step is to detect the location of the largest change in the conductance profile. Extensive experimental work has shown that this cell does not necessarily correspond to the interface position, as is currently supposed. Grégoire's method can then be applied, assuming that the interface is indeed located in this cell and then assuming that it is located in the cell immediately above. If the first assumption was true, then the calculation made under the second assumption would lead to a negligible value for the pulp level X within the interface cell. On the other hand, if the second assumption was true, then the calculation made based on the first assumption would result in an extremely high value of X (larger than the actual electrode separation). A conditional selection algorithm (IF-ELSE clause) is implemented to deal with this decision.

It was observed though, that near the electrodes, Grégoire's method leads to estimate pulp level discontinuities, in the form of a series of sudden jumps, as a result of the conditional selection algorithm. Although a maximum variation of 5 cm can be observed (for 10 cm electrode separation), a small error considering the froth-depths usually used in industry (1 m), they would be propagated to the actuator, *e.g.*, a pinch valve, when a tight level control is used, decreasing the lifetime of the actuator and increasing the variance of the control loop. To correct the above mentioned problems, a new algorithm was developed by Maldonado *et al.* [24]. While keeping the main features of Grégoire's algorithm, it substantially improves its smoothness when the interface is near an electrode or moving from one conductivity cell to another, by considering the weighted average between the two possible values of froth depth determined by the largest slope method. The steps of the algorithm are the following: (a) calculation of the conductivity profile; (b) determination of the possible cells containing the interface, *i.e.*, where the largest conductivity changes occur; (c) estimation of the X value for these two possible cases using Grégoire's algorithm; (d) calculation of the corresponding two possible froth depth values; and (e) calculation of a weighted average froth depth estimate. That is to say, the froth depth is determined as a linear combination of the two possible froth depth values, where the weighing parameters are a function of the conductivities above and below the cell containing the interface. The experimental results obtained have shown that the proposed method gives very good froth depth estimates, improving accuracy and smoothness, which makes this sensor very suitable for control and optimization purposes. The availability of a conductivity profile in the upper part of the column also opens the door to the evaluation of the bias rate as will be shown in the following section.

6.3.2 Bias Rate Sensor

Two different approaches have been proposed so far for estimating bias rate. The first one considers some sort of water mass balance on either zone of the column, above or below the interface, and makes use of measurements of external flow rates to estimate an internal variable, the bias rate. The second approach uses column internal variables, measured near the interface, for estimating the bias rate. The rationale behind the methods proposed in this category is the change in some internal variable with variation of the water content in the upper part of the column (above the interface), resulting from a change in the bias rate.

A method based on the first approach estimates bias rate (J_b) by a water mass balance in the column section below the feed port assuming a steady-state column operation, as shown in Equation 6.8 where J_{wt} and J_{wf} are the superficial velocity (volumetric flow rate divided by column cross-sectional area) of the liquid component of tails and feed streams, respectively:

$$J_b = J_{wt} - J_{wf}. \tag{6.8}$$

To implement this measurement method, four relatively expensive instruments are required: two magnetic flow meters and two gamma-ray density meters, one of each type installed in the feed and tailings streams of the column. Despite this costly set-up, the accuracy of this estimate is very poor since bias is a very small value calculated from two large quantities, J_{wt} and J_{wf}, and subjected to important error propagation (four instruments each having its own degree of uncertainty). Assuming a 3% error on each instrument reading, Finch and Dobby estimated a 75% relative error on bias estimates, which is completely unacceptable [5]. As a result of this unreliable estimation and the high cost associated with it, not a single concentrator is presently using a bias control loop in the operation of their flotation columns. In conclusion, this method can only be used for sensor calibration or for auditing purposes, and only under carefully maintained steady-state conditions.

To improve the accuracy of the bias estimation through the mass balance approach, Uribe-Salas proposed to include the liquid-phase conductivity of the tails, concentrate, feed and wash water streams [25]. This is done through the so-called "additivity rule", which states that the conductivity of a mixture is a weighed average of the component conductivities. He applied this concept to the concentrate water component J_{wc}, which is considered to be made of a wash water short-circuiting stream J_{wwc} and a water feed-entrainment stream J_{wfc}, and to the tails water-component J_{wt}, which is assumed to be composed of a wash water downward stream J_{wwt} and a water-feed downward stream J_{wft}. Then, he used these values in the bias definition given in Equation 6.8.

Uribe-Salas claimed that this approach would in principle improve the precision of the bias estimate compared with that obtained through the traditional form ($J_b = J_{wt} - J_{wf}$). However, it still requires steady-state operating conditions, otherwise the problem of time-synchronization of the various conductivity and flow rate values will reduce its reliability. The situation is particularly critical for low k_f/k_w ratio values. The extensive instrumentation required, the need for the conductivity of the liquid component of some streams, and above all, the requirement of a steady-state operation limit the practical application of this method to laboratory set-ups mostly working with two-phase systems in well controlled steady-state operation.

While running experiments to validate the use of temperature or conductivity profiles in the upper part of the column for froth depth evaluation, Moys observed that the shape of this profile changed with wash water rate variations [26]. A similar pattern was later reported by Uribe-Salas for the conductivity profile [25] but no attempt was made to practically implement this relationship.

Based on these observations, Perez-Garibay [27] managed to successfully relate the conductivity-profile change to the prevailing bias rate using an ANN. Later on, Vermette [28] did similar work using the temperature profile. These methods though, require the existence of a significant difference between wash water and feed-water conductivities or temperatures, which is usually the case, at least for northern hemisphere weather conditions. Reported experimental results (laboratory scale) are very good for two-phase and three-phase (mixture of hematite and silica) systems.

To verify the usefulness of this sensor for automatic control, dynamic tests were also conducted to evaluate the capability of the proposed approach for on-line bias estimation [27]. A 12 in diameter Plexiglas column operating with a two-phase system was submitted to a series of step-changes in wash water, tailings and air flow rates. Based on continuous conductivity measurements, the ANN provided estimates of bias and entrainment as a function of time. To verify these predictions, independent bias measurements were made through a reference method when the column was operating in steady-state conditions, before and after a given disturbance was introduced. ANN predictions for both bias and entrainment were very good.

However, the method has some drawbacks such as the need for extensive experimentation to constitute the database required for both learning and validation steps of the ANN and the need for an independent and reliable method of measuring the bias rate necessary for the ANN training database; this can easily be done on a laboratory scale by a steady-state water balance, but the method has not been attempted on an industrial scale. As a result of the above mentioned procedure, *i.e.*, ANN training using steady-state bias values, it has not yet been confirmed whether the method would satisfactorily predict transient bias values.

More recently, Aubé [29] used a simple multilinear regression technique to replace the ANN algorithm for modeling the bias rate to the conductivity profile. The steady-state validation results were quite satisfactory, eliminating at least some of the drawbacks outlined above.

Despite the above mentioned problems, both the ANN and multilinear regression approaches have been successfully used for automatic control of bias in a laboratory-scale column on various occasions [21], [30], [31].

In order to solve the remaining concerns of the previously discussed techniques, Maldonado *et al.* [32] proposed a method which combines the positive features of each of them: process phenomenology and some empirical correlations. The proposed method was developed for a water–air mixture only, although work is underway for its extension to three-phase systems. In contrast to methods based on external (water component) flow rate balances, this one gets information directly below the interface, *i.e.* where bias is defined. More precisely, it uses a linear relationship between the bias rate and the volumetric fraction of wash water (ε_{ww}) below the interface, estimated by applying the "additivity rule" (Figure 6.4(a)). The conductivity profile is obtained using the same sensor designed for froth-depth sensing [24]. Validation results are shown in Figure 6.4(b).

One important advantage of this method is the fact that it can dynamically estimate the bias rate independently of the existence of steady-state conditions. As such, the method has been successfully implemented in different control applications on a pilot scale [33, 34]. The proposed method improved the stability of the control system compared with those control strategies using bias rate estimates obtained as the difference between tails and feed rate water components [35, 36].

Figure 6.5 shows the dynamic bias estimates (smoother sequence), during step-changes in wash water rate (the variable used to control bias rate), and bias values obtained from a water balance in the collection zone (noisy sequence). During the validation test, feed rate was regulated at a constant value using a local PI controller and the froth depth was maintained at a fixed value also using a PI controller, by manipulation of tailings flow rate. Despite the fact that bias rate estimation is based on a steady-state water mass balance in the collection zone, it is intentionally plotted as a dynamic variable to emphasize the effect of the froth depth control-loop on the estimation. Figure 6.6 shows the effect of gas and wash water rate on the bias rate. It can be seen that gas rate has a strong impact on bias rate. Unfortunately, the method has never been tested in industrial columns yet. This subject will be further discussed in Section 6.4.2.

Persechini [36] also developed and tested three soft sensors to monitor froth depth, bias and air hold-up for a simplified two-phase system. The latter process variables were inferred using only two flow meters (air and wash water) and two pressure gages mounted on the column side. The proposed approach is certainly interesting and inspiring but requires further developments. The following reservations are worth mentioning:

- instrumentation would need to be adapted for a slurry application, which represents the ultimate objective; and
- validation and accuracy of the bias and gas hold-up measurements were not assessed, even for steady-state operation.

6.3.3 Collection zone Gas Hold-up

Basically three methods have been used to measure the air hold-up in flotation columns respectively based on: (a) differential hydrostatic pressure, (b) electrical conductivity, and (c) sound detection, the latter has very recently become available. Despite the fact that all three methods have been tested in industry, some more than others, and despite the importance of the measured variable for the performance of the flotation process, their continuous use for plant monitoring is extremely limited, and for control purposes, non-existent. Differential hydrostatic pressure is currently

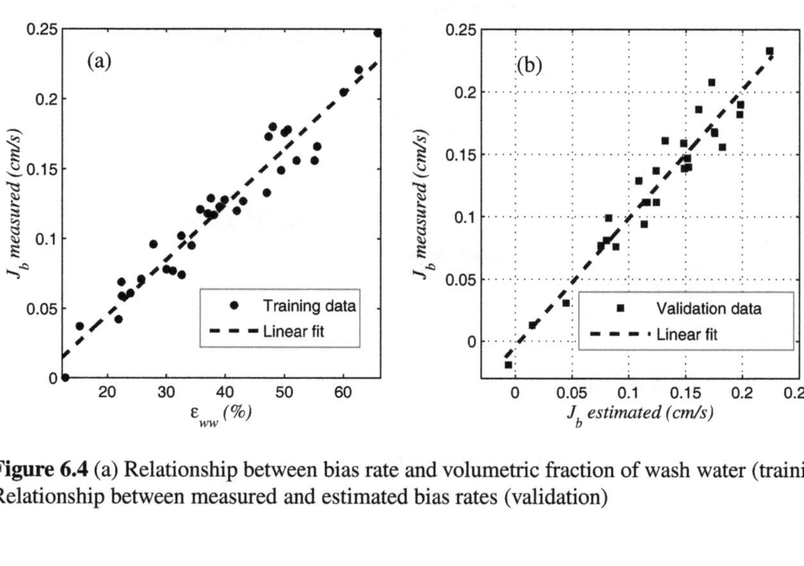

Figure 6.4 (a) Relationship between bias rate and volumetric fraction of wash water (training). (b) Relationship between measured and estimated bias rates (validation)

the oldest and simplest technique for evaluating gas hold-up. It consists of measuring the hydrostatic pressure at two points in the collection zone, distant ΔL cm. If P_1 and P_2 are the corresponding pressure values and ρ_{sl} is the average density of the pulp between the two sensors, then the gas hold-up is calculated through

$$\varepsilon_{gP} = 100\left(1 - \frac{P_1 - P_2}{\rho_{sl} \cdot g \cdot \Delta L}\right). \tag{6.9}$$

Since the slurry density value ρ_{sl} is hard to measure in industrial columns, the practical implementation of this method requires assuming a constant value, *e.g.*, feed or tail density. However, this fact could lead to erroneous estimates in practice as indicated hereafter.

When comparing this method with the conductivity-based one for two- and three-phase system in a well-controlled laboratory column, Nuñez *et al.* [37] observed that for the two-phase case, both methods gave similar results whereas for the three-phase system the pressure-based method provided continuously varying estimates as a result of the assumption of a constant slurry density used in the air hold-up calculation. This was clearly observed when the column solid content changed as the fed pulp replaced the initial water-only content, in that the ε_{gP} estimate reflected this pulp-density effect whereas the conductivity-based estimate ε_{gK} was not affected. During time periods where test conditions were stable, both methods provided consistent estimates not identical but having the same trend, as the respective sensors were installed at different column positions.

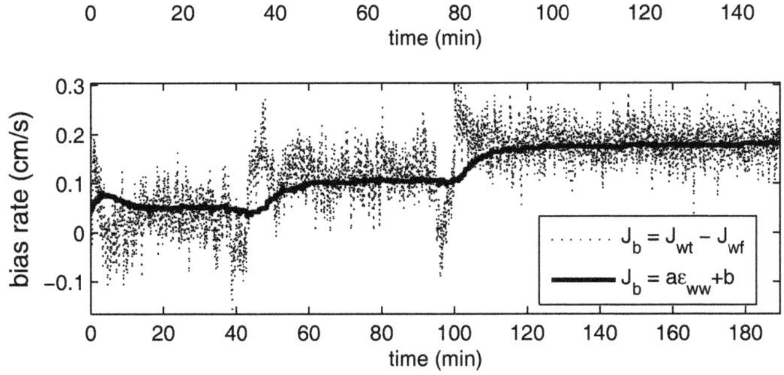

Figure 6.5 Dynamic bias estimation

A different approach for evaluating gas hold-up was developed by Uribe-Salas [38] using the Maxwell equation for a mixture of a dispersed non-conducting phase (air) in a continuous conducting phase (pulp). In this method, either phase concentration is related to its electrical conductivity. The final relation for air hold-up is given in Equation 6.10, where k_{slg} is the conductivity of the mixture pulp air and k_{sl} is the conductivity of the pulp without air. Whereas the evaluation of the former does not cause a problem, the latter (k_{sl}) represents the major challenge in using this method. In fact, this is the same problem encountered in the hydrostatic pressure method where the measurement of ρ_{sl} is required.

$$\varepsilon_{gK} = 100 \cdot \left(\frac{k_{sl} - k_{slg}}{k_{sl} + 0.5 \cdot k_{slg}} \right). \tag{6.10}$$

Various methods have been proposed for measuring k_{sl}, the simplest being the use of the tailing (or feed) stream conductivity [38, 39]. However, this estimate does not correspond to the same position where k_{slg} is measured and might lead to errors. Two other methods of evaluating k_{sl} close to the k_{slg} measurement have been proposed. The first is the "standard addition method" consisting of adding a known amount of dielectric material in a second cell and solving a system of two equations (such as Equation 6.10) and two unknowns, k_{gk} and k_{sl} [40, 41]. The second method uses an open cell for measuring k_{slg} and a parallel cell (called siphon cell), designed to avoid air bubbles suction, therefore providing an adequate measure of k_{sl} [42]. Figure 6.7 shows the concept of this method. This approach seems to provide more precise estimates of air hold-up, since the solution of the system of equations in the

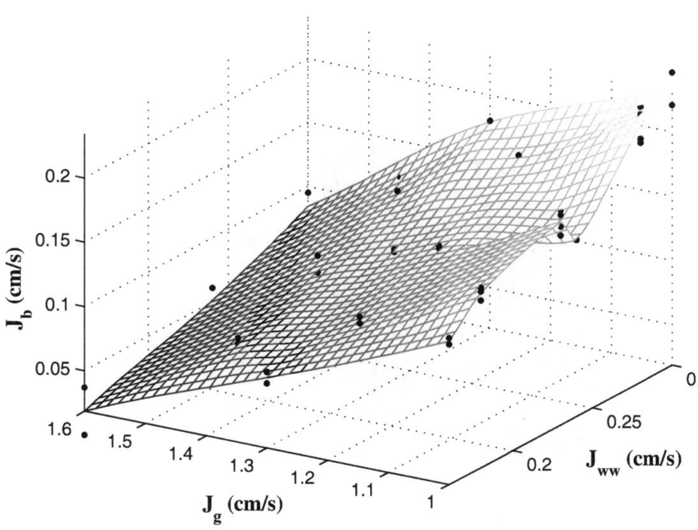

Figure 6.6 Effect of J_g and J_{ww} on bias rate

standard addition method might imply an important propagation error. However, this method might be useful for estimating the gas hold-up in the froth zone where the siphon method cannot be used.

The third method available for measuring the gas hold-up uses an array of passive sensors and sonar processing algorithms (sonar array) originally developed for fluid flow rate measurements and later adapted for measuring the amount of air entrained in a slurry pipe [43]. This sensor was developed by CIDRA Corporation (USA) and is commercialized under the name of SONARtrac®[1]. This technology can simultaneously measure the velocity of fluid flow and naturally generated acoustic waves. The gas hold-up is determined from the propagation velocity of the acoustic waves (speed of sound) via a nonlinear relationship specific to each case.

The principle of the sonar array instrument is based on the detection, tracking and measurement of the mean velocities of coherent disturbances traveling in the axial direction of a pipe. These disturbances are grouped into three major categories: disturbances conveyed by the flow, acoustic waves in the fluid, and vibrations transmitted via the pipe walls. Each disturbance class travels at a given velocity. For example, the flow will convey turbulent eddies, density variations, or other fluid characteristics at the rate of the fluid flow. Liquid-based flows rarely exceed 9 m/s. Acoustic waves in the fluid will typically have a minimum velocity of 80 m/s and a maximum velocity of 1500 m/s. The third group, pipe vibrations, travels at velocities that are several times greater than the acoustic waves. Thus each disturbance class may be clearly identified and accurately measured. The sequence of events leading to the actual gas hold-up measurement are as follows:

(a) Acoustic waves are generated by spargers, rotors, or other sources. (b) These acoustic waves enter the gas hold-up meter along with the air bubbles. (c) The

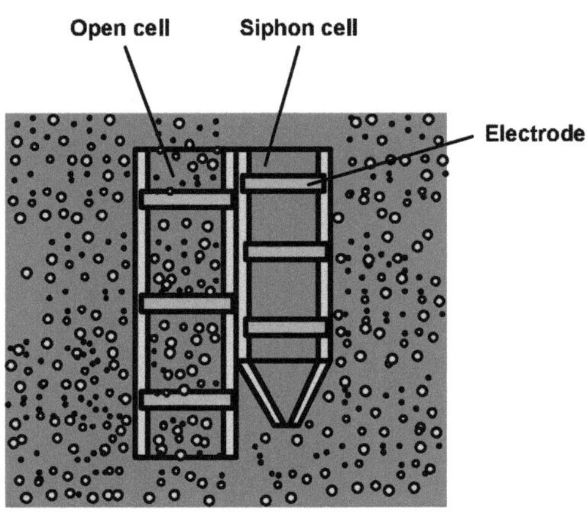

Figure 6.7 McGill two-cell sensor

[1] SONARtrac is a registered trademark of CiDRA Corporation.

passage of the acoustic waves through the pipe creates a small pressure change on the inside of the pipe wall. (d) This small pressure change results in a dynamic strain of the pipe wall itself. (e) The mechanical dynamic strain signal is converted to an electrical signal by each element of an array of passive sensors wrapped partially or fully around the pipe exterior – no coupling gels or liquids are used. These sensors are spaced at a set distance from each other along the axial direction of the pipe. (f) The resulting electrical signal from each sensor element is interpreted as a characteristic signature of the frequency and phase components of the disturbance under the sensor. (g) An array processing algorithm combines the phase and frequency information of the characteristic signature from the group of sensor array elements to calculate the velocity of the characteristic signature as it propagates under the array of sensors. (h) The velocity of the acoustic waves is used to calculate the gas hold-up percentage.

The advantage of this principle is that it is impervious to scale build-up and the acoustic velocity measurement will not drift with time or temperature. Figure 6.8 shows this device.

The physical embodiment of the gas hol-dup meter consists of a 4 in PVC pipe, open at both ends, with its axial array of sensors external to the PVC tube but encased within a sealed watertight enclosure, designed to withstand the rigors of a harsh minerals processing environment. The measurement accuracy is 5% of reading over the gas hold-up range of 0.01% to 20%. The gas hold-up meter is placed in the collection zone of the flotation device, thus allowing the free access of ascending bubbles and descending slurry through the device. The cited article reports

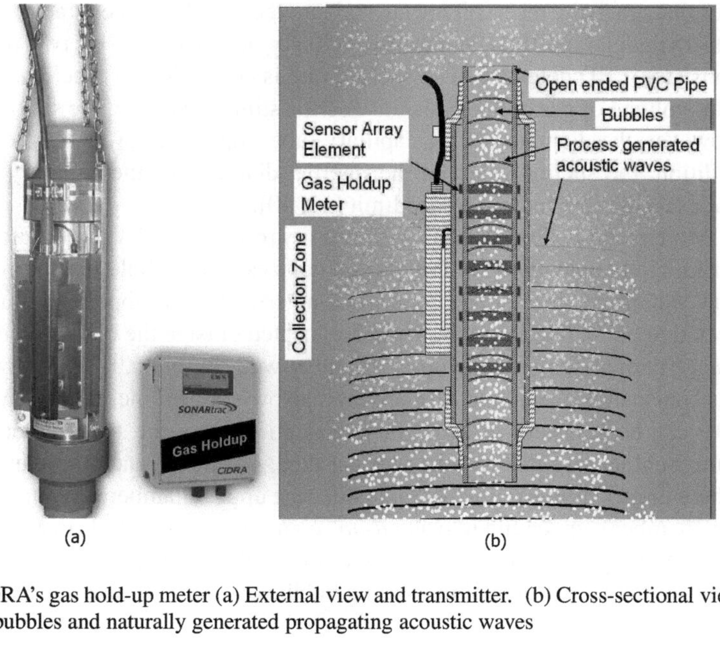

Figure 6.8 CIDRA's gas hold-up meter (a) External view and transmitter. (b) Cross-sectional view illustrating air bubbles and naturally generated propagating acoustic waves

the successful testing of these sensors in various mineral processing and tar sands plants in North America.

Other sensors have also been proposed [44–47], all consisting of the extraction of an aerated pulp sample from the flotation device and letting the two phases separate, after which the gas phase volume can be measured. Obviously, these batch methods are inappropriate for control purposes and therefore they will not be discussed here. Grau and Heiskanen proposed a semi-continuous version of this phase-separation method [48], whereby a pulp sample is continuously withdrawn from the flotation cell and sent to the separation chamber or recycled back to the cell, depending on the position of a by-pass valve. The gas measurement is made by liquid displacement in an inverted graduated cylinder, which is obviously not a convenient approach for process control purposes either.

6.3.4 Bubble Size Distribution (BSD) Sensor

For close to two decades, the practical evaluation of bubble size in flotation devices has become a privileged field of research in several universities around the world. Noteworthy research was carried out in Cape Town University (South Africa) in 1989, McGill University (Canada) in 2001, Helsinki University of Technology (Finland) in 2002, Universidad Federal do Rio Grande do Sul (Brazil) in 2003, and CETEM (Brazil) in 2009. The first BSD sensor was developed in the late 1980s at Cape Town University [49]. The UCT bubble-sizing sensor comprises a riser tube and a water reservoir where a glass capillary tube is placed. A fraction of the bubbles that reach the reservoir are sucked into the capillary where they are transformed into cylinders and their length and velocity are measured by two pairs of phototransistor–LED detectors. The suction rate is controlled by a peristaltic pump. The total volume of gas collected is utilized to estimate the absolute size of the bubbles. However, the diameter of the capillary tube imposes a lower bubble size limit. In addition, the suction rate and the internal diameter of the capillary tube seem to impose a maximum bubble size limit [45, 50].

More recently (2001), McGill University researchers developed a sensor for measuring the bubble diameter in flotation devices from bubble image processing [51]. The McGill Bubble Viewer consists of a vertical tube submerged in the pulp, connected to an inclined viewing chamber located outside the cell and a digital camera placed perpendicular to the upper chamber wall (Figure 6.9). The viewing chamber has transparent walls and a relatively narrow gap; its inclination angle (typically 20°) can be adjusted for better picture quality. Since the system works under vacuum, bubbles rise through the vertical tube, to finally reach the viewing chamber, where they continue rising, sliding under the upper chamber wall until they leave by the chamber's upper end into an *ad-hoc* container, which needs to be cleared once in a while. A light source under the bottom wall of the chamber provides the necessary light for taking pictures of the sliding bubbles.

The viewing chamber is initially filled with plant process water, to provide a clear image of the bubbles, in other words, the bubbles are washed before entering the unit. Frother can also be added to the image area to insure that the bubble size does not change while the bubbles are rising to be photographed. To obtain a bubble distribution representative of the flotation cell content, a significant number of pictures must be taken, and later processed by image analysis software. The result of the whole process is a series of pictures as shown in Figure 6.2(a), and a histogram of bubble diameters as shown in Figure 6.2(b), which is later transformed into an average diameter (usually the Sauter mean diameter). At the end of the picture taking process, the device must be emptied and refilled with clear water, since the viewing chamber must not contain a dense pulp which would prevent light from going through the chamber. Moreover, as bubbles accumulate in the container, pressure continuously increases whereas the liquid level in the chamber decreases, which poses a constraint to the time span available for continuous measurements. As a consequence, the sensor is not adequate for process control purposes in its present form. This device has been successfully used in several concentrators world-wide for diagnosis of the air injection system of mechanical cells and columns, but always using the discontinuous procedure detailed above. COREM (Consortium de Recherche en Minéralurgie, Québec, Canada) researchers have automated this device in such a way that it has been used providing semi-continuous measures for an extended period of time [52]. Despite its on-line measuring capabilities, no effort has been made so far for its use in process control applications. This point will be raised in Section 6.5.

A new sensor was proposed by Helsinki University of Technology researchers [53] based on the McGill Bubble Viewer design and including some practical fea-

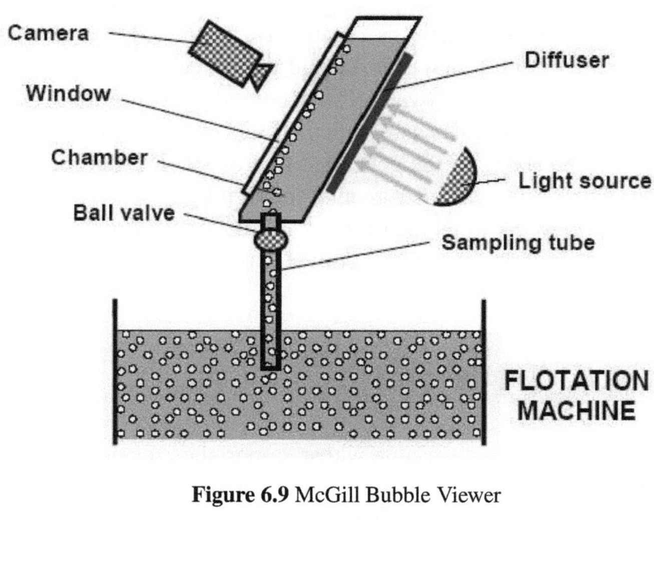

Figure 6.9 McGill Bubble Viewer

tures of the UCT bubble sampler. This bubble size analyzer (HUT-BSA) consists of a sampler tube, a viewing chamber and an image acquisition system. The sampler tube is partially immersed in the liquid phase, below the froth-liquid interface, where it is used to collect rising air bubbles, which are drawn into the viewing chamber. The function of the chamber is to expose the bubbles to a monochrome CCD camera which is connected to a personal computer, where the visual information is stored and processed. The images captured during sampling are automatically processed using commercial image analysis software and in-house visual basic user interface.

A major drawback of the present approach for analysing bubble distributions obtained from both the McGill and the HTU sensors is the fact that the use of an average bubble diameter (D_{10} and D_{32}) involves a compression of the whole BSD, where information regarding the shape of the distribution, such as tails and the presence of multiple modes, is lost. Different probability density functions all having the same D_{10} and D_{32} value can exist, which means that the use of a single average value would not allow the assessment of the multi-modal, narrowness and tails characteristics of the actual distribution.

A method to on-line estimate the whole BSD represented by its bubble size probability density function (PDF) has been recently proposed at Université Laval [54]. A Gaussian mixture model (GMM) is used to represent the bubble size PDF and a modified version of the well-known expectation-maximization algorithm (EM) is utilized to estimate the GMM parameters evolution in time. An appealing feature of this method is that the mean diameters D_{10} and D_{32} can be directly obtained from the GMM parameters, thus obtaining dynamic estimates.

Two other devices have been proposed by Brazilian researchers for the measurement of very fine bubbles. The first sensor (LTM-BSizer) was developed for two-phase systems only, but the authors mentioned future plans for three-phase system testing [55]. The sample is driven from the flotation device (a laboratory column) to a horizontal viewing chamber where bubbles decelerate and eventually stop when the downstream pinch valve is closed. Images are then taken with the aid of a microscope and a digital camera. When the pinch valve is opened the viewing chamber is purged and refilled with a new sample and the process is repeated. Images are processed using commercial software. The second device, the most recent one in the literature, was designed to measure micro-BSDs on two-phase systems at laboratory scale [56]. The sizing of the generated micro-bubbles was performed with a Mastersizer 2000SM (Malvern Instruments) through light scattering, which allows measurements down to 0.1 m. None of them seems, for the time being, to be good candidates for industrial on-line application, at least for process control purposes.

6.3.5 Superficial Gas Velocity (J_g)

Another commonly measured variable whose importance for the assessment of the hydrodynamic characteristics of flotation cells is undeniable, is the superficial gas velocity (J_g), defined as the ratio between the volumetric gas rate and the device cross-sectional area. This variable has recently attracted researchers' attention and many sensors have been developed mostly for measuring local superficial velocity in an attempt to evaluate ill gas distribution or sparger malfunctioning [45], [47, 48]. They are all based on some sort of tube submerged in the aerated pulp and a vacuuming system to drive the gas into a measuring device (typically a gas mass flow meter). Since regular commercial flow meters (*e.g.* turbines, Venturi tubes, *etc.*) are perfectly adapted for this sort of measurement in flotation columns and they are entirely adequate for process control purposes, these newly developed sensors will not be discussed here.

6.4 Automatic Control

6.4.1 Control Hierarchy

In the present context, with globalization and increasingly strong international competition, mineral processing plant managers are constantly seeking to reduce production costs while increasing throughput and product quality. One way to reach this objective is to use better equipment, *i.e.*, to replace existing parts of the plant with more efficient ones or to add new equipment. Improving plant operation can also be achieved through on-line data processing by making judicious on-line use of process sensors and large plant operation databases. Of course, this approach will only be successful if equipment is working properly.

On-line data processing consists in using measurements provided by plant sensors and knowledge of dynamic plant behavior (evaluated from measurement databases recorded during plant operation) to automatically, continuously and adequately manipulate plant actuators, such as valves and variable speed drives, in order to achieve a specific objective. The ideal on-line data processing is made up of four hierarchic layers: instrumentation and regulatory control; process observation; process control; and RTO. This scheme is depicted in Figure 6.10.

Instrumentation for the process being considered has been thoroughly discussed in the previous section. Regulatory control, the basic control layer, consists of local monovariable control loops (usually PID controllers) providing steady operation for the flotation column inputs (air, wash water, pH and reagents), but refers in general to all control elements (valves and pumps, including that of the tailings).

Process observation aims at getting useful on-line information about plant behavior, therefore allowing the "visualization" of process states and detection of operating problems, abnormal performances, *etc.* Reliable sensors are the basic equipment required for this. However, using data processing techniques such as fault detection and isolation (FDI)[57, 58], data reconciliation [59] and observers [60], may greatly improve the process "picture". FDI rapidly detects and physically localizes problems such as sensor biases and leaks, whose rapid correction will reduce detrimental effects on production. Data reconciliation techniques improve the quality of noisy measurements by forcing them to be consistent with the knowledge engineers have about the plant, such as mass and energy conservation laws. Observers are soft-sensing algorithms which infer signals that cannot be measured (not at all or not frequently enough). They generate new measurements that can be used to improve plant operation. Even if process observation is based on passive techniques, they help engineers improve their vision of the process states and therefore corrective actions can be taken. Some observation techniques are described in Chapter 2 of this book. Applications of process observation techniques to flotation columns have been reported by Bergh *et al.* [61, 62].

Once the on-line measurements have been improved by the process observation layer, they may be used to continuously adjust the regulatory controller set points to provide adequate process control actions. Through feedback and feedforward, advanced process control permits the maintenance of measured variables at selected set points despite the presence of disturbances, such as changes in ore properties. A consequence of the process stabilization around a selected operating point is a decrease in process variability usually resulting in a smoother and less costly operation. Robustness and reliability should be the primary focuses regarding any choice of control algorithms. The intermediate control level aims at maintaining process variables having a strong influence on metallurgical results (grade and recovery), such as froth depth, bias, froth parameters, air hold-up or bubble surface area flux, *etc.*, in a bounded region, *i.e.*, an acceptable operating zone where it is possible to handle the process and reach production objectives. This level is often called stabilizing control. Applications of process control to flotation columns are described in the following sections.

The fourth layer is the process optimization step [63], sometimes called supervisory control. It consists in continuously selecting set points of the control layer to optimize a control objective. The objective is usually related to profit maximization.

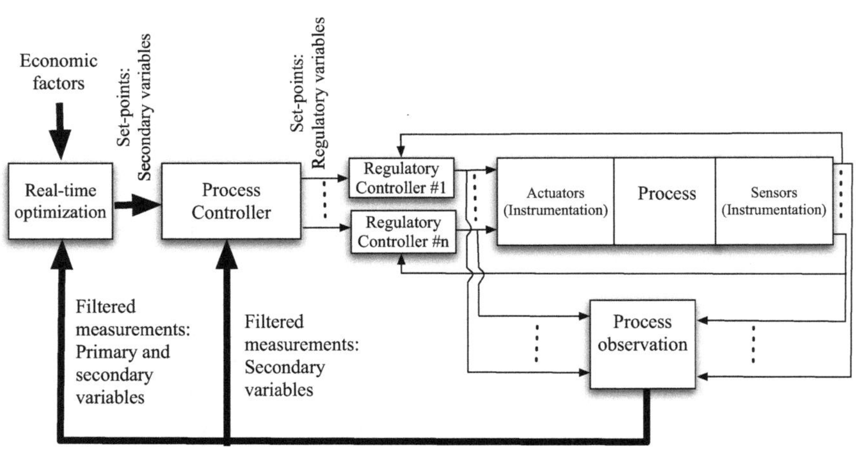

Figure 6.10 Control hierarchy

It is well-known that mineral separation performance is dictated by the so-called grade–recovery curve, which establishes a compromise between the quantity (recovery) and the quality (grade) of the device valuable product (concentrate). Usually, it is the plant metallurgist who decides which position of the curve the plant (or a section of it) should work at. This decision should, however, be dictated by the profits generated by operating at a given point on the curve and should automatically be taken by a RTO algorithm based on an economic criterion, such as the net smelter return.

6.4.2 Process Control

In this section, it is assumed that the instrumentation and regulatory control are working properly. It is also supposed that the column pulp feed has been correctly conditioned insuring adequate pH and reagent concentration (collectors, frothers, *etc.*). The design and/or tuning of the controller, model-based or not, should rely on adequate process models. Identification of such models is discussed in Chapter 4 and in Section 5.5.1, and also in several books [66, 67]. As previously explained, it is often more convenient to indirectly control the primary variables through secondary variables which can be more easily measured at a reasonable cost. A controller that adequately manipulates the regulatory controller set points (air flow rate, wash water flow rate, tailings flow rate, *etc.*) must then be designed. The secondary variable set points are the results of the RTO unit.

Undoubtedly, froth depth is the most used secondary variable, and often the only one, for control of industrial flotation columns. The position of the pulp–froth interface sets the height of the collection zone and therefore is directly related to the time available for collection of hydrophobic particles on the air bubble surface, the other factor being the tails flow rate. Consequently, it is partly responsible for the recovery of the valuable mineral. However, large froth depth set-point changes would be necessary to obtain a significant effect on recovery, seriously limiting its use for recovery optimization purposes. Nonetheless, pulp–froth interface position is an important control variable since its constancy is a guarantee of stable column operation: a continuously moving interface is certainly undesirable, since it implies the risk of feeding the column in the middle of the froth zone or deepening it so much that a risk of froth collapse exists.

To obtain an adequate cleaning action, a minimal positive value for the bias is required. On the other hand, if the bias is too large, the residence time of valuable minerals in the collection zone could decrease.

Gas is at the heart of the flotation process, therefore, controlling gas rate, gas hold-up, bubble surface area flux or BSD appears to be a way to improve metallurgical performance. The control of bubble surface area flux and BSD will be discussed in Section 6.5, where future work in this area is discussed. Since measuring gas hold-up is a more mature technology, examples of its control can be found in the literature [34–36, 68, 69], as will be illustrated in the next section.

Imaging systems can provide froth characteristics such as color, form, speed and size. Eventually, they could also be used as secondary variables for control purposes. For instance, Liu *et al.* [70] illustrate a new way for modeling and control of flotation processes based on froth appearances. Froth health monitoring is also discussed in Chapter 3. Examples where froth speed is measured to improve flotation performance in cells are described in the literature [71–73].

When controlling a multivariable process such as a flotation column, two main approaches are possible: decentralized or multivariable control. Decentralized control consists in using single-input single-output controllers. It requires a good selection of input–output pairings. Controller tuning must take into account interactions between the control loops, which may be difficult to accomplish. Despite considerable advances in multivariable control theory, decentralized PI controllers still remain the standard for most industries. They have fewer tuning parameters, are easier to understand, and are more failure tolerant. Multivariable controllers are rather complex, require more engineering manpower and have a lack of integrity, all which may result in operator non-acceptance. For the flotation column case a feasible choice of control variables is froth depth, bias rate and gas hold-up whereas the possible manipulated variables are the tailings, wash water and air flow rate set points. In the case a decentralized control approach is considered, the following pairings are suggested: froth depth – tailings flow rate, bias rate – wash water flow rate and gas hold-up – air flow rate. For a two-phase system, it has been shown that the resulting model is then an upper triangular matrix [33], indicating that decentralized control with these pairings is very appropriate. The main weakness of this approach is the difficulty of adequately taking into account constraints, as will be discussed in the next paragraph and in Section 6.4.3.

Model predictive control (MPC) is, by far, the most widely accepted multivariable control algorithm used in the process industry. Several good books describing MPC are now available [74–76]. Several commercial MPC technologies allowing model identification and control with constraints can be purchased and successful applications are found in many process areas [77]. The main advantage of MPC is its ability to handle multivariable processes while taking into account various constraints. Indeed, the plant operating point that satisfies the overall economic goals of the process, usually lies at the intersection of constraints. The main concepts of MPC are described in Sections 5.5.2 and 7.3.2. Model development and identification (Chapter 4 and Section 5.5.1) are the most difficult, critical and time consuming steps in the design of a MPC.

Two comprehensive survey papers about control have recently been presented by Bergh and Yianatos [64] and Bouchard *et al.* [65]. Table 6.1 summarizes some control applications reported in these surveys as well as more recent developments.

Table 6.1 References describing several flotation column control applications

Reference	Controlled variables	Control algorithm	Type of column
[78]	Froth depth, bias	PID	Industrial
[79]	Froth depth	Backstepping, model reference nonlinear control	Pilot, 2 phases
[61]	Grade, recovery	Fuzzy logic supervising local loops	Pilot, industrial
[62]	Grade, recovery	Expert system	Industrial
[30]	Froth depth, bias	Review of different methods	Pilot, 2 and 3 phases
[68]	Froth depth, gas hold-up	MPC	Laboratory
[4]	Froth depth	PID	Industrial
[21]	Froth depth, bias	PID	Laboratory, 2 phases
[31]	Froth depth	Gain-scheduled PI	Pilot, 2 phases
[81]	Grade	Fuzzy logic	Industrial
[82]	Column performance through pulp/froth zone densities and/or grade and recovery	Fuzzy logic	Industrial
[33]	Froth depth, bias, gas hold-up	Decentralized PI	Pilot, 2 phases
[34]	Froth depth, bias, gas hold-up	PI + MPC	Pilot, 2 phases
[83]	Column concentrate grade and tailing grade of cleaner scavenger	Expert system	Industrial
[84]	Froth depth, bias	MPC	Pilot, 2 phases
[69]	Froth depth, gas hold-up	MPC	Pilot, 3 phases
[36]	Froth depth, bias, gas hold-up	PID	Pilot, 2 phases
[85]	Froth depth, gas hold-up	MPC	Laboratory, 3 phases

6.4.3 Application Examples

Two examples are presented, the first a multivariable control (3×3) for a two-phase system and the second a multivariable control (2×2) for a three-phase system. In both cases, the MPC was selected for its ability to handle the existing operating constraints listed later on.

In the first example [34], regulatory PI controllers were implemented on an instrumented pilot flotation column (internal diameter: 5.1 cm; height: 732 cm), as illustrated in Figure 6.11. Since the primary interest was the development of a regulatory control (column hydrodynamic characteristics), a two-phase system was sufficient to illustrate the advantages of the proposed approach. Adequate frother concentration (Dowfroth 250) was used to provide proper bubbling conditions and froth stability. The controlled secondary variables were froth depth, gas hold-up and bias rate. Froth depth was measured using the weighted average method described at the end of Section 6.3.1 [24], whereas gas hold-up value was estimated from measured conductivity values using Maxwell's equation (Equation 6.10) and bias rate was estimated using the method proposed by Maldonado *et al.* [32] detailed in Section 6.3.2. The manipulated variables were tailing flow rate, wash water rate and gas rate set-points. Some controlled and manipulated variables are subjected to operating constraints which should also be taken into account:

- A low limit for bias: for an adequate cleaning action, thus reducing gangue entrainment.
- A high limit for bias: to avoid a decrease in the residence time of the valuable minerals in the collection zone [5].
- A low limit for wash water flow rate: to ensure froth stability and transport of the collected particles into the concentrate launder.
- A high limit for wash water flow rate: to avoid an increase in froth mixing and short-circuiting leading to a decrease in froth cleaning [86].
- A low limit for air flow rate: to maintain solids in suspension.
- A high limit for air flow rate: to avoid hydraulic entrainment and thus concentrate grade deterioration, interface loss and presence of large bubbles at the froth surface.

The proposed control structure is depicted in Figure 6.12. The froth depth (H) can easily be controlled with a PI controller that manipulates the tailings flow rate set point (J_t). Bias rate (J_b) and gas hold-up (ε_g) are controlled with a MPC by manipulating wash water rate (J_{ww}) and air flow rate (J_g) set points while respecting the above mentioned constraints. Since the gas hold-up is related to the total bubble surface area available for particle collection, the objective is to increase its value as much as possible while respecting operational constraints.

Experimental results are illustrated in Figure 6.13 where set points and constraints are, respectively, represented by dash and dash-dotted lines. Since a change of 50 cm in froth depth only produces a 5% change in the collection zone volume for a 10 m collection zone height, the impact of the froth depth on column optimization is not obvious. A constant 80 cm set point value was selected to damp-down the gangue entrainment [7]. The Figure 6.13 illustrates the performance of the proposed control structure when the gas hold-up set point is increased. Consequently, the air flow rate is increased which decreases bias rate until it reaches its constraint, making gas hold-up unable to attain its high value set point. Selecting such a high unreachable hold-up set point thus maximizes the bubble surface area available for particle collection while the bias low limit prevents concentrate grade deterioration due to hydraulic gangue entrainment.

The second example corresponds to a 2×2 multivariable control in a three-phase system [69]. The experimental set-up was similar to that used in the previous example. The pulp feeding the column was 15% solids, a hematite-silica ore provided by Compagnie Minière Québec-Cartier (now Arcelor-Mittal Mines Canada). Although flotation is not the most common separation technique for iron ore concentration, this mineral was retained for this work as it is reagent-free and does not oxidize, as is the case of sulphurs. Solid particles graded 66.1% iron, and have a size distribution characterized by 82.6% passing 75 m and 68.8% passing 45 m. An inverse flotation process (silica flotation) at pH of 9.5 was chosen to minimize reagent consumption. The reagents used were WW82 (starch) for iron depressing, MG83 as silica collector and MIBC as frother. Gas hold-up and froth depth were the controlled variables whereas manipulated variables were tailings and air flow rate set points. Figure 6.14 illustrates that good results can also be obtained with a constrained predictive controller for a three-phase system.

6.5 Future Industrial Applications

Previous sections have shown that a significant amount of work has already been done at the academic level to develop column flotation sensors and design suitable process control strategies. However, in spite of this, industrial applications are still rather scarce or nil. It is valid then to ask how the current industry could, from a practical point of view, benefit from these developments and what needs further investigation.

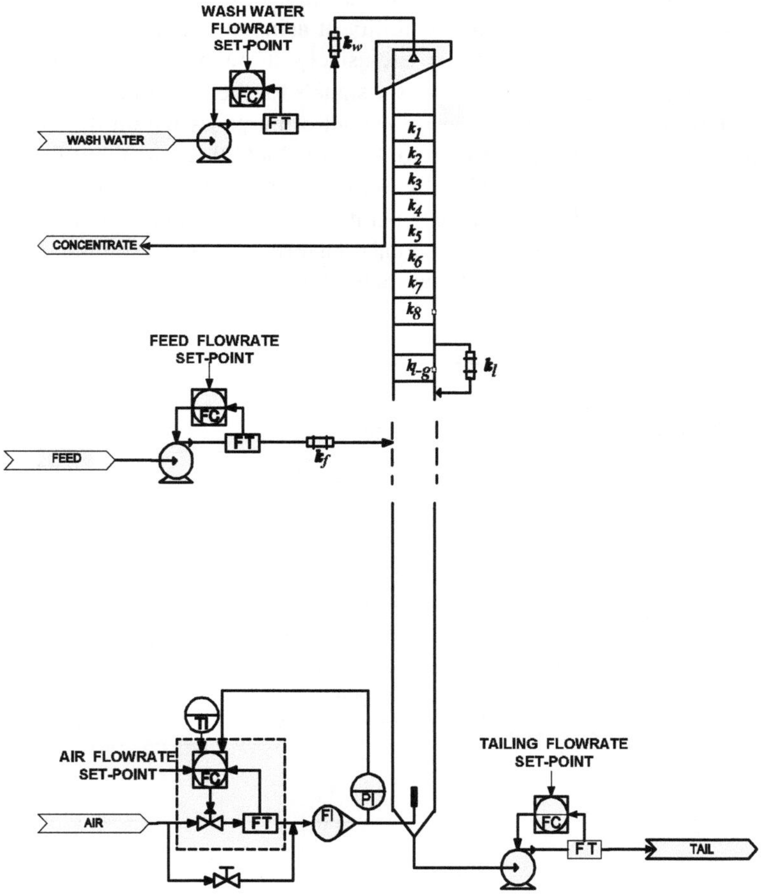

Figure 6.11 Pilot flotation column set-up

Industrial–academic research and development partnerships would simultaneously increase the number of industrial applications and improve present process supervision. Equipment and service suppliers could also take part in this venture by making column flotation sensors more robust and reliable. Through joint action nearly mature technologies like conductivity-based probes for froth depth and gas hold-up, and on-line bubble size measurements could then rapidly become standard devices for process control.

6.5.1 Sensor Development and Applications

After the industrial evaluation previously mentioned, the conductivity-based sensors for froth depth and bias rate should ideally be revisited by equipment suppliers with the aim of producing more robust and reliable commercial versions. Although commercial sensors do exist for froth depth measurement (floats), the conductivity-based probe could offer the possibility of using the same collected information for estimating the bias rate, either exclusively, through the ANN method [23], or in conjunction with gas hold-up sensor signals for the "dynamic" method [32].

Following its successful use for control purposes within the ongoing research project at the Laronde concentrator [87], the McGill conductivity-based gas hold-up sensor should also be revisited by its manufacturers to implement the required modifications for on-line and robust operation. Another interesting technique amenable to further "exploration" and eventual transformation into a commercial sensor, is the concept of the standard addition method developed by Perez-Garibay [40] for gas

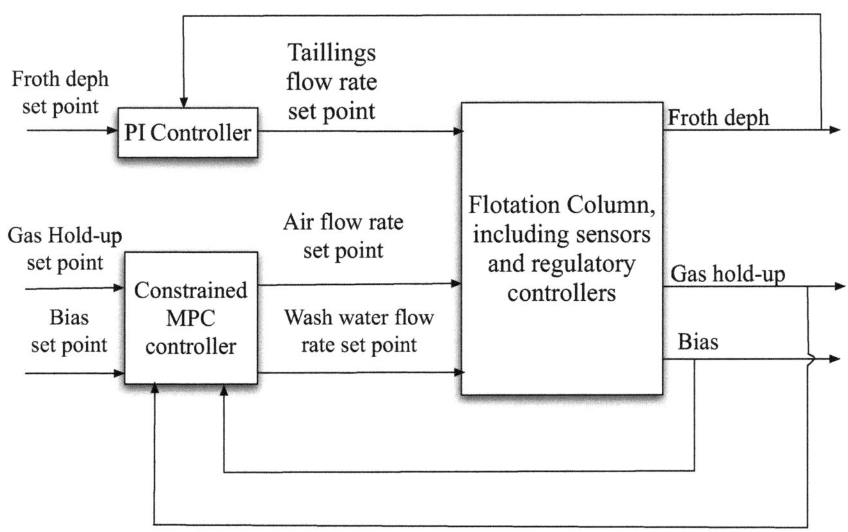

Figure 6.12 Control structure

hold-up monitoring. This technique was recently successfully tested by Arizmendi *et al.* [41] for solid hold-up measurements. Such a technique could become a low-cost and safer substitute for gamma-ray density meters. In its original application, the standard addition method remains the only available method for gas hold-up estimation in the froth zone.

Supervision methods based on multivariate image analysis (MIA) are promising options to overcome OSA limitations and to monitor froth characteristics. MIA potential applications for concentrate grade prediction have been reported by Duchesne *et al.* [88]. Liu *et al.* [89] presented a novel method to extract textural and color information related to the froth zone BSD, and the presence and amount of clear windows in flotation froths. Liu and McGregor [70] also showed that the scores of the MR-MIA (multiresolution multivariate image analysis) could be directly used in froth control. Froth image analysis based on MIA could become the alternative to some instrumentation problems of flotation columns, for instance, to monitor water entrainment. In fact, if the froth color and/or texture are related to the froth water content, it is possible to build a regression model linking froth image characteristics to concentrate water content. Like for any image analysis application, conclusions would be drawn using only surface features, therefore, even if the technique pro-

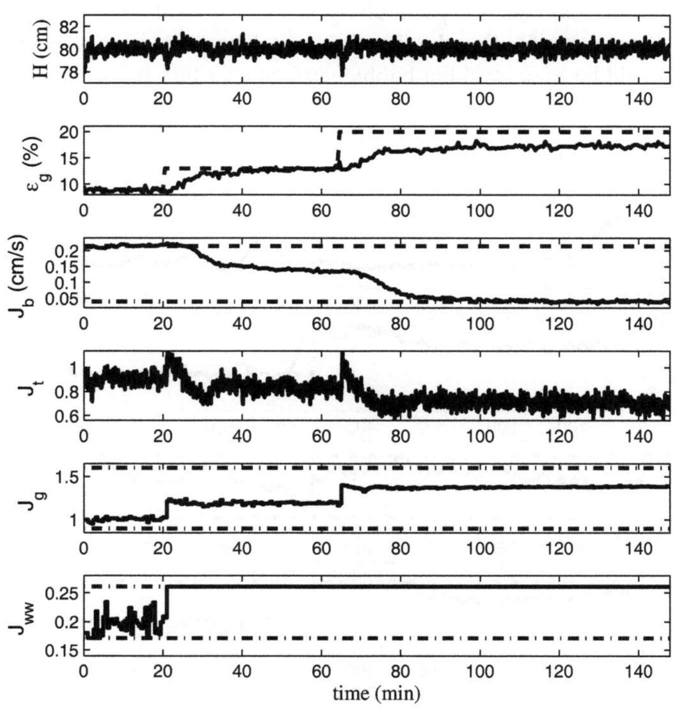

Figure 6.13 Column control – 3×3 two-phase system

vides valuable information, the local nature of the results should not be forgotten. A more detailed discussion of numerical image analysis potential for the monitoring of froth characteristics was presented by Bartolacci *et al.* [90].

Currently, commercial froth imaging systems are available to monitor different froth parameters such as velocity, bubble size, stability and color intensity. Their installation with existing or new column flotation applications is always recommended by equipment providers. Their use for more comprehensive process control strategies should be considered by researchers and practitioners.

6.5.2 Process Control and Optimization

A challenging area for future developments is the proper on-line evaluation and modeling of the BSD in the collection zone. Gas dispersion properties have proven to be a key feature of the flotation process, in particular bubble surface area flux (S_b), linearly correlated with the flotation rate constant, thus becoming a good potential control variable to achieve a target metallurgical performance. As previously explained, currently all the effort made to obtain a BSD is wasted through the calculation of an average bubble diameter. Once properly modeled, the BSD could be used for control purposes, either indirectly, through the S_b concept or directly, leading to metallurgical benefits, particularly if it can be matched to the prevailing particle size distribution feeding the flotation unit. Therefore, a reasonable control formulation would be to control the bubble surface area flux being generated from

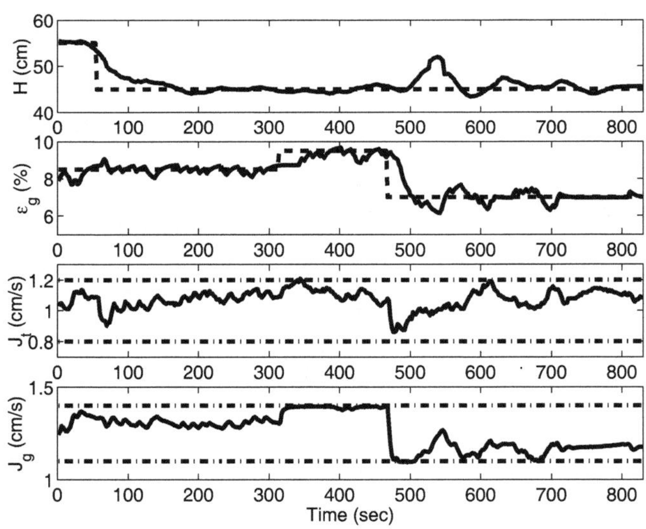

Figure 6.14 Control of the column – 2 × 2 three-phase system

a desired BSD matching a given particle size distribution. However, controlling the whole BSD is a complex problem belonging to the new research area of the control of stochastic distributions. In our view, new sparger designs will play a crucial role in increasing system controllability.

Another area which should attract research interest is the proper use of frother dosage as a control variable. So far, some reported control strategies rely on information coming either from froth characteristics (color, bubble size, froth velocity, *etc.*, usually inferred from image processing analysis) or from the collection zone (gas dispersion properties). A coordinated overall control strategy seems pertinent here. It is well known that frother content strongly affects both collection and froth zones. Indeed, frothers modify bubble size in the collection zone and stabilize the froth layer. Therefore, frother concentration may well turn out to be a relevant secondary variable to control. Frother is usually added at the head (front) of the flotation stage in a concentrator plant by implementing a ratio feedforward control based on processed tonnage, although in some places it is added at intermediate points in the flotation circuit. Nevertheless, because of the limitations of feedforward control in the presence of modeling uncertainties and unknown disturbances, such as frother persistency, evaporation rate and, most importantly, the effect of reprocessed water (containing remaining frother), frother concentration in a particular flotation machine is never precisely known, which makes its control rather complex.

Examples of column flotation control based on metallurgical objectives are rather scarce in the technical literature. What has been proposed almost exclusively relies on heuristic techniques, such as expert systems and fuzzy logic [61, 62, 80–82]. True model-based RTO, as described in Section 6.4.1, certainly deserves to be investigated. A *sine qua non* condition would be the availability of a mathematical model linking primary variables (metallurgy) to secondary (operating) variables.

As mentioned earlier in this section, the effect of froth depth on recovery is rather weak. Indeed, large froth depth set points would be required for significant effects on this primary variable, thus, its use for optimization purposes is rather limited. At this moment, there is still no consensus on the quantitative role played by bias rate on grade and recovery, owing to the lack of an appropriate sensor for this variable. The effect of S_b, ε_g and J_g on metallurgy has been studied in-depth, with each variable being studied separately. What is certainly missing is a comprehensive quantitative study aiming at determining the relationships between all of the above mentioned secondary variables and the primary variables, particularly to consider their interactions. The best way to select set points of the secondary variables as a function of the particle size distribution has not been studied either.

Reliable measurement and control of the secondary variables are required to make up for the lack of knowledge in this area. Thorough experimental campaigns consisting of a series of steady-state tests, each run at different values of the secondary set points would then be possible. During the tests, samples of feed, tail and concentrate could be taken and later processed for chemical assaying to evaluate the recovery. If available, an OSA could be used instead. These tests would allow the development of a model to predict grade and recovery from secondary variables. An economic criterion, such as the net smelter return, which is a function of grade and recovery, and therefore a function of the secondary variables, could then be maximized on-line.

Implementing a rigorous optimization plan requires investments involving financial (purchase of a simulator, hiring of a consulting firm, *etc.*), human and time resources (for implementation and minimum maintenance). However, the rewards, assessed in terms of:

- lower production costs (reagent consumption, *etc.*);
- better metallurgical performance (grade and recovery);
- lower product quality variability;
- benefits on personnel (availability, training on new technologies, *etc.*);
- fewer environmental impacts, etc;

may exceed original expectations and the payback period tends to be much shorter than in the case of major capital investments (a few weeks to a few months). This can partly be explained by the fact that tens and even hundreds of thousands of dollars of computer equipment, controllers, distributed control systems, programable logic controllers, and sensors are already available in mineral processing plants and, in most cases, only their simplest features are being used while the most powerful ones are often "left on the shelf", mainly due to the lack of human resources. In other words, many control wares have already been paid for, but their potential has not been fully exploited.

References

[1] Boutin P, Tremblay R (1964) Extraction of bitumen and oil from Athabaska tar sands. Canadian Patents CA680576, 1964-02-18
[2] Tremblay R, Boutin P (1967) Method and apparatus for the separation and recovery of ores. Canadian Patent No. 694547, 1964-09-15
[3] Wheeler DA (1985) Column Flotation – the original column. In Proceedings of 2nd Latin American Congress in Froth Flotation (S.H.Castro, J.Alvarez Eds.) 19-23 August, Chile. Also in Developments in Mineral Processing, Vol.9, Elsevier 1988, 17–40
[4] Cienski T, Coffin V (1981) Column flotation operation at Mines Gaspé molybdenum circuit. Proceedings of the 13th Annual Meeting of the Canadian Mineral Processors, CIM, Ottawa (Canada), 240–262
[5] Finch JA, Dobby G (1990) Column flotation. Pergamon Press, Oxford
[6] Clingan BV, McGregor DR (1987) Column flotation experience at Magma Copper Co. Minerals and Metallurgical Processing 3: 121–125
[7] Yianatos JB, Finch JA, Laplante AR (1987) The cleaning action in column flotation froths. Trans. of the Institution of Mining and Metallurgy, Section C96, C199–C205
[8] Gorain BK, Franzidis J-P, Manlapig EV (1997) Studies on impeller type, impeller speed and air flow rate in an industrial-scale flotation cell - Part 4: Effect of bubble surface area flux on flotation performance. Minerals Engineering 10(4):367–379
[9] Hernandez-Aguilar JR, Rao SR, Finch JA (2005) Testing the k-Sb relationship at the microscale of laboratory. Minerals Engineering 18:591–598
[10] Kratch W, Vallebuona G, Casali A (2005) Rate constant modeling for batch flotation, as a function of gas dispersion properties. Minerals Engineering 18:1067–1076

[11] Nesset JE, Hernandez-Aguilar J, Acuña C, Gomez CO, Finch, JA (2006) Some gas dispersion characteristics of mechanical flotation machines. Minerals Engineering 19:807–815
[12] Dobby GS, Finch JA (1986) Particle collection in columns - Gas rate and bubble size effects. Canadian Metallurgical Quarterly 25:9–13
[13] Finch JA, Nesset JE, Acuña C (2008) Role of frother on bubble production and behaviour in flotation. Minerals Engineering 21:949–957
[14] Laskowski JS (2003) Fundamental properties of flotation frothers. In: Lorenzen, L., Bradshaw, D.J. (Eds.), Proceedings XXIInd International Mineral Processing Congress. S. African IMM, 788–797
[15] Finch JA, Gelinas S, Moyo P (2006) Frother-related research at McGill University. Minerals Engineering, 19:726–733
[16] Azgomi F, Gomez CO, Finch JA (2007) Correspondence of gas hold-up and bubble size in presence of different frothers. International Journal of Mineral Processing 83(1–2):1–11
[17] Maldonado M, Desbiens A, del Villar R, Aguilera R (2009) On-line estimation of frother concentration in flotation processes. Proceedings of the 48th Conference of Metallurgists COM09 (C. Gomez, Eds), August 23–26, Sudbury (Canada)
[18] del Villar R, Pérez R, Díaz G (1995) Improving pulp level detection in a flotation column using a neural network algorithm. Proceedings of the 27th Annual Meeting of the Canadian Mineral Processor, CIM, Ottawa (Canada), 83–100
[19] Moys MH, Finch JA (1988) Developments in the control of flotation columns. International Journal of Mineral Processing 23:265–278
[20] Gomez CO, Uribe-Salas A, Finch JA, Huls BJ (1989) A level detection probe for industrial flotation columns. Proceedings of an International Symposium of Complex Ores (B.J. Huls, G.E. Agar and D.B. Hyma, Eds), Vol.II, Halifax (Canada), 325–334
[21] del Villar R, Grégoire M, Pomerleau A (1999) Automatic control of a laboratory flotation column. Minerals Engineering 12: 291–308
[22] Uribe-Salas A, Leroux M, Gomez CO, Finch JA, Huls BJ (1991) A conductivity technique for level detection in flotation cells. Proceedings of an International Conference on Column Flotation, Column'91. (B.J. Huls, G.E. Agar and D.B. Hyma Eds), Sudbury (Canada). Vol. II. 467–478
[23] Pérez-Garibay R, del Villar R, Flament F (1993) Level detection in a flotation column using an artificial neural network. Proceeding of the XXIV APCOM Conference (J. Elbrond, X. Tang Eds). Montreal (Canada), 174–181
[24] Maldonado M, Desbiens A, del Villar R (2008) An update on the estimation of the froth depth using conductivity measurements. Minerals Engineering 21:935–939
[25] Uribe-Salas A, Gomez CO, Finch JA (1991) Bias detection in flotation columns. Column'91-Proceedings of an International Conference on Column Flotation. Vol.2 (G.E. Agar, B.J. Huls, D.B. Hyma Eds), Canadian Institute of Mining, Metallurgy and Petroleum, Sudbury (Ontario), 391–407
[26] Moys MH, Finch JA (1988) The measurement and control of level in flotation columns. Proceedings of an International Symposium on Column Flotation, EMI Annual Meeting (K.V.S. Sastry Ed.), Phoenix (USA), 90–107
[27] Pérez-Garibay R, del Villar R (1998) Estimation of bias and entrainment in flotation columns using conductivity measurements. Canadian Metallurgical Quarterly 36(5):299–307
[28] Vermette H (1997) Mesure du biais dans une colonne de flottation par profils de température et conductivité. M.Sc. thesis, Département de mines et métallurgie, Université Laval, Québec City (Canada)
[29] Aubé V (2003) Validation semi-industrielle des capteurs de profondeur d'écume et du différentiel d'eau dans une colonne de flottation. M.Sc. thesis, Département de mines et métallurgie, Université Laval, Québec City (Canada)
[30] Bouchard J, Desbiens A, del Villar R (2005) Recent advances in bias and froth depth control in flotation columns. Minerals Engineering 18:709–720

[31] Desbiens A, del Villar R, Milot M (1998) Identification and gain-scheduled control of a pilot flotation column. 9th IFAC Symp. on Automation in Mining, Mineral & Metal Processing, Cologne (Germany), 337–342
[32] Maldonado M, Desbiens A, del Villar R, Chirinos J (2008) On-line bias estimation using conductivity measurements. Minerals Engineering 21:851–855
[33] Maldonado M, Desbiens A, del Villar R (2008) Decentralized control of a pilot flotation column: a 3X3 system. Canadian Metallurgical Quarterly 47:377–386
[34] Maldonado M, Desbiens A, del Villar R (2009) Potential use of model predictive control for optimizing the column flotation process. International Journal of Mineral Processing 93(1):26–33
[35] Carvalho MT, Durão F (2002) Control of a flotation column using fuzzy logic inference. Fuzzy Sets and Systems 125:121–133
[36] Persechini MAM, Peres AEC, Jota FG (2004) Control strategy for a column flotation process. Control Engineering Practice 12:963–976
[37] Nunez E, Desbiens A, del Villar R, Duchesne C (2006) Multivariable predictive control of a pilot flotation column. Part 1: Online sensors for froth depth and air hold-up. International Conference on Mineral Process Modelling, Simulation and Control (E. Yalcin and H. Shang, Eds). Sudbury, Ontario (Canada), 291–301
[38] Uribe-Salas A (1991) Process measurements in flotation columns using electrical conductivity. Ph.D. thesis, Department of Mining and Metallurgical Engineering, McGill University, Montreal (Canada)
[39] Marchese M, Uribe-Salas A, Finch JA (1992) Measurement of gas holdup in a three-phase concurrent down-flow column. Chemical Engineering Science 47(13–14):3475–3482
[40] Perez-Garibay R, del Villar R (1999) On-line gas hold-up measurement in flotation columns. Canadian Metallurgical Quarterly 38(2):141–148
[41] Arizmendi-Morquecho AM, Perez-Garibay R, Uribe-Salas A, Nava-Alonso F (2002) On-line solids hold-up measurement in mineral slurries by standard addition method. Minerals Engineering 15(1-2):61– 64
[42] Taveras FJ, Gomez CO, Finch JA (1996) A novel gas hold-up probe and applications in flotation columns. Trans of I.M.M., Sect. C 105:C99–C104
[43] O'Keefe C, Viega J, Fernald M (2007) Application of passive sonar technology to mineral processing and oil sands application. Proceedings of the 39th Annual Meeting of the Canadian Mineral Processors, CIM, Ottawa (Canada), 429–457
[44] Gorain BK, Franzidis J-P, Manlapig EV (1995) Studies on impeller type, impeller speed and air flow rate in an industrial-scale flotation cell – Part 2: Effect on gas hold-up. Minerals Engineering 12:1557–1570
[45] Power A, Franzidis JP, Manlapig EV (2000) The characterization of hydrodynamic conditions in industrial flotation cells. In Proceedings of AusIMM 7th Mill Operators Conference, Australasian Institute of Mining and Metallurgy, Kalgoorlie, 243–255
[46] Deglon DA, Egya-Mensah D, Franzidis, JP (2000) Review of hydrodynamics and gas dispersion properties in flotation cells in South African platinum concentrators. Minerals Engineering 13(3):235–244
[47] Yianatos JB, Bergh L, Condori P, Aguilera J (2001) Hydrodynamic and metallurgical characterization of industrial flotation banks for control purposes Minerals Engineering 14(9):1033–1046
[48] Grau R, Heiskanen K (2003) Gas dispersion measurements in flotation cell. Minerals Engineering 16:1081–1089
[49] Randall EW, Goodall CM, Fairlamb PM, Dold PL, O'Connor CT (1989) A method for measuring the sizes of bubbles in two- and three-phase systems. Journal of Physics, Section E, Scientific Instrumentation 22:827–833
[50] Tucker JP, Deglon DA, Franzidis JP, Harris MG, O'Connors CT (1994) An evolution of a direct method of bubble size distribution measurement in a laboratory batch flotation cell. Minerals Engineering 7(5-6):667–680

[51] Chen F, Gomez CO, Finch JA (2001) Bubble size measurement in flotation machines. Minerals Engineering 13(10):1049-1057
[52] Bartolacci G, Ourriban M, Lockhart A, Michaud F, Faucher A, Knutila D, Finch JA, Fortin A, Goyette G (2007) Effect of process conditions on froth stability in an industrial copper flotation circuit. In Proceedings of the Sixth International Copper Conference (R. del Villar, J.E. Nesset, C.O. Gomez, A.W. Stradling Eds), August 25–30, Toronto (Canada), Vol. 2., 221–232
[53] Grau R, Heiskanen K (2002) Visual technique for measuring bubble size in flotation machines. Minerals Engineering 15:507–513
[54] Maldonado M, Desbiens A, del Villar R, Girgin EH, Gomez CO (2008) On-line estimation of bubble size distributions using gaussian mixture models. Proceedings of V International Mineral Processing Seminar, Procemin 2008 (R. Kuyvenhoven, C.O. Gomez and A. Casali Eds), Santiago (Chile). October 22–24, 389–398
[55] Rodrigues R, Rubio J (2003) New basis for measuring the size distribution of bubbles. Minerals Engineering 16:757–765
[56] Couto HJB, Nunez DG, Neumann R, França SCA (2009) Micro-bubble size distribution measurements by laser diffraction technique. Minerals Engineering 22:330–335
[57] Berton A, Hodouin D (2003) Linear and bilinear fault detection and diagnosis based on energy and mass balance equations. Control Engineering Practice 11:103–113
[58] Gertler J (1998) Fault detection and diagnosis in engineering systems. Marcel Dekker, New York
[59] Romagnoli JA, Sanchez MC (2000) Data processing and reconciliation for chemical process operations. Academic Press
[60] Kailath T, Sayed AH, Hassibi B (2000) Linear estimation. Prentice-Hall, New Jersey
[61] Bergh LG, Yianatos JB, Cartes F (1996) Hierarchical control strategy at el Teniente flotation columns. Proceedings of International Conference on Column Flotation'96, Montreal (Canada), 369–380
[62] Bergh LG, Yianatos JB, Acuna C, Perez H, Lopez F (1999) Supervisory control at Salvador flotation columns. Minerals Engineering, 12:733–744
[63] Edgar T, Himmelblau DM (2001) Optimization of chemical processes. McGraw-Hill, 2nd edn
[64] Bergh LG, Yianatos JB (2003) Flotation column automation. Control Engineering Practice 11:67–72
[65] Bouchard J, Desbiens A, del Villar R, Nunez E (2009) Column flotation simulation and control: An overview. Minerals Engineering 22:519–522
[66] Ljung L (1999) System identification - Theory for the user. Prentice Hall, 2nd edn
[67] Verhaegen M, Verdult V (2007) Filtering and system identification - a least squares approach. Cambridge University Press, New York
[68] Chuk OD, Mut V, Núñez E, Gutierrez L (2001) Multivariable predictive control of froth depth and gas holdup in column flotation. 10th IFAC Symposium on Automation in Mining, Mineral & Metal Processing, Tokyo, Japan, 87–91
[69] Nunez E, Desbiens A, del Villar R, Duchesne C (2006) Multivariable predictive control of a pilot flotation column. Part 2: identification and control. International Conference on Mineral Process Modelling, Simulation and Control, Sudbury, Ontario, Canada, 291–301
[70] Liu JJ, MacGregor JF (2008) Froth-based modeling and control of flotation processes. Minerals Engineering 21:642–651
[71] Barria A, Valdebenito M (2008) Implementation of rougher flotation control system at Codelco Chile, Andina division. Procemin2008, V International Mineral Processing Seminar, Santiago, Chile, 215–220
[72] Cortés G, Verdugo M, Fuenzalida R, Cerda J, Cubillos E (2008) Rougher flotation multivariable predictive control; Concentrator A-1 division Codelco Norte. Procemin2008, V International Mineral Processing Seminar, Santiago, Chile, 315–325
[73] Moilanen J, Remes A (2008) Control of the flotation process. Procemin2008, V International Mineral Processing Seminar, Santiago, Chile, 305–313

[74] Camacho EF, Bordons C (2004) Model predictive control. Springer-Verlag, London
[75] Maciejowski J (2002) Predictive control with constraints. Prentice Hall, New York
[76] Rossiter JA (2003) Model-based predictive control: a practical approach. CRC Press
[77] Qin SJ, Badgwell TA (2003) A survey of industrial model predictive control technology. Control Engineering Practice 11:733–764
[78] Ameluxen R, Llerena R, Dustan P, Huls B (1988) Mechanics of column flotation operation. SME Annual Meeting Column Flotation'88, Society of Mining, Metallurgy and Exploration, Phoenix, Arizona, 149–155
[79] Barrière PA, Dumont F, Desbiens A (2001) Using semi-physical models for better control. Part 2: nonlinear control of a pilot flotation column. 10th IFAC Symposium on Automation in Mining, Mineral & Metal Processing, Tokyo, Japan, 131–136
[80] Bergh LG, Yianatos JB, Leiva C (1998) Fuzzy supervisory control of flotation columns. Minerals Engineering 11:739–748
[81] Hirajima T, Takamori T, Tsunekawa M, Matsubara T, Oshima K, Imai T (1991) The application of fuzzy logic to control concentrate grade in column flotation at Tayoha mines. Proceedings of International Conference on Column Flotation'91 vol.2, Sudbury, Canada, 375–389
[82] Kosick G, Dobby GS, Young P (1991) Columnex A powerful and affordable control system for column flotation. Proc. of Intern. Conf. on Column Flotation'91 vol. 2, Sudbury, Canada, 359–373
[83] McKay JD, Ynchausti RA (1996) Expert supervisory control of flotation columns, Proceedings of International Conference on Column Flotation'96, 353–367
[84] Milot M, Desbiens A, del Villar R, Hodouin D (2000) Identification and multivariable nonlinear predictive control of a pilot flotation column. XXI International Mineral Processing Congress, Roma (Italy), A3.120–A3.127
[85] Pu M, Gupta Y, Taweel AA (1991) Model predictive control of flotation columns. Proc. of Intern. Conf. on Column Flotation'91 vol. 2, Sudbury, Canada, 467–478
[86] Yianatos JB, Bergh LG (1995) Troubleshooting industrial flotation columns. Minerals Engineering, 8:1593–1606
[87] Desbiens A, del Villar R, Poulin E (2009) Real-time optimization of flotation column operation. Collaborative Research and Development Grant, NSERC, Canada
[88] Duchesne C, Bouajila A, Bartolacci G, Pelletier P, Breau Y, Fournier J, Girard D (2003) Application of multivariate image analysis (MIA) to predict concentrate grade in froth flotation processes. Proceedings of 35th Annual Meeting of the Canadian Mineral Processors, CIM, Ottawa, Canada, 51–526
[89] Liu J, McGregor J, Duchesne C, Bartolacci G (2005) Flotation froth monitoring using multiresolutional multivariate image analysis. Minerals Engineering 18:65–76
[90] Bartolacci G, Pelletier P, Tessier J, Duchesne C, Bossé PA, Fournier J (2006) Application of numerical image analysis to process diagnosis and physical parameter measurement in mineral processes - Part I: Flotation control based on froth textural characteristics. Minerals Engineering 19:734 –747

Chapter 7
Industrial Products for Advanced Control of Mineral Processing Plants

Aldo Cipriano

Abstract This chapter begins with some general observations on the strategic importance automation technologies have acquired in the mining business, and then reviews a series of recently available commercial advanced sensors for grinding and flotation plants. This is followed by a brief look at developments in advanced control and the main concepts involved in the solutions offered. A number of tools for advanced control development are presented and the reported benefits obtained in real-world applications are noted. The chapter ends with conclusions and a brief discussion of current tendencies.

7.1 Introduction

In recent years there has been a slow but steady growth in awareness of the contribution automation technologies can make to each of the various processes in the mining business value chain. Among producers this recognition is manifested by such trends as the creation of automation or process control management areas or units within company structures, the adoption of a vision that integrates automation with information and communication technologies, the strategic status these technologies have been given, and the formation of technology companies and research laboratories dedicated to the development of products and systems based on these technologies oriented specifically towards the mining industry [1].

This increased activity has borne fruit, as is demonstrated by the significant growth in the quantity of components, products and systems aimed at developing both basic and advanced automation functions such as those described in Chapter 1 of this volume. For each of these functions, suppliers of instrumentation equipment and automation and control systems are offering new solutions and commercial al-

Aldo Cipriano
Department of Electrical Engineering, Pontificia Universidad Católica de Chile, Santiago, Chile, e-mail: aciprian@ing.puc.cl

ternatives, some with broad-ranging industrial applications and others conceived exclusively for the mining sector. Among these solutions, particularly the ones intended for use in crushing, grinding and flotation, new sensors and analyzers, as well as environments and systems for the development and implantation of advanced control strategies are outstanding.

The rest of this chapter provides an overview of products offered in the market, focusing on the benefits they provide. Unlike other more complete analyses such as that by Hodouin *et al.* [2] and Jamsa-Jounela [3], we will confine ourselves to a limited number of products and mention research findings just to clarify major tendencies.

7.2 Advanced Sensors

The possibilities for developing control strategies in an industrial process are highly dependent on what sensors are available. In the case of mineral processing, sensors exist for measuring a variety of process variables such as those listed below:

- feeder frequency, conveyor load and crusher chute level (crushing);
- tonnage, water flow rate, mill speed, pulp level, pump speed, pulp volumetric flow rate, pulp density, cyclone and mill pressure, and the power draw of mills, screens, pumps and cyclones (grinding);
- pulp flow rate, cell and column pulp levels, air flow rate, reagent flow rate, wash water flow rate and pH (flotation).

To these instruments, which may be categorized as basic given the measurement principles they utilize, we add two more advanced ones that are also widely used: pulp particle size distribution sensors in grinding and grade analyzers in flotation.

The last few years have seen a significant increase in the availability of industrial sensors for mineral processing, some of which incorporate highly sophisticated technologies. These include units that take granulometric size measurements on crushing or grinding belt conveyors, determine mill operating states and characterize the surface froth in flotation cells.

7.2.1 Size Distribution Sensors

The first development of size distribution sensors based on image processing date back to 1985 [4]. Currently, there are various products that integrate hardware and software to determine size distribution using this technique, such as Split-Online (Split Engineering) [5] and VisioRock (Metso Minerals) [6]. KnowledgeScape [7], Tata [8] and WipWare [9] also incorporate developments in this method.

Split-Online is an automated digital image analysis system designed to operate continuously and monitor locations suitable for permanently installed cameras. This technology captures images from one or more cameras at a time interval preset by the user or in response to a trigger signal. Its software architecture is object-oriented. Split-Online applications include rock fragmentation size with an estimation for fine material, determination of agglomeration quality using texture classification, automatic image acquisition of fragmentation bucket-by-bucket, and size distribution curve averaging for each truckload on cable shovels.

VisioRock [10] is a vision technology that can determine the texture and color separately for each size class and discriminate between ore types and size distribution for different types of particles. The hardware comprises one or several cameras and their associated lighting systems, camera–computer communication hardware and one or more computers. The vision software is part of the vision module in the optimization control system (OCS©[1]) software. The OCS© vision module embeds numerous image processing algorithms.

VisioRock applications include evaluation of rock size distribution on conveyor belts, the most common situations being:

1. crusher product – automatic help for setting the crusher gap;
2. semi-autogenous (SAG) or autogenous (AG) mill feed – information on feed size to enable more highly optimized mill operation;
3. screen undersize – detecting holes in the screen or abnormal operation; and
4. feed to critical equipment – detecting wood logs or other abnormal objects on the conveyor belt that may damage equipment or disturb production.

The size distribution of grinder feed ore is one of the main disturbances affecting the mill process. The measurements provided by these sensors can therefore be usefully incorporated as a feed-forward variable in automatic mill control strategies. Since the sensors extract the data from the image analysis of surface ore, the measurements may not be sufficiently representative of the material as a whole. This factor led to the development of certain variants using, for example, 3-D laser scanning [11].

7.2.2 Mill Operating Condition Sensors

Variables such as power draw that are typically measured for supervising and controlling grinding mills only partially describe the operating condition of these installations, particularly if they are autogenous or semi-autogenous. This shortcoming explains the efforts made to develop sensors or instrumentation systems that measure other variables. The results of some of these efforts are outlined below.

The Electric Ear (Metso Minerals) [6] consists of a microphone mounted adjacent to the mill, with an amplifier and controller installed remotely. The microphone converts the sound into electrical energy and transmits it to the controller. The sound

[1] OCS is copyright of Metso Minerals Cisa BP 6009, 45060 Orlans Cedex 2, France, www.svedala-cisa.com

emitted from the mill varies in intensity as the quantity and average size of the material being ground changes. A high noise level indicates an underloaded mill and a low noise level, an overloaded mill. An underloaded mill leads to an excessive wear rate of grinding media and mill lining. As a mill becomes overloaded, movement of the grinding media is increasingly restricted and grinding diminishes.

MillSense™(Outotec) [12] is a non-contact analyzer for grinding mills providing real-time information on mill charge volume, ball charge, liner wear and grinding media trajectories. Along with mill rotation speed, charge characteristics can be used to maximize the breakage energy and therefore optimize the grinding efficiency. Charge volume measurement can also be used to avoid both overloading and underloading.

The Impactmeter [13] is an instrument comprising acoustic sensors, a processor and software that gathers and deciphers impact signals and transmits them to provide vital information relating to load impacts in real time and on-line. This information may be used either directly by the mill operator or by the mill's automatic control system to make operational adjustments that optimize impact energy and mill performance while protecting mill linings and grinding media from excessive wear, degradation and/or breakage.

The SAG Analyzer is an instrument equipped with software that identifies the charge toe position and shoulder angles on the basis of acoustic signals. Using this information together with calibrated plant data, the instrument provides on-line and real-time estimates of mill fill level as a percentage of mill volume and ball charge and ore charge as percentages of the total mill load. The SAG Analyzer also indicates the dynamic total apparent load density.

FLSmidth Minerals (formerly FFE Minerals) [14] supplies the Impactmeter and SAG Analyzer either as a combined instrument or as independent, stand-alone units.

7.2.3 Flotation Cell Froth Sensors

Operators of flotation plants frequently monitor cell surface froth visually and make adjustments based on interpretation of its appearance. They categorize froths into a number of distinct types and develop appropriate operating strategies for each one, looking for the ideal froth appearance. This is the basis for proposals to use image processing to obtain surface froth characterizations [15]. Image processing sensors provide quantitative measures of certain representative froth characteristics and offer the option of making automatic control decisions depending exclusively on them. These measures can be classified as static or dynamic features, the former including bubble size, bubble count and bubble density and the latter comprising froth color, froth speed and froth stability.

Below we briefly describe FrothMaster™ [2](Outotec) [12], JK FrothCam[3] (JK-Tech Pty. Ltd.) [16] and VisioFroth[4] (Metso Minerals Cisa) [17], three industrial products based on the foregoing principles.

Outotec's FrothMaster™ is a tool for real-time froth surface analysis. It measures several essential properties of froth appearance including froth speed and direction, bubble size distribution and froth stability, and generates a froth color histogram (RGB). The product also provides statistical data related to these variables. The froth measurements plus on-stream analyzer assays provide the information necessary for expert control of the flotation circuit by means of a hierarchical and multivariable system (FrothCon™) that controls three froth properties – speed, stability and color – and dynamically maximizes cell efficiency. FrothMaster™ also detects abnormal situations such as froth collapse. Finally, FrothMaster™ includes web server technology.

JK Frothcam employs a statistical approach known as the texture spectrum method to measure both bubble size and froth structure. A classification algorithm using patterns of texture spectra recognises froth structures that can be related to variations in flotation process performance. The height of the middle peak of the texture spectrum provides a means of estimating the average bubble size of the froth image. Froth speed is measured using an object-matching algorithm called pixel tracing. The technique assumes that both froth shape and light intensity remain unchanged during measurement.

VisioFroth, developed by Metso Minerals Cisa, furnishes mean and standard deviation measures of froth speed (in cm/s), froth stability (in %), bubble size distribution, color and brightness indexes and froth texture. It is typically installed to stabilize the froth speed by manipulating level and/or airflow set-points using the average speed signals from the OCS© statistical module. The product offers 20 methods for calculating the froth speed based on correlation, masking and fast Fourier transforms. The bubble size measurement employs a watershed-based approach.

In addition to these products, other commercial sensors based on the same technology include the PlantVision™[5] image analysis system for flotation (KnowledgeScape) [7] and WipFroth (WipWare) [9]. There has been extensive research and publications on the subject. Morar *et al.* [18] present the results of test work conducted at Rio Tinto's Kennecott Utah Copper Concentrator on the application of SmartFroth, developed at the University of Cape Town investigating the relationship between froth color and concentrate grade. Morar *et al.* show that models can be developed to predict concentrate grade based on color information and velocity and stability information. Kaartinen *et al.* [19] describe a novel approach in which

[2] FrothMaster is a registered trademark of Outotec Riihitontuntie 7 D, PO Box 86, FI-02200 Espoo, Finland, www.outotech.com

[3] FrothCam is a registered trademark of JKTechPty. Ltd. 40 Isles Road, Indooroopilly. QLD 4068 Australia, www.jktech.com.au

[4] VisioFroth is copyright of Metso Minerals Cisa BP 6009, 45060 Orlans Cedex 2, France, www.svedala-cisa.com

[5] PlantVision is a registered trademark of KnowledgeScape 254 South 600 East, Suite 202, Salt Lake City, UT, 84102, USA, www.kscape.com

a multi-camera system and expert controllers are combined to improve flotation performance. Forbes [20] shows that the Fourier ring and texture spectrum-based features perform well in classifying new images at a relatively low computational cost; results from three industrial case studies show that bubble size and texture measures can be used to identify froth classes. As described in Chapter 3, Lin *et al.* [21] present a modified texture spectrum approach and a reliable method based on binary images of bubble size estimation. Liu and MacGregor [22] use multiresolutional multivariate image analysis for numerical estimation of froth status. Núñez and Cipriano [23] introduce a new method for the characterization and recognition of visual information using dynamic texture techniques for prediction of froth speed.

Outotec has evaluated the benefits of using visual information provided by the froth characteristics sensor FrothMaster™ at Newcrest Mining's Cadia Valley Operations in New South Wales, one of Australia's largest gold producers. As Figure 7.1 shows, a first controller, using FrothMaster™ data, adjusts cell slurry level, aeration rate and reagent addition to control the froth speed set-point. A second controller uses the process data from the on-stream analyzer to manipulate the froth speed set-point so as to achieve a grade set-point specified by the plant metallurgist. For the purposes of the evaluation, three FrothMaster™ units were installed in the first three cells of line 1, enabling grade and recovery calculations to be accurately compared with their manually controlled counterparts in line 2. The results achieved in line 1 showed significant improvements in both grade control and product recovery. During a 44-day trial period the average grade error in line 1 as a percentage of target remained within 0.03%, while the average grade error recorded in line 2 was 9.99%. Copper recovery from the line 1 first rougher circuit was 5.28% higher than from the corresponding cells in line 2, while gold recovery was 5.14% higher. At the conclusion of the trial period, average frother consumption in line 1 was reduced by 7.1%, representing a significant saving in annual processing costs.

7.3 Concepts for Advanced Control

An analysis of the available information on advanced control in mineral processing reveals that despite the wide variety of existing methodologies and techniques for process control, only a handful have given rise to commercial products applied in industrial settings. These techniques may be divided into two categories:

- those we will call intelligent control, including ES and fuzzy logic;
- model predictive control, using linear or non-linear models originating in phenomenological or empirical models adjusted on the basis of operating data.

Both categories have their own underlying concepts and application methodologies, and attempts have been made to combine them into a single integrated solution. This is the case, for example, with the algorithms known as fuzzy model predictive control [24].

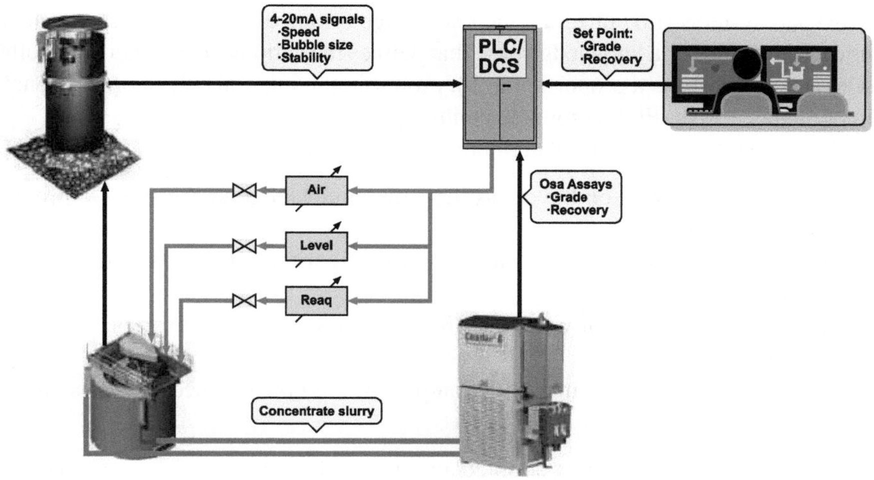

Figure 7.1 Visual information-based flotation control strategy evaluated at Newcrest Mining's Cadia Valley Operations

In crushing, grinding and flotation control there are numerous applications in each of the categories. Before presenting various industrial products that make use of these approaches, we briefly review the fundamental ideas behind them [25].

7.3.1 Intelligent Control

In engineering, knowledge is formulated in terms of mathematical models based on the laws of physics or chemistry that are generally expressed as ordinary or partial differential equations or difference equations. These models are used to quantitatively analyze the behavior of processes or systems either through analytical studies or, where that is not possible, via simulation-based numerical studies. If the models are dynamic they can be employed to develop automatic control algorithms, strategies or systems [26].

Although model-based methodologies have a solid foundation, in many cases mineral processing plant operators resort with considerable success, to decisions based on their own experience or that of colleagues or supervisors. Such empirical knowledge may be formulated as decision rules of the type

IF antecedent THEN consequent

In automatic control applications, the antecedent generally includes comparisons between variables measured by the sensors and reference values or operating limits as well as logical operations between linguistic variables defined by partitioning into sets the operating ranges of a measured variable. The consequent usually implies assigning to manipulated variables values that are sent to the actuators [27].

Expert systems (ES) integrate the knowledge of one or more process specialists into a set of rules or a knowledge base that defines the actions of an expert controller who acts similarly to a proportional (P), proportional–integral (PI) or proportional–integral–derivative (PID) automatic control algorithm.

The forms of expert control are numerous, and may:

- be based exclusively on the controller's own knowledge of the specific process being controlled;
- use rules to reproduce conventional control algorithms for P, PI, PID or other control methods;
- incorporate estimates of non-measured variables based, for example, on time-series models;
- include certainty factors in the ES control rules or take into account errors in the measurements.

One of the most frequently adopted alternatives for improving the robustness of expert control systems in handling uncertainties and errors is fuzzy logic. In this approach, a membership function μ is assigned to each linguistic variable or fuzzy set x such that $0 \leq \mu(x) \leq 1$. The most commonly used membership functions are triangular, trapezoidal or Gaussian [28, 29].

Expert systems that incorporate fuzzy logic into processing the rules are known as fuzzy ES. Since the first practical use of fuzzy logic in a control application by Mamdani [30], many advances have been made at the theoretical level as well as in applications. The wide array of variants now existing that use fuzzy logic for control include:

- those that only incorporate fuzzy sets in the rule consequents, as proposed by Mamdani [31];
- those that include mathematical expressions in the rule consequents, as suggested by Sugeno [32].

In both cases the basic concepts of automatic control can be extended through the introduction of notions such as fuzzy controller [33], fuzzy model [34, 35], fuzzy system stability [36] or fuzzy model identification [37]. By way of example, Figure 7.2 shows the modular structure of a fuzzy controller that processes signals coming from sensors (crisp inputs) to generate the signals that will be sent to the actuators (crisp outputs).

7.3.2 Model Predictive Control

Model predictive control (MPC) embraces a complete family of controllers [38] whose basic concepts are: (1) the use of an explicit dynamic model to predict process outputs at discrete future time instants over a prediction horizon; (2) computation of a sequence of future control actions through the optimization of an objective

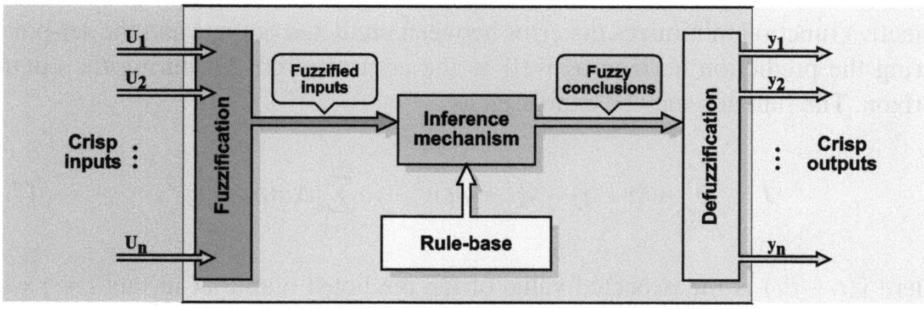

Figure 7.2 Main components of a fuzzy controller

function with given operating constraints and desired reference trajectories for process outputs; and (3) repetition of the optimization process at each sampling instant and application of the first value of the calculated control sequence (receding horizon strategy). These three characteristics allow MPC to handle multivariable, non-minimum phase, open-loop unstable and non-linear processes with a long time delay and the inclusion, if necessary, of constraints for manipulated and/or controlled variables. As shown in Figure 7.3, at instant t future process outputs are predicted over the prediction horizon N_p using an explicit model. The predictions depend on known values of the manipulated and controlled variables and future control actions $u(t+j|t)$ over the control horizon N_u. It is assumed that $N_u \leq N_p$ and that $u(t+j|t)$ remains constant for $j = N_u, ..., N_p$. The prediction horizon N_p is related to the response time of the process and N_u is usually chosen so as to be equal to the order of the model.

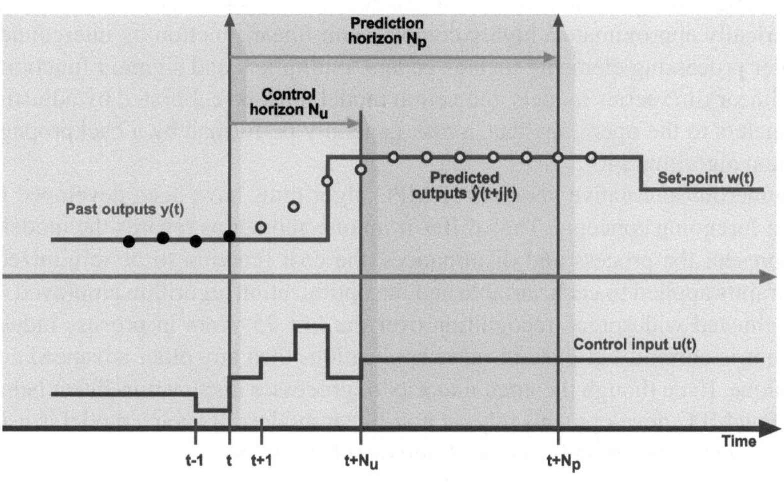

Figure 7.3 MPC strategy

The control sequence is obtained by optimizing an objective function that describes the goals the control strategy is intended to achieve. In classical MPC, an objective function minimizes the error between predicted outputs and the set-points during the prediction horizon as well as the control effort Δu during the control horizon. The function may be expressed as

$$J = \sum_{j=N_1}^{N_p} [w(t+j) - \hat{y}(t+j|t)]^2 + \lambda \sum_{j=1}^{N_u} [\Delta u(t+j|t)]^2, \tag{7.1}$$

where $\hat{y}(t+j|t)$ is the expected value of the predicted output at instant $t+j$ with known history up to time t, $w(t+j)$ is the set-point at instant $t+j$, and $\Delta u(t) = u(t) - u(t-1)$ with $u(t)$ the manipulated variable at instant t. Parameter λ weights the control in the optimization problem. N_1, N_u and N_p define the prediction and control horizons.

The optimization process may involve hard or soft constraints. For linear unconstrained systems this optimization problem is tractable and convex and can be solved analytically, but in general applications it is common to take into account constraints or non-linearities in the process, and in such cases the optimization problem must be solved using iterative numerical methods [39].

A fundamental element in MPC is the model used to characterize the dynamic behavior of the process. The origins and formulations of such models are diverse, but may be classified as follows:

- phenomenological or first principle models, in the vast majority of cases non-linear and continuous time;
- models obtained through numerical adjustments based on operating data using discrete time series, either linear or non-linear.

Notable among the non-linear models are neural networks, which are used to numerically approximate a highly complex non-linear function by interconnecting simpler processing elements such as adders, multipliers and sigmoid functions. As with linear time series models, the neural model must be calibrated by adjusting its parameters to the operating data, a task generally performed by a backpropagation gradient algorithm [40].

Numerous alternative versions of MPC algorithms have been developed based on the foregoing concepts. They differ from one another as regards the model used to represent the process and disturbances, the cost function to be minimized, the constraints applied to each variable and the optimization algorithm employed. MPC has achieved widespread recognition over the last 25 years in process industries, where it is currently utilized in more applications than any other advanced control technique. Even though the great majority of processes display non-linear behavior, very few MPC developments rely on non-linear models. If such a model is used, or if the objective function is not quadratic or subject to constraints, the optimization problem will not be convex, difficulties will arise regarding the existence of local optima and computational complexity will be much greater.

The popularity of MPC among academics is evidenced by the large number of books and articles published on the subject. Unlike many other process control techniques, however, MPC is also widely accepted in the industry. This is amply attested to by the quantity of commercial products that employ it and the 279 MPC-based patents that have been registered with the United States Patent and Trademark Office between 1976 and October 2008 [41].

In recent years, the application of MPC to hybrid dynamic systems has emerged as a significant area of research [42]. In these systems, continuous dynamic sub-processes interact with discrete event detection elements and start/stop commands. Characterizing this type of system involves combining continuous with discrete variables and differential or difference equations with finite state automata or switching theory. Although this approach increases the complexity of the model, its potential for accurately capturing the dynamic of an industrial process is much greater.

Until recently, MPC applications in mineral processing were few and far between, as was revealed in a 2003 study by Qin and Badgwell which found that only 37 of the 4,542 applications then detected were in the mining and metallurgy sector [43]. In just the last few years, however, this trend has changed markedly, as the next section will demonstrate.

7.4 Advanced Control Tools and Applications

Currently many companies working in the areas of instrumentation, automation and control are developing and offering commercial products for mineral processing plants. For reasons of space and available information we will focus here on only the following firms: ABB, Andritz, Honeywell, Invensys, Metso Minerals, Mintek and SGS. We note in passing two other suppliers of multivariable predictive control products, namely Emerson Process Management with its Delta V system [44] and Rockwell [45], provider of Pavilion Technologies [46], see Table 7.1.

Table 7.1 MPC suppliers and technologies

Company	Technology
ABB	Expert Optimizer
Andritz	BrainWave®[6]
Emerson Process Management	Delta V
Honeywell	Profit®[7] Suite
Invensys	Connoisseur
Metso Minerals	Optimizing Control System
Mintek	StarCS
Rockwell	Pavilion Technologies
SGS	MinnovEX Expert Technology

[6] BrainWave is a registered trademark of IDEAS Simulation & Control Inc.

[7] Profit is a registered trademark of Honeywell International, Inc.

ABB

ABB Expert Optimizer [47] is a hybrid intelligent system designed primarily for expert closed-loop process control and industrial process optimization, but can also be used for decision support applications. Figure 7.4 depicts the concept of the control algorithm; see European Patent EP1553470 [48]. It is based on a mathematical model of the process being controlled, for example:

$$\frac{dx}{dt} = f(t,x,u); \qquad h(t,x,u) = 0, \tag{7.2}$$

where x is the system state and u is the control variable. The algorithm uses optimal control (OC) and/or MPC techniques to select the ES targets U. ES keeps the process variables within certain bounds or close to predefined target values acting on the control variables C. OC and MPC work by minimizing the functional

$$J[x(0),u] = \int_0^T L(\tau,x(\tau),u(\tau))d\tau \tag{7.3}$$

with respect to u, subject to the constraints imposed by the mathematical model of the extended system given by the process P and the expert system ES. This model is a hybrid in the sense that both continuous dynamics (mostly process) and logical relationships (mostly ES) appear in it. A mixed logical dynamical system approach is recommended for modeling the extended plant.

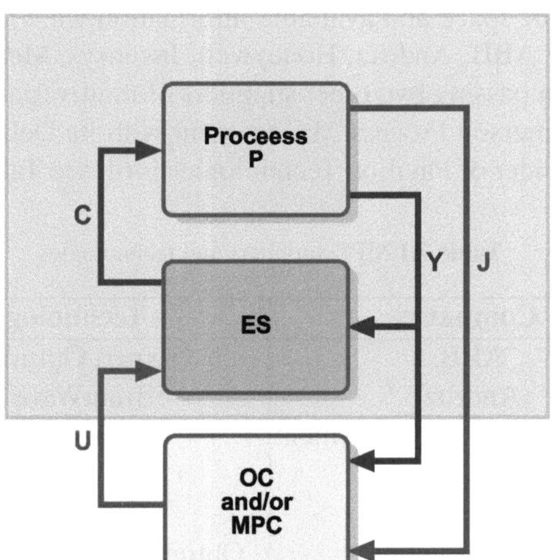

Figure 7.4 ABB control algorithm concept

In the scheme shown in Figure 7.4, ES is concerned with process stability while the OC/MPC lock takes care of optimality, that is, either OC or MPC drives processes to their economically optimal point. The performance criterion J is therefore related to some sort of economic performance measure.

In a further embodiment of the control system, several layers of the scheme described above could be put in place to optimize the process control. In such a case an ES is supervised by an OC/MPC-based controller, which in turn is supervised by another ES, *etc.* This architecture allows optimal exploitation of the advantages of each approach for each particular task.

Expert Optimizer has been applied in many cement plants. In flotation plants, a strategy has been developed that incorporates first principle modeling, ES, fuzzy logic and neural networks to generate the optimal grade–recovery curve, improve concentrate quality, stabilize flows and minimize reagent use. The manipulated variables in this strategy are reagent flow rate, cell level set-point, and air flow rates for cells, and wash water flow for columns [49].

ANDRITZ

Ideas Simulation & Control, a member of the Andritz group, markets three products with potential for application to mineral processing [50]: Ideas dynamic process simulator, ACE™[8] (Andritz Control Expert) expert control system and BrainWave® predictive control software

Ideas is a dynamic simulation environment for use in process design, control logic verification, operator training, and advanced control system design and optimization. ACE is an ES that emulates the responses of an experienced operator, and has been applied to crusher and dryer operations in combination with BrainWave®.

BrainWave® is a multivariable predictive controller based on Laguerre polynomials, a technique developed at the University of British Columbia, see U.S. Patent No. 5,335,164 [51]. It employs a model that adapts on-line and utilizes the prediction horizon as a tuning parameter. The structure of the model, built around discrete-time transfer functions (Z-transform), is shown in Figure 7.5. Once the model obtains the state equation of the dynamic system, its parameters are then identified by a recursive least-squares algorithm.

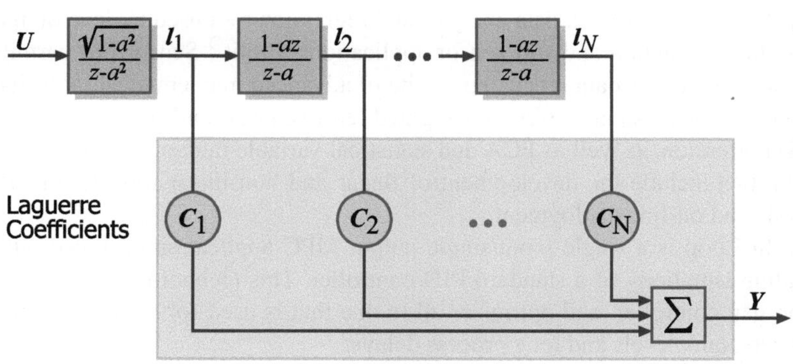

Figure 7.5 Model used by BrainWave®

[8] ACE is a registered trademark of IDEAS Simulation & Control Inc.

The application of BrainWave® has been reported in various mining processes such as level control in secondary and tertiary crushers, particle size distribution control in ball mills, cascade control of power and inventory in SAG mills, flotation cell level control, thickener density control and molybdenum concentrate dryer control.

BrainWave SAG Mill is a control package using model-based predictive control technology that accounts for changes in mill rotation, recycled pebbles and ore quality to maintain mill weight and maximize production. As the weight controller adjusts the set-point for fresh ore feed, a second BrainWave controller defines the feeder speed to ensure that the right amount of material is delivered to the mill. Additionally, BrainWave® can automatically adjust its control settings to allow for different operation due to changes in ore hardness over time. BrainWave® was installed at the Minera Los Pelambres copper concentrator site in order to tightly control the weight in the SAG mill and thereby promote optimum grinding. According to Andritz, the strategy has been demonstrated to improve mill throughput by as much as 3%.

HONEYWELL

Honeywell offers a family of products known as Profit Suite for advanced process control and optimization. Most of its components can be used in mineral processing applications. These products include Profit Suit Early Event Detection (EED) Toolkit, Profit® Sensor Pro, Profit® Loop, Profit® Controller, Profit® Optimizer, Profit® Bridge, Profit® Max and UniSim Design Suite [52].

The Profit® Suite EED Toolkit provides a statistical modeling and application environment to identify, localize and support the reduction of abnormal situations in processes and plant equipment. It uses principal component analysis (PCA) and fuzzy modeling. The Toolkit's PCA modeling interface provides intuitive controls and options to suit both novice and expert users. The Fuzzy Logic Modeling Toolkit employs an analytical method for PCA post-processing.

Profit® Sensor Pro is a data analysis and regression tool used to develop models for on-line prediction and soft sensor applications. Profit® Sensor Pro permits the analysis of complex data sets using a suite of advanced mathematical tools, including ordinary least squares (OLS), weighted least squares, and partial least squares (PLS) regression, as well as PCA and statistical variable ranking. Profit Sensor Pro capabilities include the development of linear and non-linear models for off-line analysis and on-line deployment.

Profit Loop is a single input/single output MPC application designed with the operating simplicity of a standard PID controller. This technology is a Smith predictor, gap controller, and optimizer all in one that is used for controlling discrete analyzers, tank levels and long process delays.

Profit® Controller, based on Honeywell's Robust Multivariable Predictive Control Technology (RMPCT), provides the tools for designing, implementing and maintaining multivariable control and optimization strategies to control complex and highly interactive industrial processes. The objective function employed by Profit® Controller is

$$J = \sum_i [b_i y_i(t+j) + a_i(y_i - w_i)]^2 + \sum_j [d_j u_j + c_j(u_j - m_j)]^2, \quad (7.4)$$

where y_i is a controlled variable, w_i is the corresponding reference, u_j is a manipulated variable, and a_i , b_i , c_j, d_j , and m_j are parameters. Both error and profit optimization can be performed.

Profit® Controller uses the range control algorithm, which determines the smallest process moves required to simultaneously meet control and optimization objectives, see U.S. Patent No. 5,572,420 [53]. Its handling of control through "funnels" rather than specified trajectories gives the controller additional degrees of freedom to enhance dynamic process optimization. RMPCT uses a finite impulse response (FIR) model form, one for each controlled manipulated pair; model identification is accomplished using open-loop tests on the process [54]. RMPCT provides robustness to plant–model mismatch using singular-value thresholding and a min-max design. In the latter, the optimizer determines the control effort over all possible plants realizations (worst case).

Profit ® Optimizer solves multi-unit and plant-wide optimization on a minute-by-minute basis. It leverages measured process relationships and dynamic process information residing in underlying Profit Controller applications. By accounting for constraints and economics between operating units, Profit Optimizer can implement optimal operating conditions without violating global plant constraints. In addition, it dynamically coordinates the introduction of optimal set-points for a smooth transition to optimal operation using distributed quadratic programming.

Profit® Bridge integrates non-linear process models with Profit Controller and Profit Optimizer applications. By frequently updating the linear models embedded in these applications with information obtained from non-linear process models, Profit Bridge delivers dynamic non-linear control and optimization at a fraction of the cost of a full-scale, rigorous optimization system.

Profit® Max provides a first-principle modeling and optimization system within an on-line execution environment. This solution is typically used on optimization projects requiring mixed-integer programming, or on highly non-linear processes.

UniSim™[9] Design is an interactive process modeling offering that enables engineers to create steady-state and dynamic models for plant design, performance monitoring, troubleshooting, operational improvement, business planning and asset management. With this product, steady-state models can be extended to transient models, allowing consistent model configurations to be used for process transient analysis, controllability studies, and operator training applications.

Honeywell has developed various MPC applications in mineral processing plants, one of the first being the SmartGrind ball grinding application installed at the BHP plant in Pinto Valley, Arizona (see Figure 7.6). TotalPlant SmartGrind multivariable control integrates MPC and neural networks to predict anomalous situations such as mill overloads. In this ball mill example, a projected throughput increase of 4.18% in the desired product (smaller than +65 mesh particle) was obtained. Performance

[9] UniSim is a registered trademark of Honeywell International, Inc.

data collected in December 1997 showed that control by SmartGrind rather than operator supervision brought about an increase in throughput of 2% [55].

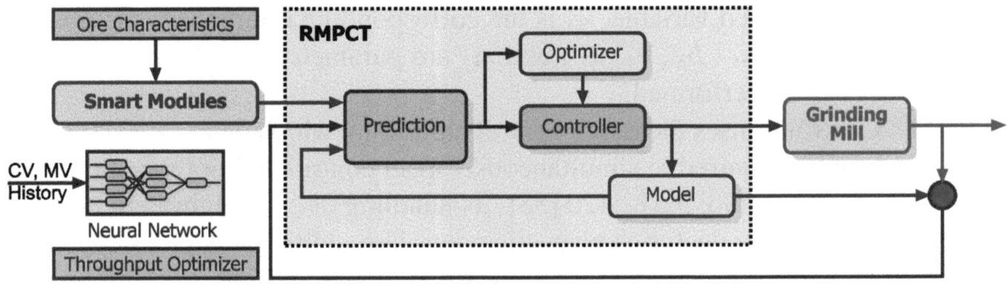

Figure 7.6 Block diagram of TotalPlant SmartGrind MPC

INVENSYS

Invensys offers a control system product known as Connoisseur that incorporates two modules, one of them a model-based predictive controller and the other a static optimizer that employs linear programming [56].

Key features of Connoisseur are the following:

- real time adaptive control, enabling the control system to be adapted to prevailing process conditions on-line;
- constrained optimization that permits operation within the physical constraints of the process, allowing the system to maximize process potential;
- application of MPC to processes that require multiple models. The product uses on-line gain scheduling and can switch automatically between multiple model sets without turning the controller off.

Connoisseur's modeling tools are cross-correlation, power spectral density and statistical techniques such as OLS and PLS. Recursive least squares is also used for on-line adaptable estimation and control with FIR and ARX models. The product enables management of hard and soft constraints including variation limits and speeds. Additionally, it offers gain scheduling functions for processes that require multimodels, neural networks (multilayer perceptron and radial basis function) for non-linear systems, and fuzzy ES to complement the MPC control.

Among other mining applications, Connoisseur has been used at Asarco Mines to improve grinding and flotation circuit performance [57]. Grinding control objectives include driving the line to its optimal operating point and maintaining control through variations in ore characteristics. The product has increased the feed rate without violating the mill's operating limits, such as maximum particle size, cyclone overflow density and cyclone pump current. It also has the ability to increase recycle mass flow for minimizing particle sizes when the feed is limited. The result has been the dual benefit of increasing recovery while minimizing wear on the ball mills. A separate Connoisseur multivariable controller is used to control each of the eight grinding circuits. Each controller has access to a number of process models which represent process dynamics under different operating regimes. The controller automatically selects the appropriate control model by determining which one provides the best fit for the prevailing conditions. The controllers also make use of Connoisseur's on-line adaptation facility to modify control models in line with longer-term drifts in operation such as mill wear. A rules-based scheme, applying fuzzy logic technology, is utilized to monitor the levels of the ore storage bins fed from the upstream ore handling facility.

METSO

Metso is a holding company active in pulp and paper (Metso Paper), mining (Metso Minerals) and automation (Metso Automation). In each of these areas it offers modeling, simulation and control products. The firm operates in the market for mining automation and control systems through Metso Minerals-Cisa, whose main control product is Optimizing Control Systems (OCS©), an expert control system based on fuzzy rules that incorporates image processing, virtual sensors, statistical models, neural networks, optimization modules and adaptive predictive models (see Figure 7.7). As with VisioRock and VisioFroth, two of the sensors described in 7.2 above, OCS© has been installed in numerous concentration plants [17].

Metso Minerals-Cisa also markets a computer-based training (CBT©[10]) system for operators based on dynamic simulation. The system permits users to visualize the effect of control actions through text, figures, animations, videos and interactive graphics, as well as providing evaluations of operator learning.

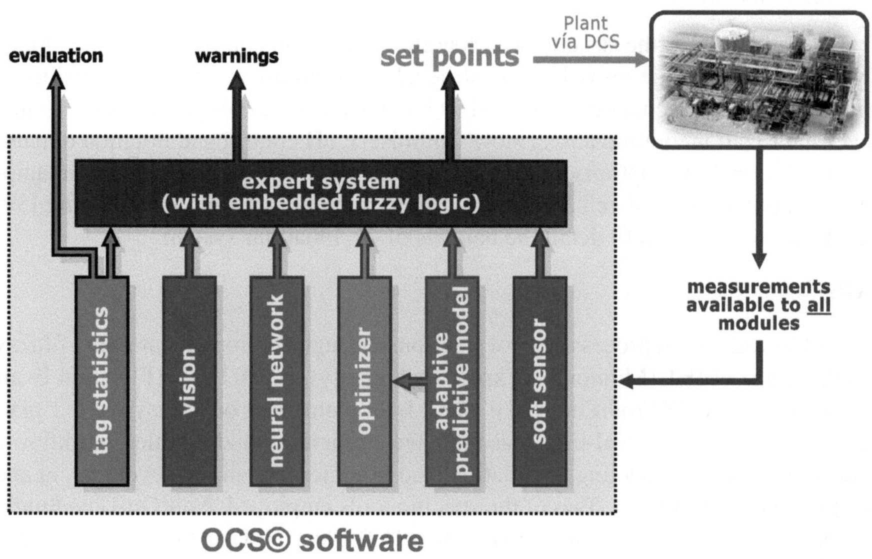

Figure 7.7 Metso Optimizing Control System

[10] CBT is copyright of Metso Minerals Cisa BP 6009, 45060 Orlans Cedex 2, France, www.svedala-cisa.com

Metso has developed control strategies for flotation plants that combine VisioFroth with feed-forward and froth speed control techniques using a cascade control system in which the controller sets the pulp and air reference levels. An assessment of the benefits obtained from these strategies at the Grasberg mine in Freeport, Indonesia, which included the installation of 172 cameras, reported a recovery increase of 2.6 to 3.9%, a reduction in frother dosage of 25% and a project payback time of 51 days.

MINTEK

Mintek offers a family of products for mineral processing control known as StarCS, an open system with OPC communication and a series of functionalities integrated in specific applications such as MillStar and FloatStar [58].

Among these functionalities are:

- StarCS Autotuner, for self-tuning of PID controllers;
- Advanced Control Package, which includes ratio controllers, advanced PID, a soft limiter, a dead time compensator, feed-forward control, advanced filters and non-linear compensators;
- StarSC MMPC, a model-based multivariable predictive controller specially adapted for processes with different numbers of controlled and manipulated variables;
- StarSC Deadbeat Controller, for processes with very long dead times;
- fuzzy logic environment for fuzzy expert control;
- dynamic simulator for continuous and discrete time systems, with non-linear blocks and white noise generators.

Figure 7.8 shows the structure of FloatStar, a flotation control system developed by Mintek for stabilizing pulp or froth levels and optimizing cell bank residence times, grades and recoveries. The product includes modules for process monitoring, level stabilization, optimization (Flow Optimiser), pH control and flotation dynamics simulation. Flow Optimiser is a multivariable predictive controller whose manipulated variables are the cell level and air flow references. Craig and Henning [59] and Houseman *et al.* [60] detail the benefits of the FloatStar system.

SGS

SGS offers advanced process control solutions through the implementation of fuzzy ES that utilize MET (MinnovEX Expert Technology), FLEET™[11] (Flotation Evaluation Tool) and JKFrothCam [61]. MET is implemented on Gensym's G2 platform, which combines real-time reasoning technologies including rules, workflows, procedures, object modeling, simulation, and graphics in a single development and deployment environment. Two of the specific tools employed: NeurOn-Line Studio and NeurOn-Line, are also Gensym products [62]. NeurOn-Line Studio is employed for viewing and cleaning training data and determining sensitivities and time lags for various model inputs. NeurOn-Line, on the other hand, is used to deploy the

[11] FLEET is a registered trademark of SGS Group, http://www.sgs.com/

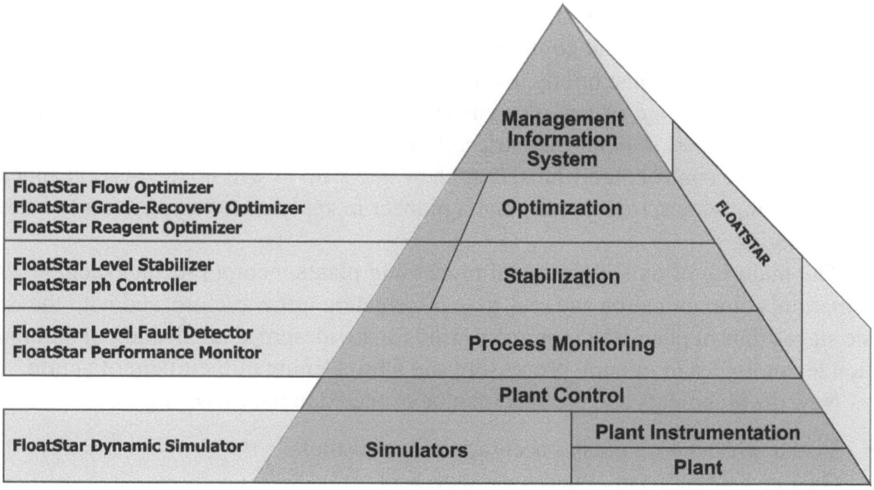

Figure 7.8 FloatStar structure

model via the ES shell. With these tools, SGS has developed soft sensors for particle size measurements in the grinding ES installed at several concentrators. As for FLEET™, it is a combination of methodology and software that analyzes the variability of flotation plant responses in order to improve their design and optimize production planning.

The introduction in 1999 of a MET ES for SAG milling at Ok Tedi brought about significant improvements in circuit throughput. In 2002 a MET neural network model was developed and deployed to enhance the existing ES solution. The main objective of this model was to predict the effect that changes in SAG tonnage and density would have on mill power draw. The result was a decrease in process variance with an associated increase in average circuit performance. The improvement in power draw was 35% over that already obtained with fuzzy expert control alone. This led to an additional 2% in calculated throughput on top of the 5.3% quoted in an earlier paper [63]. An important supplementary benefit was the reduction in variability, which ultimately benefits flotation.

As regards developments in flotation, MET and JKFrothCam were installed in 80 OK-100 copper flotation cells at Escondida to control cell level and air flow. An evaluation found that feed consistency in the cleaning circuit was improved and recovery rose by 1.3%.

7.5 Conclusions and Trends

The foregoing review has shown that a significant number of industrial products, including innovative sensors and tools for the development, implantation and evaluation of advanced control strategies, are now being applied in mineral processing plants. Other products that could also be named, such as dynamic simulators and virtual sensors, were just briefly touched on. Furthermore, research centers are perfecting a whole range of new designs and developments as well as products that are already applied in other industries. As occurred previously with visual sensors and predictive control algorithms, these new alternatives will no doubt soon appear in the mining sector, traditionally not a pioneer in applying automation and control technologies.

The majority of existing mineral processing plants incorporate an excellent automation, communication and real-time information infrastructure, and with the basic stages thus in place, the time seems ripe for a widespread application of automation technologies in mineral processing and a broadening of their field of action.

New developments and products can be expected in the following areas:

- Visual sensors with greater accuracy and robustness.
- On-line hardness and mineralogy analyzers.
- New sensors for the measurement of grinding, classification and flotation variables.
- Additional new tools for advanced control that combine ESs with model-based control and continuous with discrete control (hybrid systems).
- Dynamic optimization applications for integrated processes (*e.g.*, SAG mills and secondary grinding, cell and column flotation) and plant interconnection (crushing-grinding-flotation).

If users are to fully reap the expected benefits of all this greater technological sophistication, certain shortcomings in the training of professionals and technicians charged with supervising and operating mineral processing automation equipment will have to be resolved first [64]. Interestingly, technology can also contribute to strategies for overcoming these problems through the application of solutions already proven in other industries. These include certifications, continuing education, e-learning, project-based learning and operator training simulators.

References

[1] Barría D (2008) Digital Codelco: mining of the future (in spanish). Technological Approach Seminar, Calama (Chile)
[2] Hodouin D, Jamsa-Jounela SL, Carvalho MT, Bergh L (2001) State of the art and challenges in mineral processing control. Control Engineering Practice 9:995–1005
[3] Jamsa-Jounela SL (2001) Current status and future trends in the automation of mineral and metal processing. Control Engineering Practice 9:1021–1035
[4] Yacher L, González C, Von Borries G, Nobile R (1985) Coarse particle size distribution analyze. In H.K. Copper (Ed.), Instrumentation in the Mining and Metallurgy Industries, Instrument Society of America, 12, 71–79
[5] Split Engineering, www.spliteng.com
[6] Metso Minerals,www.metsominerals.com
[7] KnowledgeScape, www.scape.com

[8] Singh V, Rao SM (2005) Application of image processing and radial basis neural network techniques for ore sorting and ore classification. Minerals Engineering 18:1412–1420
[9] WipWare, www.wipware.com
[10] Guyot O, Monredon T, LaRosa D, Broussaud A (2004) VisioRock, an integrated vision technology for advanced control of comminution circuits. Minerals Engineering 17:1227–1235
[11] Kaartinen J, Tolonen A (2008) Utilizing 3D height measurement in particle size analysis. Proceedings of the 17th IFAC World Congress, Seoul (Korea)
[12] Outotec, www.outotec.com
[13] Valderrama WR, Pontt JO, Magne LO, Hernández JS, Salgado FI, Valenzuela JS, Poze RE (2000) The Impactmeter, a new instrument for monitoring and avoiding harmful high-energy impacts on the mill liners in SAG mills. Preprints of IFAC Workshop on Future Trend in Automation in Mineral and Metal Processing, Helsinki (Finland), 274–279
[14] FLSmidth Minerals, www.flsmidthminerals.com
[15] Cipriano A, Guarini M, Vidal R, Soto A, Sepúlveda C, Mery D, Briseño H (1998) A real time visual sensor for supervision of flotation cells. Mineral Engineering 11: 489–499
[16] JKTech Pty Ltd, www.jktech.com.au
[17] Metso Minerals Cisa, www.svedala-cisa.com
[18] Morar S, Forbes G, Heinrich G, Bradshaw D, King D, Adair B, Esdaile L (2005) The use of a colour parameter in a machine vision system, SmartFroth, to evaluate copper flotation performance at Rio Tinto's Kennecott copper concentrator. Proceedings of the Centenary of Flotation Symposium, Brisbane (Australia)
[19] Kaartinen J, Hatonen J, Hyotyniemi H, Miettunen J (2006) Machine-vision-based control of zinc flotation – A case study. Control Engineering Practice 14:1455–1466
[20] Forbes G (2007) Texture and bubble size measurements for modelling concentrate grade in flotation froth systems. PhD Thesis University of Cape Town, South Africa.
[21] Lin B, Recke B, Knudsen J, Jørgensen S (2008) Bubble size estimation for flotation processes. Minerals Engineering 21:539–548
[22] Liu J, MacGregor J (2008) Froth-based modeling and control of flotation processes. Minerals Engineering 21:642–651
[23] Núñez F, Cipriano A (2008) Visual information model based predictor for froth speed control in flotation process. Minerals Engineering 22(4):366–371
[24] Flores A, Sáez D, Araya J, Berenguel M, Cipriano A (2005). Fuzzy predictive control of a solar power plant. IEEE Transactions on Fuzzy Systems 13(1): 58–68
[25] Cipriano A, González G (2008) Advanced control for mineral processing plants. Lecture at First International Congress on Automation in the Mining Industry, Automining 2008, Santiago (Chile)
[26] Levine W (1995) The Control Handbook. CRC and IEEE Press
[27] Passino K M, Yurkovich S (1998) Fuzzy Control. Addison Wesley
[28] Driankov D, Hellendoorn H, Reinfrank M (1996) An Introduction to Fuzzy Control. Springer
[29] Driankov D, Palm R (1998) Advances in Fuzzy Control. Springer
[30] Mamdani EH (1974) Application of fuzzy algorithms for control of simple dynamics plant. Proceedings of the IEE 121:1585–1588
[31] Mamdani EH, Assilian S (1975) An experiment in linguistic synthesis with a fuzzy logic controller. International Journal of Man-Machine Studies 7(1):1–13
[32] Takagi T, Sugeno M (1985) Fuzzy identification of systems and its applications to modeling and control. IEEE Transactions on Systems, Man, and Cybernetics 15(1):116–132
[33] Tanaka K, Wang H (2001) Fuzzy Control Systems Design and Analysis. John Wiley & Sons
[34] Palm R, Driankov D, Hellendoorn H (1996) Model Based Fuzzy Control. Springer
[35] Babuska R (1998) Fuzzy Modelling for Control. Kluwer Academy Publishers
[36] Aracil J, Gordillo F (2000) Stability Issues in Fuzzy Control. Springer
[37] Abonyi J (2003) Fuzzy Model Identification for Control. Birkhaüser
[38] Camacho F, Bordons C (2005) Model Predictive Control. Springer

[39] Kouvaritakis B, Cannon M. (2001) Nonlinear Predictive Control: Theory and Practice. The Institution of Electrical Engineers, London
[40] Nørgaard M, Ravn O, Poulsen N, Hansen L (2000) Neural networks for modelling and control of dynamic systems. Springer-Verlag, London
[41] US Patent and Trademark Office, http://patft.uspto.gov
[42] Morari M, Baotic M, Borrelli F (2003) Hybrid systems modeling and control. European Journal of Control 9(2-3):177–189
[43] Qin S, Badgwell T (2003). A survey of industrial model predictive control technology. Control Engineering Practice 11:733–764
[44] Emerson Process Management, www.emersonprocess.com
[45] Rockwell, www.rockwell.com
[46] Pavilion Technologies, www.pavtech.com
[47] ABB, www.abb.com
[48] ABB Process control system patent, European Patent EP1553470, www.freepatentsonline.com
[49] Gaulocher S, Gallestey E (2008) Advanced process control of a froth flotation circuit, Proceedings of the Fifth International Seminar on Mineral Processing, Procemin 2008, Santiago (Chile)
[50] Andritz Ideas Simulation & Control, www.ideas-simulation.com
[51] Zervos C, Dumont G (1988) Deterministic adaptive control on Laguerre series representation. International Journal of Control 48(6):2333–2359
[52] Honeywell, www.honeywell.com
[53] Lu Z (1996) Method for optimal controller design for multivariable predictive control utilizing range control. US Patent No 5.572.420,
[54] Krishnan A, Kosanovich K, DeWitt M, Creec M (1998) Robust model predictive control of an industrial solid phase polymerizer. Proceedings of the American Control Conference, Philadelphia (USA)
[55] Mathur A, Parthasarathy S, Gaikwad S (1999). Hybrid neural network multivariable predictive controller for handling abnormal events in processing applications. Proceedings of the 1999 IEEE International Conference on Control Applications, Hawai (USA)
[56] Invensys, www.invensys.com
[57] Nix A, Morrow A, Gordon L (2000). Using multivariable predictive control to optimize the Asarco Mission mill. In J. A. Hebst (Ed.), Proceedings of Control 2000, SME Annual Meeting, Littleton (USA), 201–207
[58] Mintek, www.mintek.com
[59] Craig L, Henning R (2000) Evaluation of advanced industrial control projects: a framework for determining economic benefits. Control Engineering Practice 8: 769–780
[60] Roseman L., Schubert J, Jart J, Carew W (2001) Plantstar 2000: a plant – wide control platform for mineral processing, Minerals Engineering 14: 593–600
[61] SGS, www.sgs.com
[62] Gensym, www.gensym.com
[63] Keaton M, Wood M, Schaffer M (2002) Application at Ok Tedi Mining of a neural network model within the expert system for SAG mill control. SGS Internal Report, www.met.sgs.com
[64] Cipriano A, González G, Sbarbaro D (2007) Final Report Project Amira P893A, Training in Automation for optimizing the profitability of technological investments

Index

Other titles published in this series (continued):

Soft Sensors for Monitoring and Control of Industrial Processes
Luigi Fortuna, Salvatore Graziani, Alessandro Rizzo and Maria G. Xibilia

Adaptive Voltage Control in Power Systems
Giuseppe Fusco and Mario Russo

Advanced Control of Industrial Processes
Piotr Tatjewski

Process Control Performance Assessment
Andrzej W. Ordys, Damien Uduehi and Michael A. Johnson (Eds.)

Modelling and Analysis of Hybrid Supervisory Systems
Emilia Villani, Paulo E. Miyagi and Robert Valette

Process Control
Jie Bao and Peter L. Lee

Distributed Embedded Control Systems
Matjaž Colnarič, Domen Verber and Wolfgang A. Halang

Precision Motion Control (2nd Ed.)
Tan Kok Kiong, Lee Tong Heng and Huang Sunan

Optimal Control of Wind Energy Systems
Iulian Munteanu, Antoneta Iuliana Bratcu, Nicolaos-Antonio Cutululis and Emil Ceangă

Identification of Continuous-time Models from Sampled Data
Hugues Garnier and Liuping Wang (Eds.)

Model-based Process Supervision
Arun K. Samantaray and Belkacem Bouamama

Diagnosis of Process Nonlinearities and Valve Stiction
M.A.A. Shoukat Choudhury, Sirish L. Shah, and Nina F. Thornhill

Magnetic Control of Tokamak Plasmas
Marco Ariola and Alfredo Pironti

Real-time Iterative Learning Control
Jian-Xin Xu, Sanjib K. Panda and Tong H. Lee

Deadlock Resolution in Automated Manufacturing Systems
ZhiWu Li and MengChu Zhou

Model Predictive Control Design and Implementation Using MATLAB®
Liuping Wang

Fault-tolerant Flight Control and Guidance Systems
Guillaume Ducard

Predictive Functional Control
Jacques Richalet and Donal O'Donovan

Fault-tolerant Control Systems
Hassan Noura, Didier Theilliol, Jean-Christophe Ponsart and Abbas Chamseddine

Control of Ships and Underwater Vehicles
Khac Duc Do and Jie Pan

Detection and Diagnosis of Stiction in Control Loops
Mohieddine Jelali and Biao Huang

Stochastic Distribution Control System Design
Lei Guo and Hong Wang

Dry Clutch Control for Automotive Applications
Pietro J. Dolcini, Carlos Canudas-de-Wit, and Hubert Béchart

Active Braking Control Design for Road Vehicles
Sergio M. Savaresi and Mara Tanelli

Nonlinear and Adaptive Control Design for Induction Motors
Riccardo Marino, Patrizio Tomei and Cristiano M. Verrelli

MIX
Papier aus verantwortungsvollen Quellen
Paper from responsible sources
FSC® C105338

If you have any concerns about our products,
you can contact us on
ProductSafety@springernature.com

In case Publisher is established outside the EU,
the EU authorized representative is:
Springer Nature Customer Service Center GmbH
Europaplatz 3, 69115 Heidelberg, Germany

Printed by Libri Plureos GmbH
in Hamburg, Germany